# Greenhouse Gas Regulating Microorganisms in Soil Ecosystems

Santosh R. Mohanty • Bharati Kollah
Editors

# Greenhouse Gas Regulating Microorganisms in Soil Ecosystems

## Perspectives for Climate Smart Agriculture

Springer

*Editors*
Santosh R. Mohanty
Indian Institute of Soil Science
Bhopal, Madhya Pradesh, India

Bharati Kollah
ICAR Indian Institute of Soil Science
Bhopal, Madhya Pradesh, India

ISBN 978-3-031-70568-7 ISBN 978-3-031-70569-4 (eBook)
https://doi.org/10.1007/978-3-031-70569-4



This Springer imprint is published by the registered company Springer Nature Switzerland AG
The registered company address is: Gewerbestrasse 11, 6330 Cham, Switzerland

# Preface

Agriculture contributes to the global warming by acting as a source and sink for atmospheric greenhouse gases. To curb the global warming, the current agricultural practices need to be shaped in a more eco-friendly way for climate smart agriculture. Soil microbes play key role in the production and consumption of GHGs. However, there is limited comprehensive information that underpins the soil microbial processes and the microbial technologies to mitigate climate change. Few groups of soil microorganisms play key role in greenhouse gas cycling. Methanogens produce $CH_4$ while methanotrophs oxidize $CH_4$. Similarly, $N_2O$ is produced by various N cycling microbes and gets decomposed by denitrifiers. Soil heterotrophic organisms produce $CO_2$ and autotrophs fix $CO_2$. The function of these organisms depends on various factors like soil organic carbon, redox potential and nutrients. To minimize GHG emission there is need of balanced function of these organisms. Disturbance to these functional groups may alter the greenhouse gas footprint influencing the climate change. However, there is lack of comprehensive information in this aspect. Based on the above scenario, this book was framed to bring out fundamental information on microbial processes and agricultural strategies to mitigate climate change. The book unveils chapters to define the linkages among soil microbial function, crop responses, ecosystem functioning, and greenhouse gas cycling processes. Book is framed with themes including source and sink of greenhouse gas, microbial processes regulating GHG, and agricultural strategies to mitigate GHG emission. Chapters highlight the fundamentals of soil microbial diversity and interaction with climate changing factors, soil carbon dynamics in response to different agricultural practice, conservation agriculture strategies to mitigate greenhouse gas emission from agriculture, and climate change mitigation by organic farming. Chapters provide valuable information for agriculture, life science, and environmental planning modules for students, scientists, and policy makers. The book suggests different metaphors for the strategic options to make climate change work for us, rather than against us, offering practical solutions to climate change—both adaptation and mitigation.

Bhopal, Madhya Pradesh, India

Santosh R. Mohanty
Bharati Kollah

# Contents

# Sources and Sinks of Greenhouse Gasses (GHG) in Agricultural Soil Ecosystems

**Pukhrambam Helena Chanu, Aniket Singh, Sneha Bharadwaj, and Samiksha Pandey**

## Introduction

Greenhouse gases (GHGs) are gases in the Earth's atmosphere that trap heat from sunlight, contributing to climate change. Human activities have significantly increased GHG concentrations, raising concerns about climate change. Agriculture is a major source of GHG emissions, contributing to climate change. Agricultural soils can be both a source and a sink of GHGs, depending on various biogeochemical processes. These processes are influenced by factors such as soil type, land management practices, crop types, and climate conditions. GHGs are released from agricultural soils through microbe-mediated decomposition of organic matter, nutrient cycling, and fertilizer use. GHGs are sequestered or absorbed by agricultural soils through photosynthesis and nitrogen retention. The soil ecosystem is an integral part of the agroecosystem, supporting agricultural productivity, nutrient cycling, water retention, and overall ecosystem resilience (Altieri et al. 2015). Soil ecosystems are vital components of the Earth's biosphere, supporting a wide range of life

P. H. Chanu (✉)
Department of Soil Science and Agricultural Chemistry, Institute of Agricultural Sciences, Banaras Hindu University, Varanasi, UP, India
e-mail: phelena@bhu.ac.in

A. Singh
Biological Sciences, Department of Biology and Microbiology, College of Natural Sciences, South Dakota State University, Brookings, SD, USA

S. Bharadwaj
Department of Agronomy, Indian Agricultural Research Institute, New Delhi, India

S. Pandey
Department of Agronomy, National Diary Research Institute, Karnal, India

 
S. R. Mohanty, B. Kollah (eds.), *Greenhouse Gas Regulating Microorganisms in Soil Ecosystems*, https://doi.org/10.1007/978-3-031-70569-4_1

forms and playing a crucial role in agriculture and food production. Agricultural soil ecosystems encompass the dynamic interactions between soil, plants, microorganisms, and the environment, forming a complex web of relationships that shape the health and productivity of agricultural systems. Agricultural soil ecosystems provide the foundation for crop growth and serve as a reservoir for nutrients, water, and organic matter. A proper understanding of the different sources and sinks of GHGs in agricultural soil ecosystems is essential for developing effective mitigation strategies and sustainable farming practices. Greenhouse gases (GHGs), commonly carbon dioxide ($CO_2$), methane ($CH_4$), and nitrous oxide ($N_2O$), are emitted and absorbed by various processes in agricultural soils, ultimately affecting the Earth's climate. Accordingly, this chapter discusses the different sources and sinks of major GHGs in agricultural systems, the problems, anticipated solutions, and future approaches with respect to agroecosystems.

## Agroecosystem

An agroecosystem is an agricultural system that includes plants, animals, and microorganisms interacting with the environment within a specific area of cultivated land. The soil ecosystem is an integral part of the agroecosystem and plays a significant role in its functioning and in the global cycling of greenhouse gases (GHGs), which are major contributors to climate change. Understanding the sources of GHGs in agricultural soil ecosystems is crucial for developing effective mitigation strategies to minimize their impact on climate change. When discussing the manipulation of ecosystems for food security and carbon sequestration, agroecosystems offer a good scope because they are relatively small and less constrained than natural ecosystems. Thus, the agricultural landscape is a good space for implementing new technologies (Garbach et al. 2016).

### *Source of Different GHGs in Agricultural Soil Ecosystem*

Agricultural soil ecosystems emit greenhouse gases (GHGs) through various processes, including the decomposition of organic matter, respiration by plants and soil microorganisms, and burning of crop residues. Methane ($CH_4$) emissions primarily originate from anaerobic processes occurring in waterlogged or poorly drained soil conditions, where methanogenic microorganisms convert organic matter into $CH_4$. Additionally, the use of certain fertilizers, such as nitrogen-based fertilizers, can increase nitrous oxide ($N_2O$) emissions through microbial activity like nitrification and denitrification.

### *Agricultural Soil Ecosystem as a Source of Carbon Dioxide*

According to the Food and Agriculture Organization (FAO), agricultural soils store around 1500 gigatons (Gt) of carbon globally, which is about twice the amount of carbon stored in vegetation and approximately three times the amount stored in the atmosphere. Soil is the third-largest sink of $CO_2$ after the ocean and terrestrial vegetation. Various factors contribute to $CO_2$ fixation in agricultural soils.

## Contribution of $CO_2$ Resulting from Land Use Pattern Change in Soil Ecosystem

Agricultural land use can reduce the soil's ability to store carbon, leading to higher GHG emissions. Soil degradation can also increase $CO_2$ emissions in a number of ways.

Deforestation: Agricultural deforestation is the clearing of forests or other natural ecosystems for agricultural expansion or intensification. Agriculture is a major cause of deforestation, especially in tropical regions, accounting for an estimated 90–99% of deforestation, either directly or indirectly. While some of this deforestation does lead to an increase in agricultural output, only about half to two-thirds of it does (Pendrill et al. 2022). According to FAO data, deforestation for agriculture is a major source of greenhouse gas (GHG) emissions, as it releases vast amounts of carbon into the atmosphere as $CO_2$ when trees and other plants are burned or decompose. Large-scale and uncontrolled deforestation increases GHG emissions, which contribute to global warming and accelerate climate change (David et al. 2018). The conversion of natural ecosystems to agricultural land reduces soil organic carbon (SOC) levels due to increased mineralization and reduced carbon inputs (Carter et al. 1998). Normally, plants absorb extra carbon dioxide ($CO_2$), but when forests are burned or cleared for agriculture, the stored carbon is released back into the ecosystem as $CO_2$. The Intergovernmental Panel on Climate Change (IPCC) estimates that deforestation and forest degradation account for approximately 10–15% of global $CO_2$ emissions, but this estimate can vary depending on the region and the specific circumstances of deforestation. Deforestation rates vary across countries and regions, with countries with significant forest cover such as Brazil, Indonesia, and the Democratic Republic of Congo experiencing high rates of deforestation in recent decades, resulting in substantial $CO_2$ emissions.

Soil erosion: Soil erosion disproportionately removes the lighter fraction of soil organic carbon (SOC), transporting some of this carbon to depressions and water ecosystems, creating a carbon sink. Of the 1.9 billion metric tons of carbon transported from terrestrial sources to inland waters each year, about 0.8 billion metric tons are released into the atmosphere, while the rest is buried in sediment in aquatic ecosystems (Cole et al. 2007). Various factors, such as soil erosion, can affect soil $CO_2$ release (Bajracharya et al. 2000; Polyakov and Lal 2004). First, increased soil

$CO_2$ emissions at eroding sites are caused by the breakdown of soil aggregates due to direct rain and runoff shear stress, which exposes previously contained soil organic carbon (SOC) to microbial degradation (Polyakov and Lal 2008; Van Hemelryck et al. 2011). Second, because soil particles are often transported as SOC-rich aggregates, there is a greater likelihood that they will be broken down by machinery during transport from eroding sites to depositional locations (Beuselinck et al. 2000; Slattery and Burt 1997), potentially leading to $CO_2$ emissions from the soil (Lal 2003; Nadeu et al. 2011). Finally, during the erosion process, selective removal of labile SOC can lead to its accumulation in bulk in moist depositional sites, where it is vulnerable to mineralization and hence increased $CO_2$ emissions (Berhe 2012; Gregorich et al. 1998; Lal 2005; Van Hemelryck et al. 2011).

Contribution due to farming practices: Farming practices such as crop type, tillage practices, biomass burning, and grazing patterns affect soil physical and biological properties, leading to soil organic matter (SOM) degradation and $CO_2$ emissions.

Crop types: Agricultural fields are influenced by the types and productivity of crops, as the residue that remains in the fields is fixed into the soil as soil organic matter (Eghball et al. 1994). For example, continuous corn produces more residue than the corn–soybean rotation system, leading to increased accumulation of soil organic carbon (SOC) (Eghball et al. 1994). Additionally, incorporating winter cover crops or rotating crops with hay or pasture can increase overall production and carbon accumulation in soil (Bremer et al. 1994). Conversely, crops that produce relatively little residue, such as cotton (*Gossypium hirsutum* L.), result in low carbon input into the soil (Bordovsky et al. 1999).

Cropping intensity: Cropping intensity affects soil organic carbon (SOC) storage by changing the duration of soil use for crop cultivation, leading to increased annual production and deposition of carbon in the soil. Additionally, intensifying agricultural practices can reduce the decomposition rate of organic matter in the soil, which is typically high in bare areas, such as semiarid regions where fields are left bare to improve soil moisture through summer fallowing (Paustian et al. 2000). Increasing cropping intensity often involves reducing or eliminating bare summer fallowing or adding cover crops, both of which have been shown in multiple field trials to improve SOC storage. For example, Bremer et al. found that cropping systems featuring continuous wheat, hay production, or manure addition had the highest levels of total soil organic matter and labile organic carbon.

Biomass burning: Biomass burning is an important source of $CO_2$ emissions. The carbon that was taken up by organic matter, such as trees, grass, and other plant materials, is released back into the atmosphere as $CO_2$ when these items are burned. This process contributes to climate change and global warming (Smith et al. 2014). Biomass burning can occur naturally, such as during wildfires, or through human activities such as land clearing, slash-and-burn agriculture, and the burning of wood for energy. The extent of biomass burning and its contribution to atmospheric $CO_2$ levels depend on various factors, including land use practices, climate conditions, and wildfire management strategies (IPCC 2014). Crop waste burning is one of the most common biomass burning methods worldwide, accounting for 10% of all annual fire incidents worldwide (Andreae and Merlet 2001; Korontzi et al. 2006). Vegetation burning also emits aerosols (including black carbon), $CO_2$, methane ($CH_4$), nitrous oxide ($N_2O$), and ozone precursors into the atmosphere.

Tillage practices: Conventional tillage releases $CO_2$ by increasing the rate of organic matter decomposition. Tillage also increases soil porosity, which further increases $CO_2$ emissions. This phenomenon occurs more rapidly in tropical regions due to higher temperatures, which is why tropical soils have lower organic carbon content. Changes in tillage and cropping practices can affect SOC storage, with similar effects observed in both temperate and tropical regions as in long-term agricultural systems. No-till or reduced tillage practices result in the highest increase in SOC storage in tropical wet conditions, while the increase is the smallest in temperate dry conditions. Similarly, high-input cropping systems, including those that use organic amendments, have a greater increase in SOC in tropical wet conditions than in dry conditions (Ogle et al. 2005).

Grazing practice: Grazing practices can also have an impact on $CO_2$ emissions from soils. Intensive grazing practices that involve high stocking densities and overgrazing can cause soil compaction and reduced plant growth, which can decrease the organic matter content in the soil and increase $CO_2$ emissions (Conant et al. 2001). Conversely, rotational grazing practices that allow for periods of rest and recovery for pastures can promote plant growth and increase soil carbon sequestration, which can reduce soil $CO_2$ emissions (Ogle et al. 2005). Further, factors like the soil's characteristics, climate circumstances, and the type of vegetation that is being grazed influence the emission of $CO_2$ from the soil (Ogle et al. 2005).

## Contribution Due to Farm Inputs

The atmospheric emission of $CO_2$ is further increased by the use of fertilizers, manure, and other farm inputs. Fossil fuel use for powering machinery: According to Yousefi et al. (2017), electricity accounts for around 79% of carbon footprint output. In China, wheat and maize contributed 37% and 19%, respectively, to overall carbon emissions, while electrical power for rice irrigation accounted for roughly 4% (Zhang et al. 2017). Chemical and fertilizers: Fertilizers applied to agricultural fields can contribute to the release of $CO_2$ from the soil in several ways. First, some fertilizers, such as urea, undergo hydrolysis when applied to the soil, releasing $CO_2$ as a by-product. Second, fertilizers can promote plant growth, which boosts photosynthesis and the absorption of atmospheric carbon dioxide. However, when the plant matter breaks down, $CO_2$ is returned to the environment and soil. Third, excessive use of nitrogen fertilizers can acidify the soil, depleting soil organic matter and releasing $CO_2$ into the atmosphere. Overall, the impact of fertilizers on soil $CO_2$ emissions is influenced by the type and amount of fertilizer used, soil properties, and management practices. Adopting sustainable agricultural practices, such as using organic fertilizers, minimizing fertilizer use, and applying conservation tillage practices, is crucial for reducing agriculture's carbon footprint and mitigating climate change. These practices can help increase soil organic matter, improve soil health, and reduce greenhouse gas emissions from agriculture.

Agricultural soil ecosystem as a source of nitrous oxide ($N_2O$): Nitrous oxide ($N_2O$) is a potent greenhouse gas (GHG) emitted from various agricultural activities. The agriculture sector contributes approximately two-thirds of all anthropogenic $N_2O$ emissions (Janssens-Maenhout et al. 2017). According to the 2014 IPCC report, approximately 60% of global $N_2O$ emissions occur from agricultural activities (IPCC 2014). Agricultural $N_2O$ emissions can be categorized into two types: direct and indirect. Direct emissions are produced from various activities involving mineral fertilizers, biological nitrogen fixation, manure management, and cultivation of organic soils. Indirect emissions result from the warming effect of GHGs and other pollutants present in the environment, enrichment of downstream water bodies through runoff, and nitrogen accumulation on soils due to wind. The largest sources of $N_2O$ emissions in agriculture are fertilizer application, animal manure management, and cultivation of organic soils (IPCC 2014).

## Microbial Process

Nitrous oxide ($N_2O$) is produced by soil microorganisms during the processes of nitrification and denitrification, which contribute around 70% of global $N_2O$ emissions from both natural and managed soils (Syakila and Kroeze 2011). Nitrification is the conversion of ammonium to nitrate by soil bacteria, while denitrification is the process by which nitrate is converted to $N_2O$ and eventually to $N_2$ by soil bacteria under anaerobic conditions. Both processes occur naturally in the soil and are important for supplying plants with the nitrogen they need to grow, but they can also produce $N_2O$ (Zhu et al. 2014). According to Butterbach-Bahl et al. (2013), the current understanding is that the following processes are involved in the formation of $N_2O$ in the soil ecosystem. The chemical breakdown of hydroxylamine occurs during both autotrophic and heterotrophic nitrifications.

The chemodenitrification process involves the conversion of soil nitrite and the breakdown of ammonium nitrate in the presence of light, humidity, and reacting surfaces. Nitrifier denitrification refers to the occurrence of denitrification within the same nitrifying microorganisms. Combined nitrification–denitrification is carried out by distinct microorganisms. Denitrification can be performed by organisms that have the ability to utilize nitrogen oxides as alternative electron acceptors under conditions where oxygen is limited. Co-denitrification involves the simultaneous reduction of organic nitrogen compounds with nitric oxide (NO). Nitrate ammonification, also known as dissimilatory nitrate reduction to ammonium, is another process involved in nitrogen transformation.

## Livestock and Animal Manure as a Source of $N_2O$

Livestock farming and the application of animal manure on agricultural soils are major sources of $N_2O$ emissions. When animal waste is applied to the soil, the nitrogen it contains can be converted into $N_2O$ through microbe-mediated nitrification and denitrification. Additionally, manure management practices such as storage and spreading can create anaerobic conditions that favor the production of $N_2O$ by soil microbes (EPA 2021). Additionally, livestock and animal manure management practices vary across different regions and countries and by certain factors such as the type of livestock, production system, and the level of intensification.

For instance, in industrialized countries with large-scale animal production systems, manure is typically stored in lagoons or pits, where anaerobic conditions can lead to the production of $N_2O$ (FAO 2021a, b). In contrast, in smallholder systems in developing countries, manure is often used as a source of organic fertilizer, and the application rates are generally lower, which reduces the risk of emission of $N_2O$ (Hartel et al. 2019). The magnitude of $N_2O$ emissions from livestock and animal manure management practices can also vary based on the type of animal and diet. For example, ruminant animals such as cattle and sheep produce more $N_2O$ than nonruminant animals such as pigs and chickens, as their digestive system produces more nitrogen-containing waste (Reay et al. 2012). Moreover, feeding animals with high-protein diets can increase the excretion of nitrogen in manure, leading to higher $N_2O$ emissions (EPA 2021).

Understanding the factors that contribute to $N_2O$ emissions from livestock and animal manure management practices is crucial for developing effective mitigation strategies. Reducing the production and use of manure and implementing practices that promote the efficient use of nitrogen can help to reduce $N_2O$ emissions from livestock farming (FAO 2021a, b). In addition, improving manure management practices, such as reducing the storage time, using covers to prevent air exposure, and optimizing application rates, can help to reduce $N_2O$ emissions from manure (EPA 2021).

## Farming Practice as a Source of $N_2O$: Crop, Crop Residue, and Tillage Practice

Crop and crop residue management can also be a prominent source of $N_2O$ emissions in agricultural systems. The decomposition of crop residues, such as stubble and roots, in soil can result in $N_2O$ production through microbial processes like nitrification and denitrification (Chen et al. 2019). Additionally, application of crop residues as organic fertilizers can contribute to $N_2O$ emissions due to the high nitrogen content and slow release of nutrients. Agricultural practices like tillage and residue management can also impact $N_2O$ emissions from crop production. For example, the use of reduced tillage or no-till practices can reduce $N_2O$ emissions by reducing soil disturbance and promoting soil organic matter accumulation, which can enhance the capacity of soils to store carbon and nitrogen (IPCC 2019). In contrast, conventional tillage practices can result in increased soil aeration and soil moisture content, which can promote $N_2O$ production. Smith et al. (2018) found that crop production and residue management techniques are responsible for around 10% of the world's $N_2O$ emissions. The study also emphasizes how critical it is to comprehend the geographical and temporal variability of $N_2O$ emissions from crop and residue management, as well as the possibility of using management practices to reduce these emissions. For the reduction of $N_2O$ emissions from crop production systems and residue management practices, several management strategies can be implemented. These include reducing the amount of crop residue left on the soil surface, incorporating residues into the soil at a shallower depth, using cover crops to scavenge excess nitrogen, and optimizing fertilizer application rates and timing (IPCC 2019).

## Fertilizers as a Source of $N_2O$

Nitrogen fertilizers have revolutionized modern agriculture by providing the essential nutrient required for high crop yields. However, nitrogen fertilizer use also contributes significantly to $N_2O$ emissions from agricultural systems. In fact, it has been estimated that around 80% of anthropogenic emissions of $N_2O$ come from agricultural activities, with nitrogen fertilizer use being a significant contributor (Mosier et al. 2012). A number of variables, such as the quantity and timing of fertilizer application, the soil's moisture content, and soil temperature, can affect how much $N_2O$ emissions from nitrogen fertilizers are produced (IPCC 2019). In general, high nitrogen application rates and applications timed to coincide with hot, humid weather are more likely to cause an increase in emission of $N_2O$ (EPA 2021). Due to more use of nitrogenous fertilizers (NFs), agricultural soils are known to emit roughly 60% of the world's $N_2O$ as well as the nitrogenous gases NOx (NO + $NO_2$) and NH3 (Mosier et al. 2004; Galloway et al. 2014). Excessive fertilizer usage, much of which will not help the desired crop while releasing the greenhouse gas nitrous oxide into the environment, has become a significant environmental problem (McSwiney and Robertson 2005). For example, fertilizer application in the form of ammonia in rice–wheat ecosystem causes ammonia volatilization (Xue et al. 2016). Overall, the application of fertilizers is a critical aspect of modern agriculture, but it is essential to implement sustainable and efficient management practices to reduce environmental impacts such as $N_2O$ emissions and soil carbon loss.

## Agricultural Soil Ecosystem as a Source of Methane ($CH_4$)

The emission of $CH_4$ from soil ecosystems is a crucial aspect of global carbon cycle and a significant contributor to greenhouse gas emissions. Yanan et al. (2018) suggested different soil sources contribute around 15–30% of total global $CH_4$ emissions. The primary sources of $CH_4$ in soil ecosystems are anaerobic processes, such as methanogenesis, which is the production of $CH_4$ by microorganisms living under low-oxygen conditions (Conrad 2009). In addition, $CH_4$ is also produced in soil through the decomposition of organic matter by microorganisms anaerobically (Kirschbaum 1995). Methanogenesis usually occurs in different habitats, like wetlands, paddy fields, and peatlands, where the soil is saturated with water, creating an anaerobic environment suitable for microorganisms that produce $CH_4$. These are strictly anaerobic microorganisms or archaeons (Prescott et al. 2002).

## Sources of Methane Emission in Soil Ecosystem

Methanogens in soil play its role in the final step in the anaerobic decomposition of organic matter, where they convert intermediate products, such as acetate and hydrogen, into methane. This process, known as methanogenesis, occurs in anaerobic microsites within the soil, such as waterlogged or compacted areas (Wang et al. in press). Methanogens require specific environmental conditions, such as low redox potential (less than $-150$ mV) and pH, to thrive. Methanogens have a limited range of substrates that they can metabolize, including hydrogen, carbon dioxide, formate, acetate, methanol, methylamines, and methysulfides. However, their ability to utilize these substrates is dependent on their presence and interactions with a diverse array of other microorganisms (Conrad 1989; Garcia 1990). The production of methane in soil can also impact soil fertility, nutrient cycling, and plant growth. However, methanogens in soil are not only producers of methane but also important players in the regulation of methane emissions. In addition, methanogens in soil can interact with other microorganisms, like methanotrophs and sulfate-reducing bacteria, that play a role in methane consumption and sulfide production, respectively. The distribution and diversity of methanogens present in soil are affected by the availability of electron donors and the redox potential of the soil environment (Lueders and Friedrich 2003). Understanding the factors that affect the growth and activity of methanogens in the soil is essential for predicting and mitigating their impact on the environment. A proper understanding of the factors influencing the growth and activity of methanogens in the soil is crucial in order to accurately predict and effectively mitigate their environmental impact.

## *Cropping System*

The production of methane ($CH_4$) in wetland rice soils is primarily attributed to the transmethylation of acetic acid, with a minor contribution from the reduction of $CO_2$. The synthesis of methane is preceded by the production of volatile acids. Following the depletion of oxygen during flooding, there is a rapid emergence of hydrogen, which subsequently triggers the production of $CO_2$ and, consequently, the formation of methane. The delay in methane production is influenced by pH, patterns of soil reduction, substrate availability, and temperature. In tropical flooded rice soils, methane generation begins soon after the initiation of flooding in alkaline and calcareous soils. In neutral soils, there is a delay of 2–3 weeks before methane production starts, while in acidic soils, it may take up to 5 weeks or longer. There is a negative correlation between methane production and soil redox potential, meaning that higher soil redox potential reduces the production of methane. There is a positive correlation of methane production with soil temperature, soil carbon content, and the growth of rice plants. This implies that higher soil temperature, increased soil carbon content, and the presence of rice plants promote methane production. Methane production rates are typically higher during the cropping season, while the population density of methanogens in the soil remains consistent. The rate and manner in which organic matter is added to and decomposed in the soil are crucial factors that affect methane production. Initial methane production mainly stems from easily degradable crop residues, weeds during fallow periods, and soil organic matter. However, as rice plants mature, root exudates, decaying roots, and aquatic biomass become more prominent sources of methane. In addition to its nutritional values for crop, the organic amendment can influence the soil's organic carbon pool, available soil nutrients, and activity of microbes, some of which are regulating variables in the emissions of $CH_4$ and $N_2O$ into the atmosphere (Majumder et al. 2008; Ge et al. 2010).

## *Sink of Different GHGs in Agricultural Soil Ecosystem*

Agricultural soil ecosystems play a crucial role as both sources and sinks of greenhouse gases (GHGs). Different factors contribute to the removal and sequestration of GHGs from the atmosphere in agricultural soils. Carbon dioxide ($CO_2$), the most abundant GHG, can be taken up by plants through photosynthesis and stored in soil organic matter, acting as a sink for this greenhouse gas. Methane ($CH_4$), another potent GHG, can also be subject to uptake and removal in agricultural soil ecosystems by methanotrophic bacteria. Nitrous oxide ($N_2O$), a by-product of microbial processes in soil, can undergo various transformations and be either emitted to the atmosphere or sequestered in agricultural soils. Certain soil management practices can minimize $N_2O$ emissions and promote its retention in soil. By implementing sustainable land management strategies that enhance carbon sequestration, promote methane oxidation, and reduce nitrous oxide emissions, we can contribute to mitigating climate change while maintaining agricultural productivity and soil health.

## *Sink of $CO_2$*

The third-largest carbon store is soil carbon, which is comprised up of both organic and inorganic components. Soil plays an important function in capturing and storing carbon dioxide from the atmosphere, making it an important sink for greenhouse gases (Batjes 2011c; Powlson et al. 2011a, b; Banwart et al. 2013; Bhattacharyya et al. 2014a, b). The biggest reservoir of organic carbon on Earth is soil. Inorganic carbon is stored in the soil through the formation of pedogenic calcium carbonates, while soil organic carbon (SOC) and soil inorganic carbon (SIC) are the two main sinks of carbon in the soil ecosystem. Organic carbon is captured and stored by the soil through plant photosynthesis and is then transferred to the topsoil as decomposed plant materials and roots. SOC and SIC are essential for maintaining fertility, water-holding capacity, and other aspects of the soil, making them crucial for ecosystem and agroecosystem functions (Bhattacharyya et al. 2017). Moreover, carbon sequestration in soil has significant global implications, as it plays a crucial function in the global carbon cycle and can aid in reducing greenhouse gases, particularly $CO_2$. Thus, understanding the processes that influence the storage of carbon present in the soil and promoting practices that enhance soil carbon sequestration may be a successful approach reducing the impacts of climate change (Bhattacharyya et al. 2017).

## *Soil Organic Carbon*

The carbon that continues to exist in the soil after a living organism's material partially decomposes is known as soil organic carbon (SOC). Photosynthesis is the major process involved in capturing and fixing the environmental $CO_2$ in organic carbon, which improves the photosynthetic ability of the plant and leads to the removal of carbon dioxide from the atmosphere. $CO_2$ fixation by plants, including leaf litter, roots, and root exudates, is the main source of organic matter in soils, which is then broken down by soil fauna activity and converted into more robust organic molecules (collectively referred to as humus) (Trumbore 1997).

Soil organic carbon (SOC) is a crucial component of the global carbon cycle. Soils have varying SOC concentrations based on a variety of environmental factors, including regional geology, climatic conditions, land use, and management. The quality of the original plant substrate and the physical environment of the soil, such as clay content, mineralogy, pH, $O_2$ availability, and formation and disruption of soil aggregates, all affect the stabilization and fate of organic matter residues. Organic matter in soils plays an integral role in regulating soil water-holding capacity and soil structure and serves as a long-term reservoir of essential plant nutrients. According to estimates made by Kuhry et al. (2013), the northern permafrost region holds the highest level of SOC, with 190 Pg C found mostly in peat soils in the top 30 centimeters (0–30 cm) of the soil. Due to the low temperatures, limited biological activity, and slow breakdown of SOM, very large amounts of carbon accumulate in soils. Histosols have a SOC concentration ranging from 12% to 18%. On the contrary, in arid regions like Sahara, plant growth is naturally limited due to hot and dry climatic conditions, and not much carbon reaches the soil. Arenosols, the characteristic soils in these locations, often have SOC levels of less than 0.6%. Black soils, such as Chernozems, are inherently fertile because they have a relatively high SOC content (over 1%), optimal plant growth conditions in terms of nutrient exchange capacity, and a well-developed structure that enables sufficient water provision. SOC sequestration in soil can have important environmental benefits, including mitigating the impacts of climate change. Increasing SOC levels in agricultural soils has the potential to decrease atmospheric $CO_2$ concentrations, as well as improve soil health and fertility (Lal 2004). Moreover, SOC sequestration can also lead to increased water-holding capacity and improved nutrient cycling, leading to increased agricultural productivity and resilience (Lal 2010). However, SOC sequestration is not a one-size-fits-all solution, as factors such as soil type, land use, and management practices can influence the potential for SOC sequestration (Smith et al. 2016). Additionally, increasing SOC levels in agricultural soils requires careful management, as improper tillage practices or overapplication of fertilizers can release stored carbon back into the atmosphere (Paustian et al. 2016a, b).

## *Soil Inorganic Carbon*

As a crucial element of the soil ecosystem and a key player in the global carbon cycle, soil inorganic carbon (SIC) is a vital component of soil ecosystems (Banwart et al. 2013). SIC refers to the carbonates that form in soils through various processes, including mineral weathering, precipitation, and biological activity (Batjes 2011a, b). These carbonates can accumulate in the soil over time, supplying a long-term sink for atmospheric carbon dioxide ($CO_2$) (Powlson et al. 2011a, b). In addition to controlling the pH of soil and water, SIC is crucial in defining the characteristics and fertility of soil (Bhattacharyya et al. 2014a, b). In dry and semi-arid regions, which make up around 41% of the Earth's land surface (Reynolds et al. 2007; Wang et al. 2014), the majority of SIC (up to 90%) adds to the total soil C stock. The liquid phase solution including $HCO_3^-$ and $CO_3^{2-}$, the carbonate in the solid phase, and the gas phase $CO_2$ all make up the inorganic carbon in soil. In the case where pH is 0.6.5 and with good drainage conditions, the amount of inorganic carbon in the soil gas and liquid phases is minimal in comparison to the solid phase, which is the major source of soil inorganic carbon, in the event of excellent drainage and a soil pH > 6.5 (Xu et al. 2014). Soil frequently exhibits the $CO_2(g)$-$CO_2(aq)$-$HCO_3^-(aq)$-$CaCO_3(s)$ inorganic carbon balancing process. Water dissolves the $CO_2$

derived from soil respiration and the atmosphere, and at the same time, bicarbonate is hydrolyzed and transformed into pedogenic carbonates with the help of other ions including $Ca^{2+}$ and $Mg^{2+}$ (Yang and Li 2011). Soil moisture, soil type, salt level, and soil temperature all affect the reserves and distribution of inorganic carbon in the soil. It is evident that the dissolution and accumulation of inorganic carbon in time and profile space are reversible phenomena, especially in dry and semiarid regions with sparse vegetation and fluctuations in the condition of soil moisture and $CO_2$ partial pressure (Bai et al. 2017). The high temperature regimes and summer rains in dry and semiarid areas promote the mineralization of soil organic matter and increase biological respiration, which improves the partial pressure of $CO_2$ in the soil air. A solution of carbonic acid, created by the reaction of $CO_2$ with soil water, can dissolve some proportion of carbonate. When the soil water evaporates in the autumn, the bicarbonate is converted to carbonate (Pan 1999). According to US soil taxonomy (Soil Survey Staff 2006), the total SOC stock of Indian soils is predicted to be close to 30 Pg in the first 1.5 m of depth, whereas the total SIC stock is close to 34 Pg. According to data on SOC and SIC stocks in five bioclimatic zones from Bhattacharyya et al. (2008), in the top 0.3 m of soil, SOC stock is 2.5 times bigger than SIC stock, according to data on SOC and SIC stocks in five bioclimatic zones from Bhattacharyya et al. (2008). Additionally, in dry climates, the SIC stock rises with depth in all soil types (except from Ultisols), which improves the subsoil's calcareousness (Pal et al. 2003).

## *Sink of $CH_4$ in Agricultural Soil Ecosystem*

One important sink of methane in the soil is the activity of methanotrophic bacteria. Methanotrophs are microscopic organisms that use methane as a carbon source and an energy source. They consume methane from the atmosphere of soil and convert and release it in the form of carbon dioxide, a far less potent greenhouse gas.

Methanotrophs: The most reduced form of carbon, methane, cannot undergo transformation on its own without the presence of an electron acceptor. Methane-oxidizing bacteria (methanotrophs) may be isolated from all types of soils and are widespread. These microbes can oxidize methane to provide themselves with energy and carbon. Nearly all methanotrophic isolates have been found to exclusively (obligately) utilize methane and are unable to utilize other types of substrates (Topp and Hanson 1991). Certain methanotrophs are capable of autotrophic carbon dioxide fixation, while others can fix dinitrogen gas in microaerobic conditions. Methanotrophs, as opposed to methanogens, prefer a neutral pH and are mesophilic, exhibiting little to no sensitivity to temperature changes within their functioning temperature range (Topp and Pattey 1997). Methanotrophs derive energy from the oxidation of methane while utilizing the intermediate compound formaldehyde as their primary source of carbon for growth. Methane is converted to carbon dioxide via a series of intermediate steps involving methanol, formaldehyde, and formate. Since oxygen acts as a terminal electron acceptor and is required for the first stage of the methane oxidation pathway, which is aided by the enzyme methane monooxygenase (MMO), its presence is essential (Topp and Pattey 1997). This needs for oxygen to "prime" the methane oxidation pathway explains why methane is not oxidized in the soil under other redox circumstances, such as when denitrification is taking place. The MMO enzyme also needs a reducing agent, which is supplied by a cofactor that moves the electrons released by the oxidation events to the pathway's further steps. Soils that are permeable enough to allow atmospheric oxygen to diffuse through them have the ability to consume methane (Topp and Pattey 1997). Anthropogenic activities contribute to as much as 80% of total $CH_4$ emissions, while natural sources account for 20% and 47% of the emissions (Liu and Whitman 2008). According to Sabrecov et al., a significant portion of the projected yearly mean sink (570 Tg $CH_4$) of atmospheric $CH_4$ is eliminated by tropospheric hydroxyl radical (OH) oxidation. Thus, methanotrophic bacteria that live in aerated soils can oxidize the leftover $CH_4$ in the atmosphere. These bacteria have the capacity to directly consume $CH_4$ from the atmosphere, with an estimated 1–10% being oxidized through this pathway (Sabrekov et al. 2016; Sengupta and Dick 2017; Curry 2007; Feng et al. 2018). Methanotrophic activity (MTA) acts as a protective mechanism, reducing the $CH_4$ concentration in the atmosphere by up to 50% when compared to the initial levels (Sengupta et al. 2017).

### *Sink of $N_2O$ in Agricultural Soil Ecosystem*

Nitrogen dioxide ($NO_2$) is thought to be primarily produced through denitrification; however, recent studies have suggested that denitrification also serves as a sink for atmospheric $NO_2$. According to research, some haloarchaeal bacteria, such the common *Haloferax mediterranei* found in moderately and highly salinized soil and water, may be powerful denitrifiers (Torregrosa-Crespo et al. 2020; Fernández et al. 2012). Soil microbes in the presence of oxygen can consume and reduce $N_2O$, converting it back into nitrogen gas ($N_2$) or nitrate ($NO_{3-}$) through a process called nitrification and denitrification, respectively. These processes contribute to reducing $N_2O$ emissions from the soil.

## Climate Change and Effect on Agricultural Soil Ecosystem

Climate change is the term used to describe enduring alteration in temperature and weather patterns. The impact of climate change has a prominent influence on the soil ecosystem which will ultimately affect the source and sinks of GHG present in various ecosystems. Humans can benefit from a wide range of services and products that are provided by ecosystems (Costanza et al. 1997). In addition to playing significant roles in conserving biodiversity, safe energy production, and environmental protection as a vital component of terrestrial ecosystems, soil also offers concrete ecosystem services (ESs) (Pereira et al. 2018; Su et al. 2018). Global temperatures have grown in step with gradually rising $CO_2$ levels during the past century (Janet et al. 2019). According to predictions made by scientists, changes in the global climate will impair soil ESs, endangering the health of plants and animals in coming generations (Mooney et al. 2009; Nelson et al. 2013) and affecting different biogeochemical cycles.

## Effects of Climate Change on Soil Ecosystem

The soil ecosystems, which are essential for maintaining healthy and productive ecosystems, can be greatly influenced by climate change which can result in increased emissions or accumulation of GHGs in the soil ecosystem. The results are as follows:

Physical effects: Higher temperatures and $CO_2$ concentrations can lead to dryer soils and lower soil water content, which can increase the risk of wind erosion (Li et al. 2020). Climate change-driven increases in precipitation can also lead to water erosion (Nearing 2001; O'Neal et al. 2005). In addition, this can reduce the soil's capacity to store carbon in the form of soil organic matter. Altered precipitation patterns, such as increased frequency of intense rainfall events or prolonged droughts, can affect soil moisture levels, leading to soil erosion, nutrient leaching, and runoff. Runoff water from cropland containing dissolved nitrogenous fertilizer can lead to the emission of nitrogen oxides in the environment. Climate change can thaw permafrost in regions where it is present, leading to soil instability, increased soil erosion, changes in water availability and drainage patterns, and the release of stored carbon as greenhouse gases, such as methane, which further exacerbates climate change. Sea-level rise can cause saltwater intrusion into coastal soils, making them unsuitable for agriculture and affecting plant growth. Coastal erosion due to increased storm intensity and sea-level rise can also result in the loss of fertile topsoil, reducing the productivity of coastal ecosystems and agricultural lands.

Biochemical effects: Increased temperatures can accelerate the decomposition of organic materials in the soil, releasing carbon dioxide ($CO_2$) into the atmosphere. This can alter carbon storage in the soil and affect the balance between carbon sequestration and greenhouse gas emissions. Climate change can impact nutrient availability and cycling in soil ecosystems. Changes in temperature and precipitation patterns can affect microbial activity, which in turn can affect processes such as nitrogen fixation, nitrification, and denitrification, potentially increasing the emission of nitrogen oxides into the atmosphere. Alterations in nutrient cycling can impact plant growth, ecosystem productivity, and the overall balance of nutrient availability in the soil.

Soil pH and acidity: Climate change can influence soil pH and acidity levels. Increased atmospheric carbon dioxide concentrations can lead to increased dissolution of $CO_2$ in soil water, forming carbonic acid. This can contribute to soil acidification, which can have implications for nutrient availability, microbial activity, and plant nutrient uptake, and increases $CO_2$ release in the atmosphere.

Biological effects: Soil microorganisms are primarily responsible for the cycling of soil organic carbon (SOC) and other nutrients, as well as the generation and consumption of greenhouse gases including $CO_2$, $CH_4$, and $N_2O$. Climate change has an impact on the relative abundance and function of soil communities because the physiology, temperature sensitivity, and growth rates of soil community members vary (Castro et al. 2010; Gray et al. 2011; Lennon et al. 2012; Briones 2014; Delgado-Baquerizo et al. 2014; Whitaker et al. 2014). A 5 °C warming in a temperate forest altered the relative abundances of soil bacteria and increased the community's bacterium-to-fungal ratio (DeAngelis et al. 2015). The rates of microbial soil respiration are directly impacted by global causes like global warming because soil microorganisms and the activities they facilitate are temperature-sensitive. In addition, although bacterial communities remain stable, soil fungal communities can shift from one dominating member to another even when soil moisture availability is just slightly altered (30% reduction in water-holding capacity) (Kaisermann et al. 2015). Modifications to the microbial population's composition will undoubtedly have an impact on the ecosystem's performance (Schimel and Schaeffer 2012). For instance, some microbial communities control methanogenesis (Bodelier et al. 2000), nitrification (Isobe et al. 2011), denitrification (Bakken et al. 2012), and nitrogen fixation (Isobe et al. 2012). The pace of a particular process can directly impact the flux of greenhouse gases in the soil ecosystem. The impacts of climate change on soil ecosystems are diverse and related overall. It is crucial to comprehend the consequences of climate change on soil health and productivity in order to successfully prevent and respond to these effects.

## Strategies to Combat GHGs in Agricultural Ecosystem

Greenhouse gas (GHG) emissions from agroecosystems contribute significantly to climate change, making it essential to adopt strategies that effectively reduce these emissions (IPCC 2019). Conservation agricultural practices offer a promising approach to minimizing GHG emissions while ensuring sustainable food production (Lal 2018). This chapter explores various strategies that can be implemented within agroecosystems to combat GHG emissions under conservation agricultural practices. Each strategy offers unique opportunities to mitigate GHG emissions and transition toward a more sustainable future.

## Conservation Agricultural Practices

Conservation agricultural practices (CAPs) are designed to boost productivity, maintain soil health, and promote overall ecosystem sustainability, all while minimizing the adverse environmental effects of traditional farming methods (FAO 2018). Keeping soil cover intact, having minimal impact on the soil, and diversifying crops are the three primary tenets of these practices. By adopting CAPs, farmers can enhance soil structure, improve water infiltration, and optimize nutrient cycling, thereby reducing their reliance on external inputs and mitigating GHG emissions. One integral component of CAPs is conservation tillage, which involves minimizing soil disturbance by reducing or eliminating conventional plowing. This procedure is essential for maintaining organic matter, enhancing soil structure, and lowering GHG emissions. Conservation tillage promotes carbon sequestration, reduces the emission of nitrous oxide ($N_2O$), and diminishes the fossil fuel consumption associated with tillage operations (Hobbs et al. 2008).

## Reduce, Reuse, and Recycle Approach

The reduce, reuse, and recycle approach, based on circular economy principles, is critical to achieving sustainable agriculture and reducing GHG emissions (Ellen MacArthur Foundation 2015). In the context of agroecosystems under CAPs, the following are some examples of how these strategies can be applied.

(a) Reduce: Reducing GHG emissions from agricultural activities is a key priority under CAPs—adopting climate-smart agricultural practices such as integrated nutrient management, crop diversification, and water conservation practices that reduce application of synthetic fertilizers and pesticides, enhance soil health, and improve water use efficiency (Smith et al. 2007); implementing agroforestry systems that sequester carbon and provide multiple ecosystem services such as soil conservation, biodiversity conservation, and fuelwood and fodder production (Nair 2012); and using renewable energy sources such as solar pumps and biogas plants that reduce reliance on fossil fuels and lower GHG emissions (Bailis et al. 2015).

(b) Reuse: Reusing resources in agroecosystems is critical to reducing waste and improving efficiency—recycling organic waste such as crop residues, animal manure, and kitchen waste through composting, vermicomposting, or anaerobic digestion to produce biofertilizers and biogas that can be used for energy generation or soil fertility enhancement (Liu et al. 2017). Using recycled water for irrigation and adopting practices such as drip irrigation and micro-sprinklers ameliorate water use efficiency and reduce water losses (Snyder et al. 2010).

(c) Recycle: Recycling materials in agroecosystems is an essential aspect of circular economy principles—using recycled plastic and paper materials for mulching and weed suppression in crop fields (Choudhury et al. 2016); recycling agricultural derivatives such as sugarcane bagasse, rice husk, and maize stalks to produce biofuels or biodegradable materials such as paper and bioplastics; and reusing wastewater from agro-industrial processes for irrigation or fish farming, thereby reducing water pollution and conserving freshwater resources.

## Enhanced Crop Residue Management

Managing crop residues effectively can significantly impact GHG emissions. Instead of burning residues, which releases carbon dioxide ($CO_2$) and other pollutants, adopting conservation practices like residue mulching or incorporation into the soil promotes carbon sequestration. This sequesters $CO_2$ and reduces $N_2O$ emissions while improving soil health, water retention, and nutrient cycling (Kaiser et al. 2019).

Soil management by carbon farming: Carbon farming refers to a set of agricultural practices aimed at sequestering atmospheric carbon dioxide ($CO_2$) in soils. These practices enhance soil organic matter content, improve soil health, and promote carbon sequestration. Techniques such as cover cropping, conservation tillage, agroforestry, and rotational grazing help increase carbon storage in the soil, reducing $CO_2$ levels in the atmosphere (Paustian et al. 2016a, b).

Biofuels: Biofuels derived from renewable biomass sources have gained attention as a feasible substitute to fossil fuels. Ethanol and biodiesel, the most common biofuels, are derived from various biomass feedstocks, including agricultural remnants, energy crops, and algae. These biofuels have the potential to reduce GHG emissions significantly, as they generally release fewer carbon dioxide ($CO_2$) emissions when combusted compared to fossil fuels. Bioethanol production helps mitigate GHG emissions in the following ways:

Carbon neutrality: Bioethanol is considered a carbon-neutral fuel because the liberated $CO_2$ during its combustion is approximately equal to the $CO_2$ absorbed by the crops during their growth (Farrell et al. 2006). This implies that the overall release of $CO_2$ into the atmosphere is relatively low compared to fossil fuels.

Reduced fossil fuel use: Bioethanol can be blended with gasoline to create biofuel blends, such as E10 (10% ethanol) or E85 (85% ethanol), which can be used in existing vehicle engines. By displacing a portion of fossil fuel consumption, bioethanol reduces the overall GHG emissions associated with transportation (Hammerschlag 2006).

Renewable energy source: Bioethanol is derived from renewable biomass sources, unlike fossil fuels, which are finite and nonrenewable (Balat 2011). By utilizing bioethanol as a fuel, we can reduce our dependence on fossil fuels and transition toward a more sustainable energy system.

Lifecycle GHG emissions: The overall GHG emissions associated with bioethanol production can differ relying upon factors such as the feedstock, agricultural practices, and energy sources used during production. However, several studies have shown that bioethanol production can result in lower lifecycle GHG emissions compared to gasoline, even when considering emissions from land use changes and indirect effects (Searchinger et al. 2008).

Co-products and waste utilization: Bioethanol production often generates valuable co-products, viz., animal feed and industrial chemicals, helping in maximizing resource utilization and minimizing waste (Wang et al. 2007). This further contributes to the sustainability and environmental benefits of bioethanol production.

Additionally, bioethanol alone may not be sufficient to achieve substantial reductions in GHG emissions. It should be seen as part of a broader strategy that includes other renewable energy sources, energy efficiency improvements, and sustainable transportation systems to effectively combat climate change.

Precision nutrient management: Efficient nutrient management is crucial for reducing GHG emissions. Precision nutrient management techniques, such as site-specific nutrient application and balanced fertilization, help optimize nutrient use efficiency, minimizing nitrogen losses and subsequent $N_2O$ emissions (Robertson et al. 2009). The application of organic fertilizers, cover crops, and crop rotations also enhances soil fertility and reduces reliance on synthetic fertilizers, further reducing GHG emissions (Fageria et al. 2011).

Integrated pest and nutrient management (IPNM): IPNM is a holistic approach that aims to minimize the reliance on synthetic pesticides while also reducing the associated greenhouse gas (GHG) emissions. By implementing IPM practices, the use of biological control agents, crop rotation, habitat manipulation, and pest monitoring are promoted to effectively manage pests (Pedigo and Rice 2009). This reduction in pesticide application not only helps protect beneficial insects and maintain ecological balance but also mitigates environmental contamination (Pretty et al. 2006). Furthermore, the implementation of precision nutrient management (INM) practices within IPNM plays a crucial role in optimizing the use of nitrogen-based fertilizers and minimizing the potential for $N_2O$ emissions. This involves employing appropriate quantities and timing of fertilizer application, utilizing slow-release fertilizers, and incorporating nitrification inhibitors. By adopting these practices, farmers can enhance nutrient efficiency, reduce nitrogen losses, and significantly minimize GHG emissions associated with agricultural activities.

Renewable energy integration: Integrating renewable energy sources into agroecosystems can offset GHG emissions associated with energy consumption. Utilizing solar panels, wind turbines, or biogas systems can provide clean energy for irrigation, machinery, and farm operations (Liu et al. 2013). Dependency to fossil fuels is drastically minimized with the implementation of these resources, thereby mitigating GHG emissions and contributing to sustainable farming practices (Kumar et al. 2020).

Rice biomass as a renewable resource: Rice biomass refers to various plant residues generated during rice cultivation, including straw, husks, and bran. Traditionally, these residues were often burned in the field, leading to the release of carbon dioxide ($CO_2$) and other pollutants. However, alternative approaches can convert rice biomass into a valuable resource with multiple environmental benefits.

Energy generation: Incorporation of rice biomass as a renewable energy source through various technologies such as combustion, gasification, and anaerobic digestion (Nguyen et al. 2013). Burning rice straw or husks in controlled conditions can produce heat and electricity, displacing the need for fossil fuel-based energy sources (Parikh et al. 2005). Additionally, anaerobic digestion of rice biomass can generate biogas, primarily composed of methane, which can be used for cooking, heating, or electricity production (Sialve et al. 2009).

Biochar production: Biochar are produced after rice biomass undergoes pyrolysis process, where organic matter is heated without oxygen (Lehmann et al. 2011). Biochar is a stable carbon-rich material that can be incorporated into soils. Its addition enhances soil fertility water retention, and carbon sequestration capacity, thereby mitigating GHG emissions and improving soil health (Jeffery et al. 2011).

Mitigating methane emissions: Rice paddy fields are known to be significant sources of methane emissions due to anaerobic conditions during cultivation. However, the incorporation of rice biomass into the soil can contribute to reducing methane (Yao et al. 2019) emissions through the following mechanisms:

Reduced methane production: Incorporating rice straw or husks into the soil increases carbon availability, stimulating the growth of beneficial microbes that consume methane. This process reduces methane emissions from paddy fields, as the methane-consuming microorganisms outcompete methane-producing microbes.

Improved soil aeration: Rice biomass added to the soil enhances its physical structure, promoting better aeration and reducing waterlogging. Adequate aeration limits the anaerobic conditions that favor methane production, effectively mitigating emissions (Wassmann et al. 2000).

Challenges and considerations: While utilizing rice biomass presents significant potential for mitigating GHG emissions, some challenges need to be addressed.

Harvesting and handling: Efficient collection and handling of rice biomass require appropriate machinery and logistical planning. Ensuring cost-effective and sustainable methods for biomass collection is essential.

Nutrient management: Removal of rice biomass from the fields can lead to nutrient depletion. Implementing nutrient management strategies, such as balanced fertilization or returning nutrient-rich ash from biomass combustion, can mitigate potential nutrient deficiencies.

Socioeconomic implications: Engaging farmers and local communities in adopting biomass utilization practices is crucial. Educating farmers about the benefits, providing access to appropriate technologies, and ensuring economic incentives are essential for successful implementation.

## Future Scenario and Cleaner Production Technology

In anticipation of the future, the adoption of cleaner production technologies plays a pivotal part in the reduction of GHG emissions across diverse sectors. These technologies are geared toward optimizing resource efficiency, minimizing waste generation, and reducing energy consumption. Here are some examples:

Renewable energy integration: The integration of renewable energy sources like solar, wind, and hydroelectric power is crucial in reducing dependence on fossil fuels (IPCC 2011). By embracing these sustainable alternatives, the generation of $CO_2$ emissions associated with energy production can be mitigated.

Energy-efficient manufacturing processes: The implementation of energy-efficient practices and technologies within the manufacturing sector yields significant reductions in energy consumption and the accompanying GHG emissions (UNIDO 2016). This involves adopting advanced process controls, utilizing energy recovery systems, and improving insulation to enhance overall energy efficiency.

Circular economy: Embracing the principles of a circular economy model holds immense potential for resource conservation and waste minimization (Ellen MacArthur Foundation 2012). By maximizing product reuse, recycling, and remanufacturing, the circular economy approach minimizes the need for raw material extraction and reduces the GHG emissions associated with such processes.

By prioritizing these cleaner production technologies, the way for a sustainable future can be paved that is characterized by reduced GHG emissions, efficient resource utilization, and minimized environmental impact.

## Conclusion

In conclusion, the agricultural soil ecosystem can as act as both source and a potential sink of numerous greenhouse gases. The relationships between soil, plants, and microbial communities in agricultural systems contribute to the emission and sequestration of greenhouse gases, ultimately influencing climate change. $CH_4$ is a dominant greenhouse gas with a major source in agricultural soil ecosystems. It is primarily produced through anaerobic microbial processes in waterlogged or poorly drained soils, such as paddy fields and wetlands. Rice cultivation, livestock farming, and organic waste decomposition are significant contributors to methane emissions. However, certain agricultural practices, such as improved water management and the use of methane capture technologies, can help mitigate methane emissions from agricultural soils. $N_2O$ is another important greenhouse gas emitted from agricultural soil ecosystems. It primarily originates from microbial processes in the soil, particularly during the nitrification and denitrification processes. Nitrogen fertilizers, manure application, and crop residue decomposition contribute to $N_2O$ emissions. Implementing best management practices, such as optimized fertilizer application, cover cropping, and soil conservation techniques, can help reduce $N_2O$ emissions from agricultural soils. $CO_2$ emissions from agricultural soil ecosystems are mainly associated with land use changes, such as deforestation and conversion of natural habitats to agricultural land. These activities release large amounts of carbon stored in vegetation and soil organic matter, contributing to increased atmospheric $CO_2$ concentrations. On the other hand, agricultural soils also have the potential to sequester carbon through practices like conservation tillage, cover cropping, and agroforestry systems. These practices enhance soil organic carbon content, promoting carbon sequestration and mitigating $CO_2$emissions. Overall, managing greenhouse gas emissions from agricultural soil ecosystems is crucial for climate change mitigation efforts. It requires a comprehensive approach that combines sustainable agricultural practices, improved soil management techniques, and the adoption of innovative technologies. By reducing emissions of methane, nitrous oxide, and carbon dioxide, while simultaneously enhancing carbon sequestration, we can contribute to the long-term sustainability of agricultural systems and help mitigate climate change impacts.

## References

Altieri MA, Nicholls CI, Henao A, Lana MA (2015) Agroecology and the design of climate change-resilient farming systems. Agronomy for Sustainable Development 35(3):869–890. https://doi.org/10.1007/s13593-015-0285-2

Andreae MO, Merlet P (2001) Emission of trace gases and aerosols from biomass burning. Global Biogeochem Cycles 15(4):955–966. https://doi.org/10.1029/2000GB001382

Bai SG, Jiao Y, Yang WZ et al (2017) Review of progress in soil inorganic carbon research. IOP Conference Series: EES 100(1):012129. https://doi.org/10.1088/1755-1315/100/1/012129

Bailis R, Drigo R, Ghilardi A, Masera O (2015) The carbon footprint of traditional woodfuels. Nature Climate Change 5(3):266–272. https://doi.org/10.1038/nclimate2491

Bajracharya RM, Lal R, Kimble JM (2000) Erosion effects on carbon dioxide concentration and carbon flux from an Ohio Alfisol. SSSAJ 64(2):694–700. https://doi.org/10.2136/sssaj2000.642694x

Bakken LR, Bergaust B, Liu L et al (2012) Regulation of denitrification at the cellular level: a clue to the understanding of $N_2O$ emissions from soils. Philos Trans R Soc Lond B 367:1226–1234

Balat M (2011) Production of bioethanol from lignocellulosic materials via the biochemical pathway: a review. Energy Convers Manag 52(2):858–875. https://doi.org/10.1016/j.enconman.2010.08.013

Banwart S, Black H, Cai Z et al (2013) Benefits of soil carbon: report on the outcomes of an international scientific committee on problems of the environment rapid assessment workshop. Carbon Manag 4(4):435–447. https://doi.org/10.1080/17583004.2013.846514

Batjes NH (2011a) Global status of soil degradation: soil carbon. In: Earthscan soils for life: humans and soils, partnering for the future, pp 23–34

Batjes NH (2011b) Soil organic carbon stocks under native vegetation—revised estimates for use with the simple assessment option of the carbon benefits project system. Agric Ecosyst Environ 142:365–373. https://doi.org/10.1016/j.agee.2011.06.007

Batjes NH (2011c) Soil carbon stocks, changes, and the role of soils in the global carbon cycle. Proceedings of the 19th World Congress of Soil Science, Soil Solutions for a Changing World, Brisbane, Australia. Available at: http://www.iuss.org/19th%20WCSS/symposium/pdf/WG2.3%20Batjes.pdf

Berhe AA (2012) Decomposition of organic substrates at eroding vs depositional landform positions. Plant Soil 350:261–280. https://doi.org/10.1007/s11104-011-0902-z

Beuselinck L, Steegen A, Govers G, Nachtergaele J, Takken I, Poesen J (2000) Characteristics of sediment deposits formed by intense rainfall events in small catchments in the Belgian Loam Belt. Geomorphology 32(1–2):69–82. https://doi.org/10.1016/S0169-555X(99)00068-9

Bhattacharyya T, Pal DK, Chandran P, Ray SK, Mandal C, Telpande B (2008) Soil carbon storage capacity as a tool to prioritize areas for carbon sequestration. Curr Sci 95:482–494

Bhattacharyya P, Chandran P, Ray SK, Mandal C, Mandal B, Pal DK, Sarkar D (2014a) Soil organic carbon pools in relation to land use and soil depth in tropical soils of Tripura, India. Geoderma Reg 2-3:14–22. https://doi.org/10.1016/j.geodrs.2014.05.002

Bhattacharyya T, Sarkar D, Ray SK, Chandran P, Pal DK, Mandal DK et al (2014b) Georeferenced soil information system: assessment of database. Curr Sci 107(9):1400–1419

Bhattacharyya T, Wani SP, Pal DK, Sahrawat KL (2017) Soil as source and sink for atmospheric $CO_2$. In: Goel M, Sudhakar M (eds) Carbon utilization green energy and technology. Springer Singapore. https://doi.org/10.1007/978-981-10-3352-0_4
Bodelier PLEP, Roslev T, Henckel P, Frenzel P (2000) Stimulation by ammonium-based fertilizers of methane oxidation in soil around rice roots. Nature 403:421–424
Bordovsky DG, Choudhary M, Gerard CJ (1999) Effect of tillage, cropping, and residue management on soil properties in the Texas rolling plains. Soil Sci 164:331–340
Bremer E et al (1994) Sensitivity of total, light fraction and mineralizable organic matter to management practices in a Lethbridge soil Can. J Soil Sci.
Briones MJI (2014) Soil fauna and soil functions: a jigsaw puzzle. Front Environ Sci 2:7
Butterbach-Bahl K, Baggs EM, Dannenmann M, Kiese R, Zechmeister-Boltenstern S (2013) Nitrous oxide emissions from soils: how well do we understand the processes and their controls? Philos Trans R Soc Lond Ser B Biol Sci 368(1621):20130122. https://doi.org/10.1098/rstb.2013.0122
Carter MR, Gregorich EG, Angers DA, Donald RG, Bolinder MA (1998) Organic C and N storage, and organic fractions, in adjacent cultivated and forested soils of eastern Canada. Soil Tillage Res 47:253–261
Castro HF, Classen AT, Austin EE, Norby RJ, Schadt CW (2010) Soil microbial community responses to multiple experimental climate change drivers. AEM 76:999–1007
Chen H, Li L, Luo X, Li Y, Liu DL, Zhao Y, Feng H, Deng J (2019) Modeling impacts of mulching and climate change on crop production and $N_2O$ emission in the Loess Plateau of China. Agric For Meteorol 268:86–97. https://doi.org/10.1016/j.agrformet.2019.01.002
Choudhary M, Sharma PC, Jat HS, Nehra V, McDonald AJ, Garg N (2016) Crop residue degradation by fungi isolated from conservation agriculture fields under rice–wheat system of North-West India. Int J Recycl Org Waste Agric 5:349–360
Conant RT, Paustian K, Elliott ET (2001) Grassland management and conversion into grassland: effects on soil carbon. Ecol Appl 11(2):343–355
Conrad R (1989) Control of methane production in terrestrial ecosystems. In: Schimmel DS (ed) Andrae MO. Exchange of trace gases between terrestrial ecosystems and the atmosphere, JWS, pp 39–58
Conrad R (2009) The global methane cycle: recent advances in understanding the microbial processes involved. Environ Microbiol Rep 1:285–292. https://doi.org/10.1111/j.1758-2229.2009.00038.x
Costanza R, Arge G, Groot R, Farberk S, Grasso M, Hannon B et al (1997) The value of the world's ecosystem services and natural capital. Nature 387:253–260
Curry CL (2007) Modeling the soil consumption of atmospheric methane at the global scale. Glob Biogeochem Cycles 21:GB4012. https://doi.org/10.1029/2006GB002818
David C, Lemke R, Helgason W, Farrell RE (2018) Current inventory approach overestimates the effect of irrigated crop management on soil-derived greenhouse gas emissions in the semi-arid Canadian Prairies. Agric Water Manag 208:19–32. https://doi.org/10.1016/j.agwat.2018.06.006
DeAngelis KM, Pold G, Topcuoglu BD, van Diepen LTA, Varney RM, Blanchard JL, Melillo J, Frey SD (2015) Long-term forest soil warming alters microbial communities in temperate forest soils. Front Microbiol 6:104
Delgado-Baquerizo M, Maestre FT, Escolar C, Gallardo A, Ochoa V, Gozalo B, Prado-Comesaña A (2014) Direct and indirect impacts of climate change on microbial and biocrust communities alter the resistance of the N cycle in a semiarid grassland. J Ecol 102:1592–1605
Eghball B, Mielke LN, McCallister DL, Doran JW (1994) Distribution of organic carbon and inorganic nitrogen in a soil under various tillage and crop sequences. JSWC 49:201–205
Ellen MacArthur Foundation (2012) Towards the circular economy. Retrieved from https://www.ellenmacarthurfoundation.org/assets/downloads/publications/Ellen-MacArthur-Foundation-Towards-the-Circular-Economy-vol.1.pdf
Ellen MacArthur Foundation (2015) "What is a circular economy?", Ellen MacArthur Foundation. Available at: https://www.ellenmacarthurfoundation.org/circular-economy. Accessed 6 July 2018

Environmental Protection Agency (EPA) (2021) Nitrous oxide emissions. Environmental Protection Agency. Retrieved from https://www.epa.gov/ghgemissions/nitrous-oxide-emissions
Fageria NK, Baligar VC, Clark RB (2011) Nutrient management in plants. CRC Press
FAO (2018) Circular economy in agriculture: closing the loop. Food and Agriculture Organization of the United Nations
Farrell AE, Plevin RJ, Turner BT, Jones AD, O'Hare M, Kammen DM (2006) Ethanol can contribute to energy and environmental goals. Science 311(5760):506–508. https://doi.org/10.1126/science.1121416
Feng J, Li F, Zhou X, Xu C, Ji L, Chen Z, Fang F (2018) Impact of agronomy practices on the effects of reduced tillage systems on $CH_4$ and $N_2O$ emissions from agricultural fields: a global meta-analysis. PLoS One 13(5):e0196703
Fernández I, Cabaneiro A, Carballas T (2012) Carbon mineralization dynamics in soils after wildfires in a shrubland ecosystem of NW Spain. Sci Total Environ 430:87–95. https://doi.org/10.1016/j.scitotenv.2012.04.059
Food and Agriculture Organization (FAO) (2021a) Emissions from agriculture and forest land. Global, regional, and country trends 1990–2019. FAOSTAT Analytical Brief Series, 25
Food and Agriculture Organization (FAO) (2021b) Livestock and the environment. Food and Agriculture Organization of the United Nations. Retrieved from https://www.fao.org/livestock-environment/en/
Galloway JN, Leach AM, Bleeker A, Erisman JW (2014) Nitrogen footprints: past, present, and future. Environ Res Lett 9(11):115003. https://doi.org/10.1088/1748-9326/9/11/115003
Garbach K, Milder JC, DeClerck FAJ, Driscoll L, Montenegro M, Herren B (2016) Examining multi-functionality for crop yield and ecosystem services in five systems of agroecological intensification. Int J Agric Sustain. https://doi.org/10.1080/14735903.2016.1174810
Garcia JL (1990) Taxonomy and ecology of methanogens. FEMS Microbiol Rev 87:297–308
Ge H, Zhang F, Xu M, Gu B, Chen X (2010) Nitrous oxide and methane emissions from a rice-wheat rotation ecosystem in southeast China. Soil Biol Biochem 42(9):1549–1557. https://doi.org/10.1016/j.soilbio.2010.05.026
Gray SB, Classen AT, Kardol P, Yermakov Z, Miller RM (2011) Multiple climate change factors interact to alter soil microbial community structure in an old-field ecosystem. SSSAJ 75:2217–2226
Gregorich EG, Greer KJ, Anderson DW, Liang BC (1998) Carbon distribution and losses: erosion and deposition effects. Soil Tillage Res 47(3–4):291–302
Hammerschlag R (2006) Ethanol's energy return on investment: a survey of the literature 1990-present. Environ Sci Technol 40(6):1744–1750. https://doi.org/10.1021/es052024h
Hartel M, Mader P, Bauer M, Scheu S (2019) Manure management practices in smallholder dairy farming systems in the East African highlands: a case study from western Uganda. Sci Total Environ 659:887–898
Hemelryck HV, Govers G, Oost KV, Merckx R (2011) Evaluating the impact of soil redistribution on the in situ mineralization of soil organic carbon. Earth Surf Process Landf 36(4):427–438
Hobbs PR, Sayre K, Gupta R (2008) The role of conservation agriculture in sustainable agriculture. Philos Trans R Soc 363:543–555. https://www.fao.org/policy-support/tools-and-publications/resources-details/en/c/1382716/
IPCC (2011) IPCC special report on renewable energy sources and climate change mitigation. Retrieved from https://www.ipcc.ch/site/assets/uploads/2019/10/SRREN-Chapter-7_FINAL.pdf
IPCC (2014) Climate change 2014: synthesis report. In: Pachauri RK, Meyer LA (eds) Contribution of working groups I, II and III to the Fifth Assessment Report of the Intergovernmental Panel on Climate Change. IPCC, Geneva
IPCC (2019) Climate change and land: an IPCC Special Report on climate change, desertification, land degradation, sustainable land management, food security, and greenhouse gas fluxes in terrestrial ecosystems. Intergovernmental Panel on Climate Change

Isobe K, Koba S, Otsuka S, Senoo K (2011) Nitrification and nitrifying microbial communities in forest soils. J For Res 16:351–362

Isobe K, Ohte N (2012) Ecological perspectives on microbes involved in N-cycling. Microbes and Environments 27(2):1–11. https://doi.org/10.1264/jsme2.ME12003

Janssens-Maenhout G, Crippa M, Guizzardi D, Muntean M, Schaaf E, Olivier JGJ, Peters JAHW, Schure KM (2017) EDGAR v4.3.2 Global Atlas of the three major Greenhouse Gas Emissions for the period 1970–2012. J Earth Syst Sci 9(2):693–697. https://doi.org/10.5194/essd-9-693-2017

Janet J, Sheehy J, Sparkes DL (2019) Evaluation of nitrogen management strategies for reducing greenhouse gas emissions in arable crop production systems. Agric For Meteorol 276–277:107617. https://doi.org/10.1016/j.agrformet.2019.107617

Jeffery S, Verheijen FG, van der Velde M, Bastos AC (2011) A quantitative review of the effects of biochar application to soils on crop productivity using meta-analysis. Agric Ecosyst Environ 144(1):175–187

Kaiser M, Devkota KP, Sharma B, Maharjan S, Sapkota TB, Rana RB (2019) Impact of tillage and crop residue management on greenhouse gas emissions in agricultural systems: a review. Sustain For 11(2):527. https://doi.org/10.3390/su11020527

Kaisermann APA, Maron L, Beaumelle L, Lata JC (2015) Fungal communities are more sensitive indicators to non-extreme soil moisture variations than bacterial communities. Appl Soil Ecol 86:158–164

Kirschbaum MU (1995) The temperature dependence of soil organic matter decomposition, and the effect of global warming on soil organic C storage. Soil Biol Biochem 27(6):753–760

Korontzi S, McCarty J, Loboda T, Kumar S, Justice C (2006) Global distribution of agricultural fires in croplands from 3 years of Moderate Resolution Imaging Spectroradiometer (MODIS) data. Glob Biogeochem Cycles 20. https://doi.org/10.1029/2005GB002529

Kuhry P, Grosse G, Harden JW, Hugelius G, Koven CD, Ping C, Schirrmeister L, Tarnocai C (2013) Characterization of the Permafrost Carbon Pool. Permafr Periglac Process 24(2):146–155. https://doi.org/10.1002/ppp.1782

Kumar D, Sahoo SK, Sahoo BB (2020) Sustainable agriculture using renewable energy sources: a review. Renew Energy 148:1392–1411

Lal R (2003) Soil erosion and the global carbon budget. Environ Int 29(4):437–450. https://doi.org/10.1016/S0160-4120(02)00192-7

Lal R (2004) Soil carbon sequestration to mitigate climate change. Geoderma 123(1–2):1–2

Lal R (2005) World crop residues production and implications of its use as a biofuel. Environ Int 31(4):575–584

Lal R (2010) Beyond Copenhagen: mitigating climate change and achieving food security through soil carbon sequestration. Food Secur 2(2):169–177

Lal R (2018) Sustainable intensification of China's agroecosystems by conservation agriculture. Int Soil Water Conserv Res 6(1):1–12

Lehmann J, Gaunt J, Rondon M (2011) Bio-char sequestration in terrestrial ecosystems—a review. Mitig Adapt Strateg Glob Chang 11(2):403–427

Lennon JT, Aanderud ZT, Lehmkuhl BK, Schoolmaster DR Jr (2012) Mapping the niche space of soil microorganisms using taxonomy and traits. Ecology 93(8):1867–1879. https://doi.org/10.1890/11-1745.1

Li J, Ma X, Zhang C (2020) Predicting the spatiotemporal variation in soil wind erosion across Central Asia in response to climate change in the 21st century. Sci Total Environ 709:136060. https://doi.org/10.1016/j.scitotenv.2019.136060

Liu Y, Whitman WB (2008) Metabolic, phylogenetic, and ecological diversity of the methanogenic archaea. Ann N Y Acad Sci 1125:171–189. https://doi.org/10.1196/annals.1419.019

Liu X, Zhang A, Ji C, Joseph S, Bian R, Li L, Pan G, Paz-Ferreiro J (2013) Biochar's effect on crop productivity and the dependence on experimental conditions—A meta-analysis of literature data. Plant Soil 373(1–2):583–594

Liu C, Jiang L, Li X, Hou H, Zhang J (2017) Recycling of organic wastes in agriculture for sustainable crop production and environmental protection: a review. Agron Sustain Dev 37(1):10. https://doi.org/10.1007/s13593-016-0386-5

Lueders T, Friedrich M (2003) Archaeal abundance and diversity pattern in a peat soil ecosystem. Environ Microbiol 5(12):1121–1127

Majumder B, Mandal B, Bandyopadhyay PK (2008) Soil organic carbon pools and productivity in relation to nutrient management in a 20-year-old rice–berseem agroecosystem. Biol Fertil Soils 44:451–461. https://doi.org/10.1007/s00374-007-0226-6

McSwiney CP, Robertson GP (2005) Nonlinear response of $N_2O$ flux to incremental fertilizer addition in a continuous maize (*Zea mays* L) cropping system. Glob Change Biol 11(10):1712–1719. https://doi.org/10.1111/j.1365-2486.2005.01040.x

Mooney H, Larigauderie A, Cesario M, Elmquist T, Hoegh-Guldberg O, Lavorel S et al (2009) Biodiversity, climate change, and ecosystem services. COSUST 1:46–54

Mosier AR, Halvorson AD, Reule CA, Liu XJ (2004) Net global warming potential and greenhouse gas intensity in irrigated cropping systems in northeastern Colorado. Journal of Environmental Quality 35(4):1584–1598. https://doi.org/10.2134/jeq2005.0157

Mosier AR, Kroeze C, Nevison C, Oenema O, Seitzinger S, Van Cleemput O (2012) Closing the global $N_2O$ budget: nitrous oxide emissions through the agricultural nitrogen cycle. Nutr Cycl Agroecosyst 74(3):157–172. https://doi.org/10.1007/s10705-012-9460-7

Nadeu E, de Vente J, Martínez-Mena M, Boix-Fayos C (2011) Exploring particle size distribution and organic carbon pools mobilized by different erosion processes at the catchment scale. J Soil Sci 11:667–678

Nair PKR (2012) Climate change mitigation and adaptation: a low hanging fruit of agroforestry. In: Nair PKR, Garrity DP (eds) Agroforestry: the future of global land use. Springer, Dordrecht, pp 31–67

Nearing MA (2001) Potential changes in rainfall erosivity in the U.S. with climate change during the 21st century. JSWC 56(3):229–232

Nelson EJ, Kareiva P, Ruckelshaus M, Arkema K, Geller G, Girvetz E et al (2013) Climate change's impact on key ecosystem services and the human well-being they support in the US. Front Ecol Environ 11(9):483–893

Nguyen LX, Stichnothe H, Hoang THA (2013) Rice straw for biogas production: availability, potential, and impact on soil amelioration and carbon footprint. Bioresour Technol 129:281–292

Ogle SM, Breidt FJ, Paustian K (2005) Agricultural management impacts on soil organic carbon storage under moist and dry climatic conditions of temperate and tropical regions. Biogeochemistry 72(1):87–121

O'Neal MR, Nearing MA, Vining RC, Southworth J, Pfeifer RA (2005) Climate change impacts on soil erosion in Midwest United States with changes in crop management. Catena 61(2–3):165–184

Pal DK, Srivastava P, Bhattacharyya T (2003) Clay illuviation in calcareous soils of the semiarid part of the Indo-Gangetic Plains, India. Geoderma 115(3–4):177–192

Pan GX (1999) The significance of the occurrence of soil carbonates and the carbon transfer in the land system in arid regions of China. J Nanjing Agric Univ 22:51–57

Parikh J, Channiwala SA, Ghosal GK (2005) A correlation for calculating HHV from proximate analysis of solid fuels. Fuel 84(5):487–494

Paustian K, Six J, Elliott ET, Hunt HW (2000) Management options for reducing CO2 emissions from agricultural soils. Biogeochemistry 48:147–163

Paustian K, Lehmann J, Ogle S, Reay D, Robertson GP (2016a) Soil carbon sequestration for climate food security. Science 352(6290):685–686. https://doi.org/10.1126/science.aad5761

Paustian K, Lehmann J, Ogle S, Reay D, Robertson GP, Smith P (2016b) Climate-smart soils. Nature 532(7597):49–57

Pedigo LP, Rice ME (2009) Entomology and pest management. Pearson Education

Pendrill F et al (2022) Disentangling the numbers behind agriculture-driven tropical deforestation. Science 377:eabm9267. https://doi.org/10.1126/science.abm9267

Pereira P, Bogunovic I, Muñoz-Rojas M, Brevik EC (2018) Soil ecosystem services, sustainability, valuation, and management. CurrOpin Environ Sci Health 5:7–13

Polyakov VO, Lal R (2004) Soil erosion and carbon dynamics under simulated rainfall. Soil Sci 169(8):590–599. https://doi.org/10.1097/01.ss.0000138414.84427.40

Polyakov V, Lal R (2008) Soil organic matter and $CO_2$ emission as affected by water erosion on field runoff plots. Geoderma 143(1–2):216–222. https://doi.org/10.1016/j.geoderma.2007.11.005

Powlson DS, Gregory PJ, Whalley WR, Quinton JN, Hopkins DW, Whitmore AP (2011a) Soil management in relation to sustainable agriculture and ecosystem services. Food Policy 36(Suppl 1):S72–S87. https://doi.org/10.1016/j.foodpol.2010.11.025

Powlson DS, Whitmore AP, Goulding KWT (2011b) Soil carbon sequestration to mitigate climate change: a critical re-examination to identify the true and the false. Eur J Soil Sci 62(1):42–55. https://doi.org/10.1111/j.1365-2389.2010.01342.x

Prescott LM, Harley JP, Klein DA (2002) Microbiology, 5th edn. McGraw-Hill Higher Education, Boston

Pretty J, Bharucha ZP, Poulton C (2006) Pesticide use and ecological change in Indonesian vegetable farming systems: a case study of IPM adoption among market-oriented farmers. Int J Agric Sustain 4(1):5–20

Reay DS, Davidson EA, Smith KA, Smith P, Melillo JM, Dentener FJ (2012) Global agriculture and nitrous oxide emissions. Nat Clim Chang 2(6):410–416

Reynolds JF, Smith DMS, Lambin EF, Turner BL, Mortimore M, Batterbury SPJ et al (2007) Global desertification: building a science for dryland development. Science 316(5826):847–851

Robertson GP, Vitousek PM, Tiedje JM (2009) Nitrogen in agriculture: balancing the cost of an essential resource. Annu Rev Environ Resour 34:97–125

Sabrekov AF, Glagolev M, Alekseychik PK, Smolentsev BA, Terentieva IE, Krivenok LA, Maksyutov SS (2016) A process-based model of methane consumption by upland soils. Environ Res Lett 11(7):075001. https://doi.org/10.1088/1748-9326/11/7/075001

Schimel JP, Schaeffer SM (2012) Microbial control over carbon cycling in soil. Front Microbiol 3. https://doi.org/10.3389/fmicb.2012.00348

Searchinger T, Heimlich R, Houghton RA, Dong F, Elobeid A, Fabiosa J, Tokgoz S, Hayes D, Yu TH (2008) Use of U.S. croplands for biofuels increases greenhouse gases through emissions from land-use change. Science 319(5867):1238–1240. https://doi.org/10.1126/science.1151861

Sengupta A, Dick WA (2017) Methanotrophic bacterial diversity in two diverse soils under varying land-use practices as determined by high-throughput sequencing of the pmoA gene. Appl Soil Ecol 119:35–45. https://doi.org/10.1016/j.apsoil.2017.05.031

Sengupta A, Majumder AL, Roy A (2017) Integrated nutrient management for sustaining the productivity of rice (Oryza sativa L.)–wheat (Triticum aestivum L.) cropping system under Indo-Gangetic Plains. Arch Agron Soil Sci 63(12):1694–1708. https://doi.org/10.1080/03650340.2017.1314825

Sialve B, Bernet N, Bernard O (2009) Anaerobic digestion of microalgae as a necessary step to make microalgal biodiesel sustainable. Biotechnol Adv 27(4):409–416

Slattery MC, Burt TP (1997) Particle size characteristics of suspended sediment in hillslope runoff and stream flow. Earth Surf Process Landf 22:705–719

Smith P, Martino D, Cai Z, Gwary D, Janzen H, Kumar P, McCarl B, Ogle S, O'Mara F, Rice C, Scholes B, Sirotenko O (2007) Agriculture. In: Climate change 2007: mitigation. Contribution of Working Group III to the Fourth Assessment Report of the Intergovernmental Panel on Climate Change. Cambridge University Press, pp 497–540

Smith P, Bustamante M, Ahammad H et al (2014) Agriculture, forestry and other land use (AFOLU). In: Edenhofer O, Pichs-Madruga R, Sokona Y, Farahani E, Kadner S, Seyboth K, Adler A et al (eds) Climate change 2014: mitigation of climate change, Contribution of Working Group III to the Fifth Assessment Report of the Intergovernmental Panel on Climate Change. Cambridge University Press, Cambridge

Smith P, Davies CA, Ogle SM et al (2016) Towards an integrated global framework to assess the impacts of land use and management change on soil carbon: current capability and future vision. Glob Chang Biol 22(3):1008–1028

Smith P, Bustamante M, Ahammad H, Clark H, Dong H, Elsiddig EA, Haberl H, Harper R, House J, Jafari M, Masera O, Mbow C, Ravindranath NH, Rice CW, Robledo Abad C, Romanovskaya A, Sperling F, Tubiello F, Tubiello FN (2018) Agriculture, Forestry and Other Land Use (AFOLU). In: Climate Change 2014: Mitigation of Climate Change. Contribution of Working Group III to the Fifth Assessment Report of the Intergovernmental Panel on Climate Change. Cambridge University Press, pp 811–922

Snyder RL, Orang M, Matyac JC (2010) Microirrigation technology: advances in agriculture. CRC Press

Soil Survey Staff (2006) Keys to Soil Taxonomy (10th ed.). USDA-Natural Resources Conservation Service, Washington, DC

Su C, Liu H, Wang S (2018) A process-based framework for soil ecosystem services study and management. Sci Total Environ 627:282–289

Syakila A, Kroeze C (2011) The global nitrogen budget revisited. GHG Measure Manage 1:17–26. https://doi.org/10.3763/ghgmm.2010.0007

Topp E, Hanson RS (1991) Metabolism of radiatively important trace gases by methane-oxidizing bacteria. In: Rogers JE, Whitman WB (eds) Microbial production and consumption of greenhouse gases: methane, nitrogen oxides, and halomethanes. ASM, pp 71–90

Topp E, Pattey E (1997) Soils as sources and sinks for atmospheric methane. Can J Soil Sci 77:167–178

Torregrosa-Crespo J, Pire C, Richardson DJ, Martínez-Espinosa RM (2020) Exploring the molecular machinery of denitrification in *Haloferaxmediterranei* through proteomics. Front Microbiol 11:605859. https://doi.org/10.3389/fmicb.2020.605859. PMID: 33363526; PMCID: PMC7754194

Trumbore SE (1997) Potential responses of soil organic carbon to global environmental change. PNAS 94(16):8284–8291

United Nations Industrial Development Organization (UNIDO) (2016) Resource efficiency and cleaner production: an assessment of the global status of resource efficiency and cleaner production. Retrieved from https://www.unido.org/sites/default/files/2017-09/rcp_global_status_report_2016_web_0.pdf

Wang, X., Li, Y., Zhang, Z., & Chen, J. (in press). Methanogenesis in waterlogged soils: Mechanisms and environmental implications. J Soil Sci Plant Nutr.

Wang M, Wu M, Huo H, Liu J (2007) Life-cycle energy and greenhouse gas emission impacts of different corn ethanol plant types. Environ Res Lett 2(2):024001. https://doi.org/10.1088/1748-9326/2/2/024001

Wang C, Wang X, Liu D et al (2014) Aridity threshold in controlling ecosystem nitrogen cycling in arid and semi-arid grasslands. Nat Commun 5:4799. https://doi.org/10.1038/ncomms5799

Wassmann R, Neue HU, Lantin RS, Buendia LV, Corton TM, Lu Y, Rennenberg H (2000) Characterization of methane emissions from rice fields in Asia. III Mitigation options and future research needs. Nutr Cycling Agroecosyst 58(1–3):23–36

Whitaker JN, Ostle AT, Nottingham A, Ccahuana N, Salinas RD, Bardgett P, Meir P, McNamara NP (2014) Microbial community composition explains soil respiration responses to changing carbon inputs along an Andes-to-Amazon elevation gradient. J Ecol 102:1058–1071

Xu WQ, Chen X, Luo GP et al (2014) Progress of soil carbon cycle research based on stable isotope technology. Arid Land Geography 37:980–987

Xue L, Yu Y, Yang L, Liu J, Chen F (2016) Mitigating ammonia volatilization from paddy fields under the rice-wheat rotation system in the Taihu Lake Region, China. Agric Ecosyst Environ 233:55–66. https://doi.org/10.1016/j.agee.2016.08.028

Yanan X, Yang SH, Xu J, Ding J, Sun X, Jiang Z (2018) Effect of biochar amendment on methane emissions from paddy field under water-saving irrigation. Sustain For 10:1371. https://doi.org/10.3390/su10051371

Yang LF, Li GT (2011) Progress in soil inorganic carbon research. Chinese J Soil Sci 42:986–990
Yao Z, Yan G, Zheng X, Wang R, Liu C (2019) Crop straw management alters soil microbial community composition and carbon metabolism associated with greenhouse gas emissions. Soil Biol Biochem 129:50–61
Yousefi M, Khoramivafa M, Damghani AM (2017) Water footprint and carbon footprint of the energy consumption in sunflower agroecosystems. Environ Sci Pollut Res Int 24:19827–19834
Zhang Y, Peng C, Li W, Tian L, Zhu Q (2017) Multiple interactions of soil organic carbon, nutrients, and microbial activity as affected by the conversion of forest to agricultural land. For Ecol Manag 391:316–323. https://doi.org/10.1016/j.foreco.2017.02.006
Zhu YG, Wang XH, Yang XR, Xu HJ, Jia Y (2014) Key microbial processes in nitrous oxide emissions of agricultural soil and mitigation strategies. Huan Jing KeXue 35(2):792–800

# Soil and Environmental Variables Influencing Greenhouse Gas Cycling in an Agroecosystem

**Sunil Kumar Srivastava**

## Introduction

As world scientific communities reported, climate change is one of human civilization's biggest challenges. Few researchers suspected that the impact of climate change has already been initiated, and some significant impacts will be observed shortly. The contribution of microorganisms available in the soil for releasing greenhouse gases (GHGs) during the decomposition of organic matter cannot be ignored. Few microorganisms play a significant role in the biogeochemical cycle of nutrients and also facilitate nutrient availability in soil. The importance of nitrogen-fixing bacteria in leguminous plants in an agroecosystem is well-known and reported by various researchers (Forster et al. 2007; Srivastava 2020). Achaean bacteria contribute a significant proportion of GHGs to the environment by decomposing organic material in an anaerobic environment. The increase of organic materials in landfill areas, which increases the $CH_4$ in the atmosphere through anaerobic decomposition, was reported recently.

The study of reports published by various researchers indicate that approximately 60–80% of total $N_2O$ released from agroecosystems is mainly contributed through chemical fertilizers (Davidson 2009; Maraseni and Qu 2016). Nitrous oxide is a more dangerous GHG in comparison to $CO_2$ (~298 times the global warming potential (GWP) of $CO_2$ over a 100-year timescale) (Forster et al. 2007). In India, the emission of GHGs from agricultural activities has reported significant growth of ~161% in the last 50 years from 14.81 TgCE/year (0.12 tCE $ha^{-1}$ $yr^{-1}$) in 1960 to 38.71 TgCE/year (0.28 tCE $ha^{-1}$ $yr^{-1}$) in 2010 mainly due to the application of chemical fertilizer in the recent past (Sah and Devakumar 2018). Researchers reported increases in GHGs in the agroecosystem due to changes in

S. K. Srivastava (✉)
Department of Chemistry, Siddharth University Kapilvastu, Siddharth Nagar, UP, India
e-mail: sunil44@suksn.edu.in



S. R. Mohanty, B. Kollah (eds.), *Greenhouse Gas Regulating Microorganisms in Soil Ecosystems*, https://doi.org/10.1007/978-3-031-70569-4_2

cultivation patterns from fewer GHG-producing crops (coarse cereals) to more GHG-producing crops (paddy). The recent study reported a decrease in the cultivation area/land by ~16% for cereal crops and an increase in the cultivation area/land for paddy crops by ~22%. The maximum GHG emission was observed through paddy crops (~23.75 Tg CE/ha) and the least reported from *Cajanus cajan* (~2.98 Tg CE/ha).

A few researchers have reported a sharp rise in the human population, reaching 9.7 billion in 2050. This will also affect the agroecosystem by consuming more food than Earth's carrying capacity (Garlapati et al. 2021; Yang et al. 2021; Anonymous 2019). Furthermore, the sharp rise in food demand associated with the ever-increasing human population raises concerns for global food security (Hodson 2017); air, soil, and water pollution (Hou and Ok 2019; Tollefson 2019; Xu et al. 2019); and environmental sustainability (Puma 2019; Srivastava and Dhaker 2024; Tilman and Clark 2014).

## Agroecosystem

Agroecosystems are communities of plants, animals, and other living beings interacting with their physical and chemical environments. Recently, people have modified agroecosystems significantly to produce food, fiber, fuel, and other needs for human consumption and processing (Maes et al. 2018). Li et al. (2019) reported that crop plants could release ~21% of their total fixed carbon as root exudates in the rhizosphere, dramatically impacting soil microorganisms. However, more data may facilitate an understanding of the behavior of crop diversification, which can impact agroecosystem and available soil microorganisms and health in long-term rotations.

Agroecosystem contains numerous living (biotic) and nonliving (abiotic) components. The abiotic components of an agroecosystem contain soil (micro- and macronutrients), temperature, water, pH, temperature, climatic conditions (rainfall, humidity, evaporation, wind velocity), and chemical and physical characteristics of soils. The biotic component in the agroecosystem involves numerous multicellular species of microorganisms and plants. They involve the emission and fixation of GHGs through various biological activities like respiration, photosynthesis, and decomposition (Srivastava 2021; Parashar et al. 2018). The statistics indicate that the agroecosystem significantly influences the global biogeochemical cycle, which has recently caused a massive impact on GHG emissions. Statistics also justify it by indicating that soil is a significant carbon sink. Scientists are trying to develop plant statistics based on their carbon-fixing capacity. It was observed that carbon fixation varies in C-3 and C-4 type plants. The refusal of agro-residue is also responsible for the emission of GHGs through decomposition in anaerobic conditions (preferably $CH_4$) or oxidation during pyro-treatment (preferably $CO_2$) (Parashar et al. 2019a, b).

## Soil Chemistry

Soil is essential to life. All life-supporting components derive, either directly or indirectly, from the soil. Plants grow in soil and utilize the available nutrients, which are used for creating biomass. All living beings, including animals and humans, use this biomass as part of one or another food chain. The role of soil chemistry in the nutrient cycle and its further contribution to the existence of living beings on the earth system is substantial. The primary soil components are sand, silt, clay, organic matter, peds, profiles, pedons, and landscape. Peds are formed by aggregation of sand, silt, and clay particles to form larger (secondary) soil structures that result from the action of soil-forming structures (Conklin 2005).

To delineate the role of the agroecosystem, it is necessary to understand soil chemistry since soil works as sinks for GHGs. To reveal the broad prospect of soil storage and emission capacities, precise quantifications are needed to obtain reliable global budgets for better land use. The soil formation is initiated during rock weathering through runoff water of the surface water. Many sediment rivers are deposited in the delta region, and organic materials in the sediment propagate soil formation. These organic materials also support numerous microorganisms in freshly formed sediments. Organic load and climate conditions facilitate soil formation and influence its chemistry (Srivastava 2019).

The chemistry of soil is influenced by sediment composition, organic load, and climate conditions. It was observed that the high precipitation rate and organic load support more living beings who can participate in soil formation. Thus, soil formation is more active in climatic conditions, which supports chemical weathering and the presence of biodegradable organic materials. Soil features, textures, peds, profiles, and other material properties are combined to form different soils. Scientists have classified soil into 12 orders (Conklin 2005). It depends on factors like climate conditions, parent material, topography, biota, and time duration under which all these interact together. Dead plant leaf decomposition (by microorganisms) causes complex cellular materials and biochemicals, forming simpler intermediate bioorganic and organic compounds (Srivastava and Ramanathan 2018a).

The essential soil components are minerals, organic matter, water, and air. The typical soil consists of ~45% minerals, ~5% organic matter, 20–30% water, and 20–30% air (Fig. 1). The upper layer of soil composition fluctuates and influences the basis of water supply, cultivation practices, and soil type. Soil organic matter contains a C/N ratio of ~10–12. In the case of a high C/N ratio, microorganisms accelerate the process of decomposition and cause a reduction in the proportion of inorganic nitrogen (Conklin 2005; Prasad et al. 2006; Ranjan et al. 2017; Srivastava and Ramanathan 2008).

The global impact of soil degassing is staggering. It contributes approximately 35% of total $CO_2$, 47% of total $CH_4$, 53% of total $N_2O$, and 21% of total NO annual emissions. These statistics, related to soil degassing (IPCC 2007), underscore the urgent need for further research and action to mitigate this significant source of greenhouse gas emissions. However, the potential for further research gives us hope

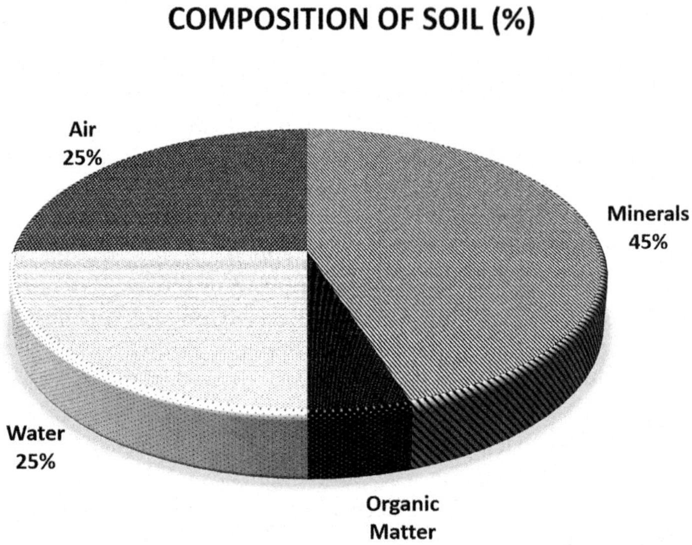


**Fig. 1** Approximate composition of soil

and should motivate us to act. Methane is one of the most dangerous GHGs. It is produced in soil by methanogenesis under anaerobic conditions. Other important GHG $N_2O$ and NO are released in the atmosphere from the agroecosystem during the process of nitrification (nitrification process is defined as oxidation of ammonium ion $NH_4^+$ to nitrate $NO_3^-$ via $NO_2^-$) and denitrification (denitrification process is defined as a reduction of $NO_3^-$ to $N_2O$ and $N_2$). It was observed that $N_2O$ production accelerated through denitrification in the anaerobic condition in those areas where water-filled pore space (WFPS) is more than 50%. In comparison, NO emission is reduced in soil that possesses WFPS of less than 10% because of inhibition in nutrient flow (Ussiri and Lal 2013; Brummer et al. 2008; Srivastava and Ramanathan 2018b). It has recently been a concern to worldwide communities due to the excessive utilization of chemical fertilizers and pesticides.

## Microorganisms in Soil

Soil microbiology is one of the branches of soil science that deals with studying microorganisms in the soil, their function, and their understanding of how they affect soil nature (Tayyab et al. 2021). Scientific communities have reported five different types of microorganisms in soil. These are bacteria, actinomycetes, fungi, protozoa, and nematodes. Bacteria play a significant role by releasing nutrients from dead organic material into the soil. Actinomycetes were once classified as a fungus and acted similarly to release nutrients in soil. However, it was observed that some actinomycetes are predators and cause harm to the plant, while others living in soil can act as antibiotics for the plant. Like bacteria, fungi can also live in the root zone and facilitate nutrients available to plants. Mycorrhizae, as fungi, are key players in the ecosystem. They efficiently facilitate the transfer of water and nutrients from the soil to the plants' roots. In addition, they provide essential nutrients such as sugar, amino acids, and other elements, thereby enhancing plant growth and health. Protozoa are also important microbes that love to consume and are surrounded mainly by bacteria. Nutrients eaten by bacteria are released when protozoa, in turn, eat the bacteria. Finally, nematodes are microscopic worms that live around or inside the plant. Some nematodes are predators, while others are beneficial, eating pathogenic nematodes and secreting plant nutrients.

Soil is essential for all living plants. Plants utilize the soil's micronutrients, including water, during photosynthesis and synthesizing a primary biomass. All living beings, including microorganisms, consume these plant products for survival and release greenhouse gases (GHGs) in the atmosphere during respiration and other microbial activities. Further, decomposers (microbes) release nutrients in the environment at the end of the food chain. Hence, the role of microorganisms in the biogeochemical cycle is quite significant. Microbes influence and control the atmospheric carbon in various manners by participating in the carbon, nitrogen, phosphorus, and sulfur cycle. Microbes contributed significantly to fixing/releasing atmospheric carbon.

## Methodology

This study was performed to understand the agroecosystem's role in greenhouse gas emissions. A systematic and organized literature review of more than 125 research papers was conducted to understand the correlation between an agroecosystem and GHG emissions. Further, in this work, few field studies were performed to develop/evaluate a relationship between an agroecosystem and its impact on total GHG emissions. This chapter compiles all this information, which facilitates understanding the soil, agroecosystem, and environmental variables influencing GHG cycling in an environment.

## Results and Discussion

Plant roots and tops behave differently with the soil because the roots are always in intimate contact, extracting water and nutrients and exuding material into the soil (Conklin 2005). Meanwhile, the plant's top portion (stem and leaves) depends on the nutrient supply of the root. The plant's top portion (dead leaves) indirectly enriches the soil's organic load, which serves as food for soil microorganisms. The decomposition of these organic materials in soil releases GHGs in the atmosphere. There are various natural and anthropogenic sources of GHGs. The data released by the USA's

Environmental Protection Agency (EPA) suggested that $CO_2$ contributed ~80% of total GHG production in the USA. Approximately 35% of total $CO_2$ is contributed by transportation, ~31% by electricity production, ~16% by industries, ~11% by residential and commercial, and the remaining 8% by other nonfossil combustion.

Production of the microbial enzyme through utilizing agro-residue can minimize specific level load in GHG emissions, which takes place during the combustion of this agro-residue (Garlapati et al. 2021; Srivastava and Ramanathan 2012). The proper management of agro-residue for producing microbial enzymes and organic manure can reduce specific proportions of $CH_4$ and $CO_2$, which will be released during the anaerobic and aerobic decomposition of organic materials. The various cities in India, including Delhi, are suffering from smog/fog due to the burning of agro-residue by the farmers in the winter season. The burning of agro-residue is increasing the air quality index (AQI) value of Delhi and other cities of NCR up to 500.

The studies indicate that $CO_2$ contributes ~80% of total GHGs, as reported by EPA. The maximum quantity of $CO_2$ is produced during vehicle transportation through fossil fuel combustion.

The statistics indicate ~35% of total $CO_2$ is contributed by "transportation," ~30% of total $CO_2$ is contributed by "thermal power plants," ~15% of total $CO_2$ is contributed by "industries," ~10% by "residential and commercial," and ~7.5% of total $CO_2$ contributed by "nonfossil fuel combustion." The details of individual section contributions are shown in Fig. 2. Developing and developed countries' role in

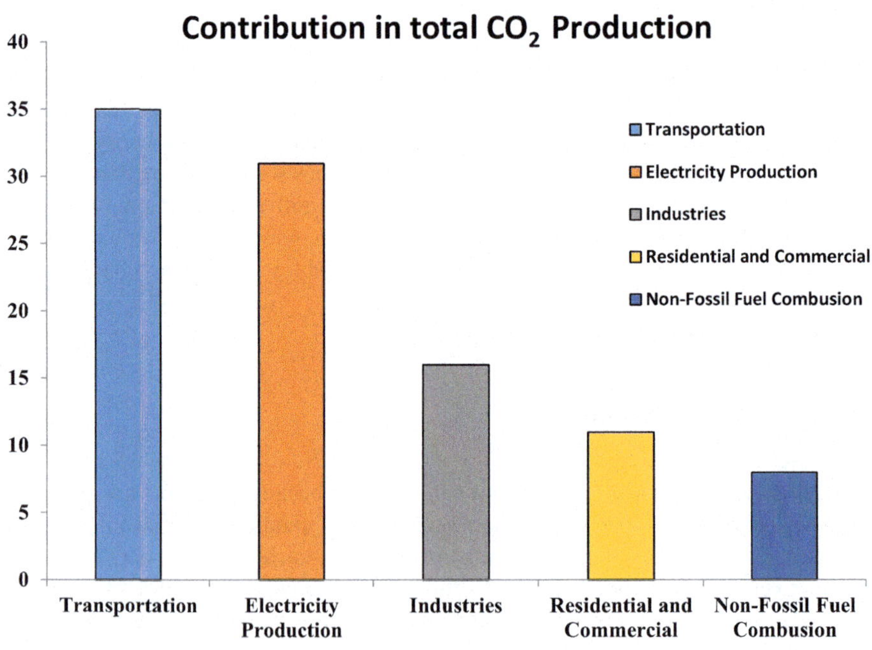


**Fig. 2** Contribution of total $CO_2$ production by various sources

releasing GHGs is crucial because the primary industries responsible for $CO_2$ production are thermal power plants and the cement industry. The consumption of electricity and cement in the last few decades has increased significantly, creating a challenge to the environment.

## Crop Rotation Impact on GHG Emission and Cycling in Agroecosystem

The rotation of crops is quite an old tradition in Asia and other countries, and this popular method is still quite effective in maintaining balance in agroecosystems. Scientists worldwide collected some data to understand the impact of crop rotation on the emission of GHGs from the agroecosystem. Temporal and spatial rotation of crops enhanced the soil's fertility by improving the soil's nutrient availability and water-holding capacity and reducing water logging problems. It enhances the fertility of the soil.

The agroecosystem plays an essential role in the cycling of GHGs in the atmosphere in various manners by fixing the excess proportion of $CO_2$ and other gases. The rotation of crops may yield a certain level of stability in GHG proportion in the atmosphere. Various researchers worldwide reported significant observations through rotations of crops. Scientific communities performed crop rotation on various soils under different climatic conditions, which also influenced the emission of GHGs from the crops, e.g., Behnke et al. (2018) reported that soybeans in crop rotation reduced daily and cumulative GHG emissions, especially $N_2O$, in silty clay loam soil. Soy–corn had much lower GHG emissions. Rotation combined with tillage significantly increased daily $CO_2$ emissions but not cumulative GHG emissions.

Similarly, Jarecki et al. (2018) reported that gray–brown lavish crop diversification in rotations increased observed and predicted soil carbon sink capabilities. Salton et al. (2014) performed crop rotation and concluded that soy/oat–pasture grass without tillage is more efficient in reducing the GHGs in agroecosystems. Cai et al. (2018) performed a crop rotation experiment with rice wheat and rice-rapeseed. They observed that crop rotation along with leguminous plants reduces GHGs in the atmosphere and further helps reduce carbon footprint by up to 37–50%. Ma et al. (2012) reported a reduction of GHGs in the atmosphere in a significant proportion of leguminous plants used in crop rotation. He has suggested a quantitative benefit in fixing atmospheric carbon and nitrogen during crop rotation.

The biological nitrogen fixation and other soil nutrient changes caused by legume crops can modify soil microbial community composition. The contributions of other nutrients (C, S, P, etc.) to the cycle cannot be avoided, e.g., it was observed that total global nitrogen fixation was ~400 Tg $year^{-1}$ from terrestrial and marine ecosystem systems. At the same time, agroecosystems contributed ~70 Tg of available nitrogen annually (Fowler et al. 2013; Herridge et al. 2008).

A recent study reported an increase in carbon footprint in agroecosystems due to the high load of food grains for fulfilling growing population needs. The carbon footprint of crop production per unit of farmland is the balance between the equivalent carbon emissions and carbon sequestration. The most anthropogenic interventions in agroecosystems, such as fertilization, pesticide applications, and farming implementation operations, contribute to the carbon footprint of crop production. Mineral fertilization is a primary anthropogenic source of the carbon footprint of the cropping system. The study performed in Canada reported that nitrogen fertilization contributed ~50% of the total $CO_2$ equivalent of GHG emissions in wheat production (of which ~22% from $N_2O$) (Gan et al. 2014).

Agricultural activities contributed ~16% of total GHGs nationwide in Australia (Young et al. 2009). The conversion of $NH_3$ to $NO_3^-$ and $N_2O$ and NO gases preferably depends on various soil microbes involved in nitrification and denitrification, e.g., *Nitrosomonas*, *Nitrobacter*, *Pseudomonas*, and *Clostridium*, and Archaea phylum *Thaumarchaeota*. The activities of these microorganisms increase $N_2O$ emissions and amend the carbon footprint. In addition, excess application of nitrogen-based chemical fertilizer may shift the C/N ratio in soil and cause denitrification.

However, due to carbon resource limitations in nitrogen-rich soil, *nir*- and *nor*-gene clusters showed immense relative abundance than *nos*-gene clusters, which could preferentially increase $N_2O$ and NO emissions over $N_2$ emissions in the soil (Yang et al. 2015), aggravating the negative effect of N fertilization to some extent. A few bacteria play a significant role in the formation of $CH_4$ in soil, which is the second most abundant GHG after $CO_2$. Methane is more potent than $CO_2$ and less potent than $N_2O$ in causing global warming. Since soils work as carbon sinks, methane consumption occurs due to interaction among soil bacteria, including methanotrophic archaea, $NH_3$-oxidizing bacteria, and $SO_4^{2-}$ reducing bacteria. In their study, Singh and Gupta (2016) reported that the application of bio-fertilizer reduces the emission of $CH_4$ from the soil. They also suggested long-term chemical fertilizer (nitrogen) utilization could inhibit $CH_4$ oxidation due to modification in the methanotrophic Archean communities. Systematic and planned agriculture activities through the rotation of crops may provide a balance in an agroecosystem. The crop rotations also provide ample opportunities for the interaction of microorganisms to contribute to the nutrient cycle more effectively.

The average $CO_2$ emission in an irrigated wheat agroecosystem was ~3184.4 kg $CO_2$-eq $ha^{-1}$ and ~680.4 kg $CO_2$-eq $t^{-1}$. In contrast, it was reported to be ~553.1 kg $CO_2$ equivalent per hectare and ~381.3 kg $CO_2$ equivalent per ton in a dryland wheat agroecosystem (Mondani et al. 2017).

Higher $CO_2$ emissions in irrigated ecosystems indicate more energy consumption in an irrigated agroecosystem compared to a dryland agroecosystem. Hence, suitable climatic conditions will reduce water consumption during irrigation and the emission of GHGs. Similarly, crop rotation will also enhance the soil's nutrient load and water-holding capacity, reducing the stress of water irrigation and reducing the emission of GHGs into the atmosphere.

## Conclusion

Agroecosystems contribute significant amounts of GHGs ($CH_4$, $NO_2$, $N_2O$, NO) into the atmosphere annually worldwide. These influenced the agroecosystem's chemistry due to the application of chemical fertilizers and pesticides during farming activities. It observed that the chemistry of the agroecosystem also depends on climatic conditions (temperature, rainfall, humidity, wind) and soil characteristics, as well as the presence of microorganisms in the soil. These microorganisms participate in various biogeochemical activities and contribute a significant role in the biogeochemical cycle of nutrients. The average greenhouse gas emission through agroecosystems is comparatively low compared to other anthropogenic sources. Agricultural soils are responsible for approximately 60–80% of anthropogenic $N_2O$ sources, mainly from synthetic fertilizers, manure applications, and farm crop residues.

Total greenhouse gas (GHG) emissions globally contributed by soil degassing are ~35% $CO_2$, ~47% $CH_4$, ~53% $N_2O$, and ~21% NO of the respective total annual emissions. The leading GHG emission was observed through paddy crops (~23.75 Tg $CEha^{-1}$) and the least reported from *Cajanus cajan* (~2.98 Tg $CEha^{-1}$). The Environmental Protection Agency (2021) states that methane accounts for ~10% of all GHGs emitted by anthropogenic activity. Methane also contributes to the atmosphere through natural activities like the decomposition of organic material in an anaerobic environment. The natural processes occurring within soil molecules and further chemical reactions in the atmosphere help to remove methane from the atmosphere. Methane's ($CH_4$) lifetime in the atmosphere was much shorter than that of $CO_2$, but $CH_4$ is more dangerous in the environment for trapping radiation than $CO_2$. The comparative impact of $CH_4$ is ~25 times stronger than $CO_2$ over 100 years (https://www.epa.gov/ghgemissions/overview-greenhouse-gases).

Crop rotation helped resolve certain levels of GHG emissions by controlling soil chemistry and improving organic load in the soil. It also improves soil nutrient load and retains soil fertility. The study indicates that the problem of salinity hazard and sodium hazards could be minimized up to a certain level by rotating the crops according to climatic conditions. It has recently been a concern to worldwide communities due to the excessive utilization of chemical fertilizers and pesticides. An unbalanced agroecosystem may cause various issues, but global warming is more concerning due to the emission of GHGs into the atmosphere. Hence, it is necessary to coordinate an agroecosystem and its environment through crop rotation and further relevant information for policy and decision-makers to implement for a sustainable response.

## References

Anonymous (2019) Sustainable development goals, United Nations

Behnke GD, Zuber SM, Pittelkow CM, Nafziger ED, Villamil MB (2018) Long-term crop rotation and tillage effects on soil greenhouse gas emissions and crop production in Illinois. USA Agric Ecosyst Environ 261:62–70

Brummer C, Bruggemann B-BK, Falk U, Szarzynski J, Vielhauer K, Wassmann R, Papen H (2008) Soil-atmosphere exchange of N2O and NO in near-natural savanna and agricultural land in Burkina Faso (W. Africa). Ecosystems 11:582–600

Cai S, Pittelkow CM, Zhao X, Wang S (2018) Winter legume-rice rotations can reduce nitrogen pollution and carbon footprint while maintaining net economic benefits. J Clean Prod 195:289–300

Conklin AR (2005) Introduction to soil chemistry: analysis and instrumentation. Wiley, pp 1–2009

Davidson EA (2009) The contribution of manure and fertilizer nitrogen to atmospheric nitrous oxide since 1860. Nat Geosci 2:659

Environmental Protection Agency EPA, USA, https://www.epa.gov/ghgemissions/overview-greenhouse-gases (access on 1st Dec. 2021)

Forster P, Ramaswamy V, Artaxo P, Berntsen T, Betts R, Fahey DW, Haywood J, Lean J, Lowe DC, Myhre G et al (2007) Changes in atmospheric constituents and radiative forcing. In: Solomon S, Qin D, Manning M, Chen Z, Marquis M, Averyt KB, Tignor M, Miller HL (eds) Climate change 2007: the physical science basis. Contribution of working group I to the fourth assessment report of the Intergovernmental Panel on Climate Change. Cambridge University Press, Cambridge/New York, pp 129–234

Fowler D, Coyle M, Skiba U, Sutton MA, Neil Cape J, Reis S, Sheppard LJ, Jenkins A, Grizzetti B, Galloway JN, Vitousek P, Leach A, Bouwman AF, Butterbach-Bahl K, Dentener F, Stevenson D, Amann M, Voss M (2013) The global nitrogen cycle in the twenty-first century. Philos Trans R Soc B Biol Sci 368:20130164

Gan Y, Liang C, Chai Q, Lemke RL, Campbell CA, Zentner RP (2014) Improving farming practices reduces the carbon footprint of spring wheat production. Nat Commun 5:5012

Garlapati VK, Parashar SK, Klykov S, Vundavilli PR, Sevda S, Srivastava SK, Taherzadeh MJ (2021) Invasive weed optimization coupled biomass and product dynamics of tuning soybean husk towards lipolytic enzyme. Bioresour Technol 344:126254. https://doi.org/10.1016/j.biortech.2021.126254

Herridge DF, Peoples MB, Boddey RM (2008) Global inputs of biological nitrogen fixation in agricultural systems. Plant Soil 311:1–18

Hodson R (2017) Food security. Nature 544:S5

Hou D, Ok YS (2019) Soil pollution: speed up global mapping. Nature 566:455

IPCC (2007) In: Solomon S, Qin D, Manning M, Chen Z, Marquis M, Averyt KB, Tignor M, Miller HL (eds) Climate change 2007: the physical science basis. Contribution of working group I to the fourth assessment report of the intergovernmental panel on climate change. Cambridge University Press, Cambridge/New York

Jarecki M, Grant B, Smith W, Deen B, Drury C, VanderZaag A, Qian B, Yang J, Wagner-Riddle C (2018) Long-term trends in corn yields and soil carbon under diversified crop rotations. J Environ Qual 47:635–643

Li X, Jousset A, De Boer W, Carrión VJ, Zhang T, Wang X, Kuramae EE (2019) Legacy of land-use history determines reprogramming plant physiology by soil microbiome. ISME J 13:738–751

Ma BL, Liang BC, Biswas DK, Morrison MJ, McLaughlin NB (2012) The carbon footprint of maize production as affected by nitrogen fertilizer and maize-legume rotations. Nutr Cycl Agroecosyst 94:15–31

Maes J, Teller A, Erhard M, Grizzetti B, Barredo JI, Paracchini ML, Condé S, Somma F, Orgiazzi A, Jones A, Zulian A, Vallecilo S, Petersen JE, Marquardt D, Kovacevic V, Abdul Malak D, Marin AI, Czúcz B, Mauri A, Loffler P, BastrupBirk A, Biala K, Christiansen T, Werner B (2018) Mapping and assessment of ecosystems and their services: an analytical framework for ecosystem condition. Publications office of the European Union, Luxembourg

Maraseni TN, Qu J (2016) An international comparison of agricultural nitrous oxide emissions. J Clean Prod 135:1256–1266

Mondani F, Aleagha S, Khoramivafa M, Ghobadi R (2017) Evaluation of greenhouse gas emission based on energy consumption in wheat agroecosystems. Energy Rep 3:37–45

Parashar SK, Srivastava SK, Dutta NN, Garlapati VK (2018) Engineering aspects of immobilized lipases on esterification: a special emphasis on crowding, confinement, and diffusion effects. Eng Life Sci 18:308–316. https://doi.org/10.1002/elsc.201700082

Parashar SK, Srivastava SK, Garlapati VK (2019a) The mathematical modeling for optimizing triacylglycerol acylhydrolases production through artificial neural network and genetic algorithm. Int J Pharm Bio Sci 10(3):135–143. https://doi.org/10.22376/ijpbs.2019.10.3.b135-143

Parashar SK, Srivastava SK, Garlapati VK, Dutta NN (2019b) Production of microbial enzyme triacylglycerol Acylhydrolases by ASPERGILLUS SYDOWII JPG01 by in submerged fermentation using agro residues. Asian J Microbiol Biotechnol Environ Sci 21(4):1076–1079. https://www.envirobiotechjournals.com/article_abstract.php?aid=10217&iid=295&jid=1

Prasad MBK, Ramanathan AL, Srivastava SK, Anshumali, Saxena R (2006) Metal fractionation studies in surfacial and core sediments in the Achankovil River basin, India. Environ Monit Assess 121:77–102. https://doi.org/10.1007/s10661-005-9108-2

Puma MJ (2019) Resilience of the global food system. Nat Sustain 2:260–261

Ranjan R, Srivastava SK, Ramanathan AL (2017) Assessment of hydro-geochemistry of two wetlands located in Bihar State in the subtropical climatic zone of India. Environ Earth Sci 76(16):1–17. https://doi.org/10.1007/s12665-016-6330-x

Sah D, Devakumar AS (2018) The carbon footprint of agricultural crop cultivation in India. Carbon Manag 9(3):213–225

Salton JC, Mercante FM, Tomazi M, Zanatta JA, Concenço G, Silva WM, Retore M (2014) Integrated crop-livestock system in tropical Brazil: toward a sustainable production system. Agric Ecosyst Environ 190:70–79

Singh JS, Gupta VK (2016) Degraded land restoration in reinstating $CH_4$ sink. Front Microbiol 7. https://doi.org/10.3389/fmicb.2016.00923

Srivastava SK (2019) Assessment of groundwater quality for irrigation suitability and its impacts on crop yields in the Guna district, India. Agric Water Manag 216:224–241. https://doi.org/10.1016/j.agwat.2019.02.005

Srivastava SK (2020) Advancement in biogas production from solid waste by optimizing anaerobic digestion. Waste Dispos Sustain Energy 2:85–102. https://doi.org/10.1007/s42768-020-00036-x

Srivastava SK (2021) New challenges on natural resources and their impact on climate change in the Indian context. In: Islam MN, van Amstel A (eds) India: climate change impacts, mitigation, and adaptation in developing countries, vol 1. Springer Climate, Cham, pp 1–15. https://doi.org/10.1007/978-3-030-67865-4_1

Srivastava SK, Dhaker KL (2024) Data-driven approach for copper recovery from hazardous e-waste. Process Saf Environ Prot 183:665–675

Srivastava SK, Ramanathan AL (2008) Geochemical assessment of groundwater quality in the vicinity of Bhalswa landfill, Delhi, India, using graphical and multivariate statistical methods. Environ Geol 53:1509–1528. https://doi.org/10.1007/s00254-007-0762-2

Srivastava SK, Ramanathan AL (2012) Groundwater near the landfill. Application of the graphical and multivariate statistical method for hydrogeochemical characterization of groundwater. Lambert Publication, Germany, p 1–380. ISBN-13:978-3847328858, ISBN-10:3847328859

Srivastava SK, Ramanathan AL (2018a) Assessment of landfills vulnerability on the groundwater quality near floodplain of the perennial river and simulation of contaminant transport. Model Earth Syst Environ 4(2):729–752. https://doi.org/10.1007/s40808-018-0464-7

Srivastava SK, Ramanathan AL (2018b) Geochemical assessment of fluoride enrichment and nitrate contamination in groundwater in hard rock aquifer by using graphical and statistical methods. J Earth Syst Sci 127(104):1–23. https://doi.org/10.1007/s12040-018-1006-4

Tayyab M, Yang Z, Zhang C, Islam W, Lin W, Zhang H (2021) Sugarcane monoculture drives microbial community composition, activity, and abundance of agricultural-related microorganisms. Environ Sci Pollut Res 28(35):48080–48096

Tilman D, Clark M (2014) Global diets link environmental sustainability and human health. Nature 515:518–522

Tollefson J (2019) Air pollution science under siege at US environment agency. Nature 568:15–16
Ussiri D, Lal R (2013) Soil emission of nitrous oxide and its mitigation. Springer, Dordrecht, p 378
Xu Z, Xu J, Yin H, Jin W, Li H, He Z (2019) Urban river pollution control in developing countries. Nat Sustain 2:158–160
Yang C, Hamel C, Gan Y (2015) Incongruous variation of denitrifying bacterial communities as soil N level rises in Canadian canola fields. Appl Soil Ecol 89:93–101
Yang T, Lupwayi N, St-A M, Siddique KHM, Bainarde LD (2021) Anthropogenic drivers of soil microbial communities and impacts on soil biological functions in agroecosystems. Glob Ecol Conserv 27:1521
Young R, Wilson B, Harden S, Bernardi A (2009) Accumulation of soil carbon under zero tillage cropping and perennial vegetation on the Liverpool Plains, eastern Australia. Aust J Soil Res 47:273–285

# Greenhouse Gas Cycling and Ecosystem Services: Organic Amendments Under Rice-Rice-Rice Cropping System and Net Ecosystem Carbon Status

**MdMozammel Haque, Jatish C. Biswas, M. Maniruzzaman, M. B. Hossain, and U. A. Naher**

## Introduction

Rice is a staple food crop for approximately half of the world's population and serves around 150 million people in Bangladesh. Rice in Bangladesh occupies 74.64% of the total cropped land area (BBS 2017). In contrast, an additional 44.6 million tons (MT) of milled rice will be needed in 2050 to feed about 215.4 million people (Kabir et al. 2015). Organic nutrient sources and management will be essential measures to improve rice productivity and soil quality (Timsina et al. 2006; Sihi et al. 2017). Incorporating organic materials can increase soil carbon sequestration and facilitate greenhouse gas (GHG) emissions under intensive cropping systems (Haque et al. 2021; Zhao et al. 2016). The emissions of greenhouse gases (GHGs) from crop fields contributes to the increase in atmospheric carbon dioxide ($CO_2$). These gases are considered the main drivers of global warming (Haque et al. 2015a). Globally, anthropogenic GHGs are primarily produced from agricultural fields. Their contribution to national GHG emissions could be 10–12% (Hou et al. 2012).

Improvements to paddy soils increase $CH_4$ emissions and global warming potential (GWP) (Haque et al. 2013, 2015a, b; Ibrahim et al. 2015). Management strategies for reducing GHG emissions include the use of straw biochar (Dong et al. 2013), reduced tillage practices (Lu and Lu 2017), selecting suitable rice cultivars, and changing the cropping regime (Setyanto et al. 2000; Haque et al. 2017c). Although these measures reduced GHG emissions and the overall greenhouse

M. Haque (✉) · M. Maniruzzaman · M. B. Hossain · U. A. Naher
Bangladesh Rice Research Institute, Gazipur, Bangladesh

J. C. Biswas
Krishi Gobeshona Foundation, Farmgate, Dhaka, Bangladesh



S. R. Mohanty, B. Kollah (eds.), *Greenhouse Gas Regulating Microorganisms in Soil Ecosystems*, https://doi.org/10.1007/978-3-031-70569-4_3

effect, the application of these technologies was not convenient for many farmers in Bangladesh. However, organic amendments like as vermicompost and cow dung are good management strategies for enhancing rice productivity and soil organic carbon storage.

The net ecosystem carbon balance is considered based on carbon inputs and outputs.

Carbon inputs into the soil include adding decomposable organic sources, crop biomass, litter, etc. Carbon outputs are considered as carbon losses through $CH_4$-C, $CO_2$-C, removal of crop biomass, etc. Existing literature has shown carbon emission losses and net carbon balance under double-rice cropping systems, single rice systems, rice–wheat rotation systems, and other cropping systems (Haque et al. 2015a, 2017a; Xie et al. 2017).

## *Study Location and Initial Soil Characteristics*

The field experiment took place at the Bangladesh Rice Research Institute (BRRI) experimental farm (23°85.9′N, 90°82.4′E, elevation 12 m). Initial soil sampling and analysis revealed a pH of 7.0 (1:2.5 water extraction), 1.3 g kg$^{-1}$ organic matter content, 1.2 g kg$^{-1}$ total nitrogen, 0.02 g kg$^{-1}$ available phosphorus, 0.08 g kg$^{-1}$ available potassium, 0.010 g kg$^{-1}$ available sulfur, and 0.004 g kg$^{-1}$ available zinc.

## Experimental Treatments, Design, and Management Practices

The treatments consisted of chemical fertilizers (RCF), vermicompost (VC), and cow dung (CD) as integrated plant nutrient system (IPNS)-based inorganic fertilizations. Rice cultivars were grown in the experimental field under fully irrigated conditions. Rice seedlings were transplanted in the first week of January and harvested in the third week of May for the Boro season, in the last week of May and harvested in the last week of July for the Aus season, and in the first week of August and harvested in November for the T. Aman season. Two to three rice seedlings (25–35 days old in wet season, 45–50 days old in dry season, and 20–25 days old in pre-monsoon season) were transplanted at a spacing of 20 × 20 cm. After harvest, the crop root biomass was incorporated into the soil. The recommended sole chemical fertilizer (RCF) rates applied were 100-10-80-5-5, 138-10-80-5-5, and 60-10-80-5-5 kg ha$^{-1}$ of N-P-K-S-Zn for the T. Aman, Boro, and Aus seasons, respectively.

The nutrient sources were urea, triple superphosphate, muriate of potash, gypsum, and zinc sulfate. Cow dung (CD) and vermicompost (VC) were applied at 2 t ha$^{-1}$ along with IPNS-based inorganic fertilizers as a basal dose. Nitrogen was split into three equal applications at final land preparation before rice transplanting, active tillering stage, and 1 week before the panicle initiation stage. Total phosphorus, potassium, sulfur, and zinc were applied as basal fertilizers before rice transplanting. The rice cultivars 'BRRI dhan49', 'BRRI dhan29', and 'BRRI dhan48' were used as indicator crops in the wet, dry, and pre-monsoon seasons, respectively.

## Rice Yield and Yield Component Measurement

Grain yield was determined at physiological maturity from three 5 $m^2$ areas within each plot, followed by bagging and threshing. The samples were sun-dried for 3 days, cleaned, and weighed using a digital balance. Moisture content was measured using a digital grain moisture meter. Grain yield was recorded at approximately 14% moisture content (wet weight basis).

### *Soil Chemical Analysis*

Initial soil samples were collected from the surface layer (0–15 cm) and air-dried. The sieved soils (<2 mm) were analyzed for pH (1:2.5 water extraction), soil organic carbon content (Walkley and Black method 1934; Allison 1965), total nitrogen (Kjeldahl method; Bremner and Mulvaney 1982), available phosphorus (Olsen method; Fox et al. 1964), 0.16 M calcium dihydrogen phosphate-extractable sulfur (Fox et al. 1964), and DTPA-extractable zinc (Olsen and Ellis 1982).

### *Gas Sampling and Analysis*

The static closed-chamber method (Haque et al. 2013, 2016a, b, c) was used to measure methane ($CH_4$) and carbon dioxide ($CO_2$) emission rates during the Boro, T. Aman, and Aus rice seasons in 2018–2019. Transparent glass chambers with a surface area of 0.62 × 0.62 m and a height of 1.12 m were permanently placed in the flooded soil after rice transplanting for monitoring $CH_4$. Nine rice plants were enclosed in each chamber, with two holes in the bottom to maintain a water level 5–7 cm above the soil–water interface. Smaller closed chambers (20 × 50 cm) were placed between rice plants to measure $CO_2$ emission rates. All chambers were equipped with a circulating fan for gas mixing and an internal thermometer to monitor temperature during sampling.

Chambers were kept open except during gas sampling. Air gas samples were collected using 50 ml gas-tight syringes at 0 and 30 min after chamber closing. Gas sampling was carried out three times (8:00, 12:00, and 16:00) in a day to measure the average greenhouse gas (GHG) emission rate. Three gas samples were drawn from the chamber headspace equipped with a three-way stopcock for each replicate of a treatment. Collected gas samples were immediately transferred into 20 ml air-evacuated glass vials sealed with a butyl rubber septum for analysis.

Collected samples were analyzed using gas chromatography (Shimadzu, GC-2014, Japan) with a Porapak NQ column (Q 80–100 mesh). A flame ionization detector (FID) and a thermal conductivity detector (TCD) were used for quantifying $CH_4$ and $CO_2$, respectively. The column, injector, and detector temperatures were set at 100 °C, 60 °C, and 100 °C for $CH_4$ and 45 °C, 75 °C, and 270 °C for $CO_2$, respectively. Helium and hydrogen gases were used as the carrier and burning gases, respectively. $CH_4$ and $CO_2$ emission rates were calculated from the increase in their concentrations per unit surface area of the chamber over a specific time interval. A closed-chamber equation (Lou et al. 2004) was used to estimate seasonal fluxes from each treatment as follows:

$$F = \rho \times (V / A) \times (\Delta c / \Delta t) \times (273 / T)$$

where $F$ is the $CH_4$ and $CO_2$ emission rate (mg m$^{-2}$ h$^{-1}$), $\rho$ is the gas density of $CH_4$ and $CO_2$ under a standardized state (mg cm$^{-3}$), $V$ is the chamber volume (m$^3$), $A$ is the chamber surface area (m$^2$), $\Delta c/\Delta t$ is the rate of increase of $CH_4$ and $CO_2$ gas concentrations in the chamber (mg m$^{-3}$ h$^{-1}$), and $T$ (absolute temperature) is 273 + mean temperature (°C) inside the chamber.

The seasonal $CH_4$ and $CO_2$ flux for the entire crop period was computed as reported by Singh et al. (1999):

$$\text{Seasonal}\, CH_4 \text{ and } CO_2 \text{ flux} = \Sigma (R_i \times D_i),$$

where $R_i$ is the rate of $CH_4$ and $CO_2$ flux (g m$^{-2}$ d$^{-1}$) in the $i$th sampling interval, $D_i$ is the number of days in the sampling interval, and $n$ is the number of samplings.

## Estimation of Net Ecosystem Carbon Balance (NECB)

The findings of Jia et al. (2012), Ma et al. (2013), Haque et al. (2015a), and Haque et al. (2020) were summarized for the determination of NECB as follows:

$$\text{NECB} = \text{GPP} - \text{Re} - \text{Harvest} - CH_4 + \text{organic manure}, \quad (1)$$

$$\text{GPP} = \text{NPP} + \text{Ra}, \quad (2)$$

$$\text{Re} = \text{Ra} + \text{Rh}, \quad (3)$$

where GPP is gross primary production and NPP is net primary production.

The harvest term includes rice straw and grains, and organic manure C inputs were calculated from the CD and VC application rates.

Equation (1) was converted to Eq. (4) by using Eqs. (2) and (3):

$$NECB = NPP - Rh - Harvest - CH_4 + organic\ manure, \quad (4)$$

$$NPP = aboveground\ biomass + belowground\ biomass + litter + rhizodeposits.$$

## Statistical Analysis

Statistical analyses were performed using the SAS package, version 9.1 (SAS Institute 2003). Tukey's test was used for comparisons, and differences were considered significant at $p \leq 0.05$.

## Results

### *Biomass Productivity in the Boro Season*

The vermicompost (VC) treatment showed significantly higher belowground and aboveground biomass productivity compared to the other treatments (Table 1). In all three rice growing seasons, the sole recommended chemical fertilizer (RCF) treatment had the lowest aboveground and belowground productivity compared to the organic amendments. Rice grain yields were significantly higher (28–47%) under the organic amendments than the sole chemical fertilization treatment. However, grain yields also varied significantly between the cow dung (CD) and VC organic amendments under the triple rice cropping system. The application of VC resulted in a 4–9% increase in grain yield compared to the CD treatment. Grain yields were significantly higher in the Boro season than in the T. Aman and Aus seasons.

### *$CH_4$-C Emission Patterns During Boro Season Rice Cultivation*

The vermicompost (VC) and cow dung (CD) treatments showed varying $CH_4$-C emission patterns during Boro rice cultivation (Fig. 1). The highest $CH_4$-C emission peak was observed at 28 days after transplanting (DAT). The CD treatment showed significantly higher $CH_4$-C emission (16.9 mg $m^{-2}$ $hr^{-1}$) than the VC treatment (14.8 mg $m^{-2}$ $hr^{-1}$). The lowest $CH_4$-C emission (6.16 mg $m^{-2}$ $hr^{-1}$) was observed with the sole recommended chemical fertilizer (RCF) treatment.

**Table 1** Ground biomass as influenced by fertilizer packages under Boro–T. Aman–Aus season rice cultivation

| Seasons (C) | Year (B) | Treatment (A) | Biomass productivity (kg $ha^{-1}$) Belowground (root) | Aboveground (straw) | Aboveground (grain) |
|---|---|---|---|---|---|
| Boro | 2018 | RCF | 1038 | 5230 | 5150 |
| | | CD | 1442 | 7230 | 7190 |
| | | VC | 1514 | 7690 | 7450 |
| | 2019 | RCF | 1054 | 5320 | 5220 |
| | | CD | 1467 | 7421 | 7250 |
| | | VC | 1523 | 7680 | 7550 |
| T. Aman | 2018 | RCF | 804 | 4050 | 3990 |
| | | CD | 1039 | 5282 | 5110 |
| | | VC | 1102 | 5570 | 5450 |
| | 2019 | RCF | 860 | 4500 | 4100 |
| | | CD | 1064 | 5400 | 5240 |
| | | VC | 1105 | 5610 | 5550 |
| Aus | 2018 | RCF | 750 | 3885 | 3700 |
| | | CD | 858 | 4350 | 4230 |
| | | VC | 941 | 4830 | 4580 |
| | 2019 | RCF | 770 | 3920 | 3850 |
| | | CD | 875 | 4410 | 4320 |
| | | VC | 964 | 4920 | 4720 |
| Statistical analysis | | | | | |
| A | | | *** | *** | |
| B | | | ns | ns | |
| C | | | *** | *** | |
| A*B | | | ns | ns | |
| A*C | | | *** | *** | |
| A*B*C | | | ns | ns | |

ns non-significant; *** significant at $P < 0.001\%$ levels

## $CH_4$-C Emission Patterns During T. Aman Season Rice Cultivation

The $CH_4$-C emission patterns varied with different fertilizer treatments (Fig. 1). The emission rates were low at the early rice growing stage and gradually increased with the development of soil reductive conditions and plant growth. The highest peak of $CH_4$-C was observed around 35 DAT and then gradually decreased to the initial levels at the grain filling stage. Among the treatments, the CD treatment showed the highest $CH_4$-C emission peak. The organic amendment treatments increased $CH_4$-C emission by about 32–58% compared to the RCF treatment. However, the VC treatment decreased $CH_4$-C emission by about 11% compared to the CD treatment. The

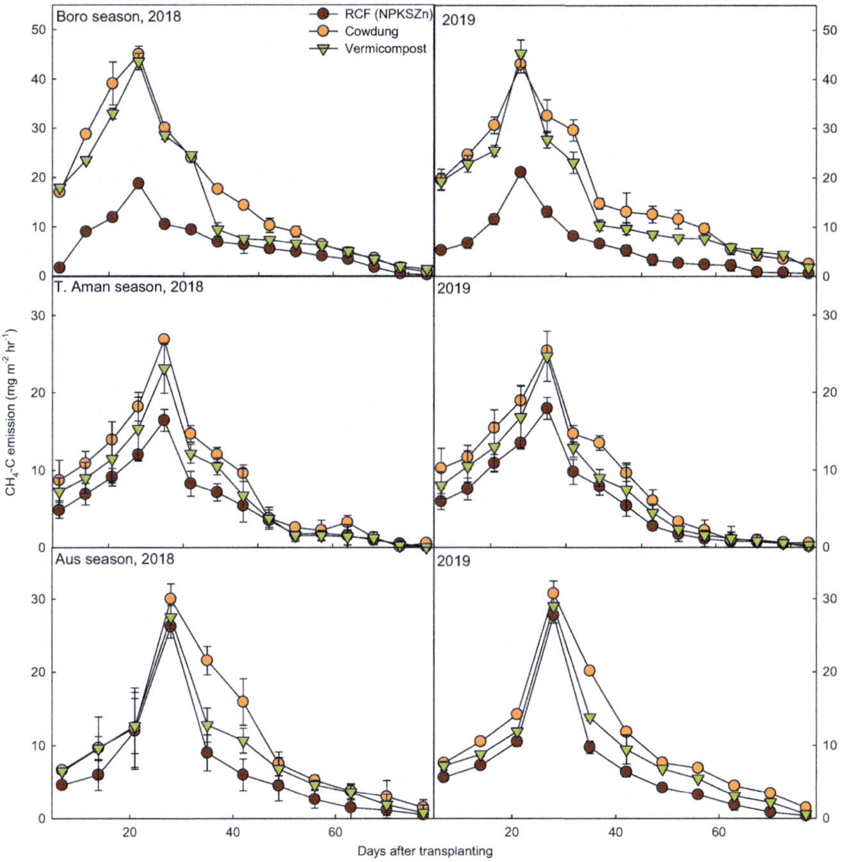


**Fig. 1** $CH_4$-C emission as influenced by fertilizer management under Boro–T. Aman–Aus season rice cultivation during 2018–2019

mean $CH_4$-C emission rates in the VC plots were about 7.28 mg $m^{-2}$ $hr^{-1}$, which was comparable to 8.75 mg $m^{-2}$ $hr^{-1}$ in the CD plots (Fig. 1).

## *$CH_4$-C Emission Patterns During Aus Season Rice Cultivation*

In the Aus season, the $CH_4$-C emission peak was highest at 28 DAT and then decreased thereafter for different treatments. The CD treatment had a $CH_4$-C emission rate of about 17.17 mg $m^{-2}$ $hr^{-1}$, which was about 30% higher than the VC treatment. The organic amendment treatments increased $CH_4$ emission by about 35–74% compared to the sole RCF treatment (Fig. 1).

## $CO_2$-C Emission Patterns During Boro Season Rice Cultivation

The highest $CO_2$-C emission peaks occurred around 42–56 days after transplanting (DAT) during the Boro rice cultivation period. However, at the initial and later crop growth stages, the emission peaks were significantly lower (Fig. 2). Among the treatments, the cow dung (CD) application resulted in higher $CO_2$-C emissions (10.00 mg $m^{-2}$ $hr^{-1}$) compared to the vermicompost (VC) amendment (8.34 mg $m^{-2}$ $hr^{-1}$). The sole recommended chemical fertilizer (RCF) treatment exhibited the lowest $CO_2$-C emissions (6.49 mg $m^{-2}$ $hr^{-1}$) throughout the study period.

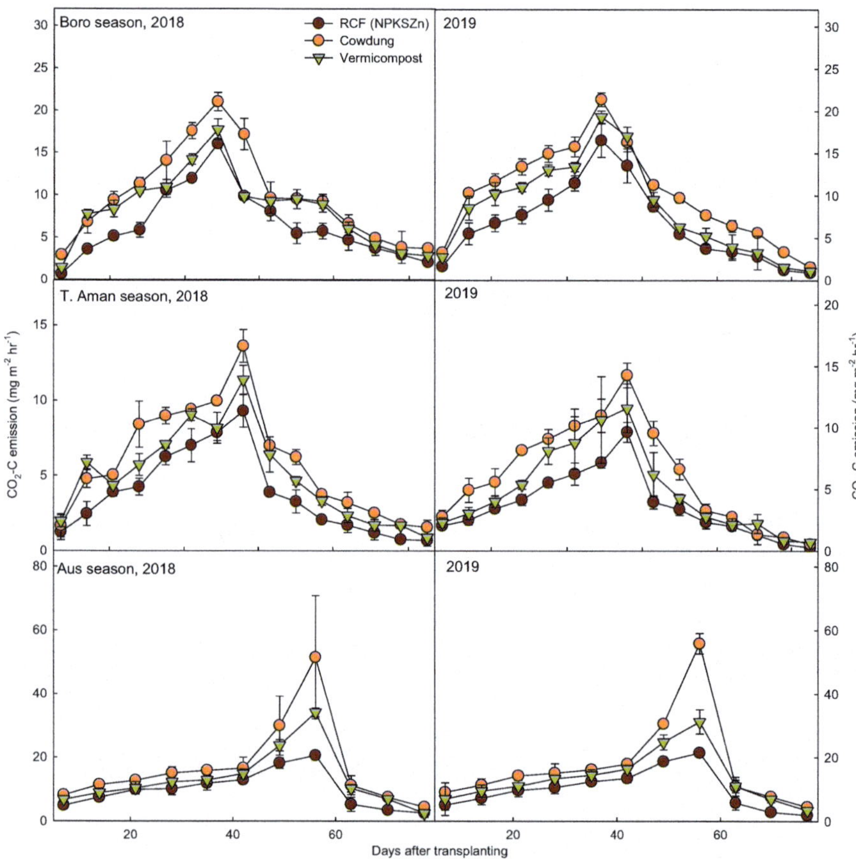


**Fig. 2** $CO_2$-C emissions as influenced by fertilizer packages during Boro–T. Aman–Aus season rice cultivation during 2018–2019

### *$CO_2$-C Emission Patterns During T. Aman Season Cultivation*

In the T. Aman season, the highest $CO_2$-C emission peak was observed around 49–56 DAT, depending on the treatments (Fig. 2). The CD treatment displayed the highest $CO_2$-C emission peak throughout the study period. However, the lowest emissions were observed under the sole RCF treatment (3.62–3.72 mg $m^{-2}$ $hr^{-1}$) compared to the other treatments. $CO_2$-C emissions of 4.86–4.96 mg $m^{-2}$ $hr^{-1}$ were found under the VC treatment, whereas emissions of 5.85–6.04 mg $m^{-2}$ $hr^{-1}$ were observed under the CD amendments.

### *$CO_2$-C Emission Patterns During Aus Season Rice Cultivation*

The highest $CO_2$-C emission peak was recorded at 49–56 DAT during the study period.

The $CO_2$-C emissions were 4.36–13.53 mg $m^{-2}$ $hr^{-1}$ under the VC treatment and 5.85–17.57 mg $m^{-2}$ $hr^{-1}$ under the CD treatment. Under the sole RCF treatment, $CO_2$-C emissions of 3.95–4.00 mg $m^{-2}$ $hr^{-1}$ were found (Fig. 2). The VC treatment reduced $CO_2$-C emission loss by approximately 15–17% compared to the CD treatment.

## Net Primary Production Input Carbon (NPPIC) During Boro–T. Aman–Aus Season Rice Cultivation

The seasonal totals NPPIC were 5476–8050, 4246–5901, and 4006–5109 kgC $ha^{-1}$ during Boro–T. Aman–Aus rice cultivation (Table 2). In the Boro season, the NPPIC was significantly higher than other seasons. The Aus season showed lower NPPIC than Boro and T. Aman seasons. Among the treatments, VC significantly increased total NPPIC by 4%, 5%, and 10% over CD under Boro–T. Aman–Aus cropping. However, only sole RCF treatment showed significantly the lowest total NPPIC during year-round Boro–T. Aman–Aus rice cultivation (Table 2).

## Total Output Carbon During Boro–T. Aman–Aus Season Rice Cultivation

In the Boro season, the total output carbon was 4557–6736 kg C $ha^{-1}$, which was about 1.3 and 1.5 times higher than the T. Aman and Aus seasons, respectively. Interestingly, the $CH_4$-C loss was 1.5–1.5–1.9 times greater than the $CO_2$-C loss under the Boro–T. Aman–Aus cropping system. As a result, the vermicompost (VC)

**Table 2** Net primary production input carbon as influenced by fertilizer packages under Boro–T. Aman–Aus season rice cultivation

| Seasons (C) | Year (B) | Treatment (A) | Net primary production input carbon (kg C ha$^{-1}$) Straw C | Grain C | Root C | Litter | Rhizodeposits | Manure C | Total NPPIC (kg C ha$^{-1}$) |
|---|---|---|---|---|---|---|---|---|---|
| Boro | 2018 | RCF | 2197 | 2009 | 390 | 230 | 631 | 20 | 5476 |
| | | CD | 3037 | 2804 | 542 | 319 | 876 | 45 | 7623 |
| | | VC | 3230 | 2906 | 569 | 335 | 920 | 45 | 8005 |
| | 2019 | RCF | 2234 | 2036 | 396 | 233 | 641 | 20 | 5560 |
| | | CD | 3117 | 2828 | 552 | 325 | 892 | 45 | 7757 |
| | | VC | 3226 | 2945 | 573 | 337 | 926 | 45 | 8050 |
| T. Aman | 2018 | RCF | 1701 | 1556 | 302 | 178 | 489 | 20 | 4246 |
| | | CD | 2218 | 1993 | 391 | 230 | 632 | 40 | 5504 |
| | | VC | 2520 | 2126 | 414 | 244 | 670 | 40 | 5833 |
| | 2019 | RCF | 1890 | 1599 | 323 | 191 | 523 | 20 | 4546 |
| | | CD | 2268 | 2044 | 400 | 236 | 647 | 40 | 5634 |
| | | VC | 2356 | 2165 | 415 | 244 | 671 | 40 | 5901 |
| Aus | 2018 | RCF | 1632 | 1443 | 282 | 168 | 461 | 20 | 4006 |
| | | CD | 1827 | 1650 | 323 | 190 | 522 | 40 | 4551 |
| | | VC | 2029 | 1786 | 354 | 208 | 572 | 40 | 4989 |
| | 2019 | RCF | 1646 | 1502 | 290 | 172 | 472 | 20 | 4101 |
| | | CD | 1852 | 1685 | 329 | 193 | 531 | 40 | 4630 |
| | | VC | 2066 | 1841 | 362 | 213 | 586 | 40 | 5109 |
| Statistical analysis | | | | | | | | | |
| A | | | | | | *** | *** | *** | |
| B | | | | | | ns | ns | ns | |
| C | | | | | | *** | *** | *** | |
| A*B | | | | | | ns | ns | ns | |
| A*C | | | | | | *** | *** | *** | |
| A*B*C | | | | | | ns | ns | ns | |

ns non-significant; *** significant at $P < 0.001\%$ levels

treatment showed 3%–4%–8% higher carbon loss than the cow dung (CD) treatment under the tested cropping system. In the Aus season, the output carbon was significantly lower than the other seasons (Table 3).

This means that the carbon loss was much smaller during Aus rice cultivation compared to the Boro and T. Aman seasons. The higher total output carbon and greater relative contribution of $CH_4$-C losses in the Boro season indicate different carbon dynamics compared to the other seasons. Additionally, VC resulted in slightly higher carbon losses than CD across the cropping system. The Aus season had the lowest overall carbon outputs and losses among the three seasons.

**Table 3** Total output carbon as influenced by fertilizer packages under Boro–T. Aman–Aus season rice cultivation

| | | | Output carbon (kg C ha$^{-1}$) | | | | |
|---|---|---|---|---|---|---|---|
| Seasons (C) | Year (B) | Treatment (A) | Grain biomass carbon | Straw biomass carbon | $CH_4$-C | $CO_2$-C | Total output C (kg C ha$^{-1}$) |
| Boro | 2018 | RCF | 2009 | 2197 | 159 | 162 | 4526 |
| | | CD | 2804 | 3037 | 423 | 248 | 6512 |
| | | VC | 2906 | 3230 | 368 | 209 | 6712 |
| | 2019 | RCF | 2036 | 2234 | 152 | 166 | 4588 |
| | | CD | 2828 | 3117 | 433 | 259 | 6636 |
| | | VC | 2945 | 3226 | 378 | 212 | 6760 |
| T. Aman | 2018 | RCF | 1556 | 1701 | 137 | 94 | 3488 |
| | | CD | 1993 | 2218 | 216 | 148 | 4575 |
| | | VC | 2126 | 2520 | 176 | 125 | 4766 |
| | 2019 | RCF | 1599 | 1890 | 146 | 91 | 3726 |
| | | CD | 2044 | 2268 | 225 | 152 | 4689 |
| | | VC | 2165 | 2356 | 181 | 122 | 4824 |
| Aus | 2018 | RCF | 1443 | 1632 | 125 | 73 | 3273 |
| | | CD | 1650 | 1827 | 197 | 103 | 3777 |
| | | VC | 1786 | 2029 | 164 | 86 | 4065 |
| | 2019 | RCF | 1502 | 1646 | 131 | 74 | 3353 |
| | | CD | 1685 | 1852 | 200 | 101 | 3838 |
| | | VC | 1841 | 2066 | 165 | 86 | 4158 |
| Statistical analysis | | | | | | | |
| A | | | *** | *** | *** | | |
| B | | | ns | ns | ns | | |
| C | | | *** | *** | *** | | |
| A*B | | | ns | ns | ns | | |
| A*C | | | *** | *** | *** | | |
| A*B*C | | | ns | ns | ns | | |

ns non-significant; *** significant at $P < 0.001\%$ levels

# Net Ecosystem Carbon Budget (NECB)

To estimate the impact of soil management practices on $CO_2$ emissions, the net ecosystem carbon balance (NECB) was calculated for year-round Boro–T. Aman–Aus rice cultivation (Table 4). The total organic carbon inputs varied from 5476 to 8050 kg ha$^{-1}$.

During the study period, a total of 4526–6760 kg C ha$^{-1}$ was lost through $CH_4$-C and $CO_2$-C emissions and the removal of rice grains and straw. As a result, the NECB ranged from 950 to 1293 kg C ha$^{-1}$ during the Boro season, indicating that this amount of carbon was sequestered during rice cultivation. In the T. Aman season, the NECB was 758–1077 kg C ha$^{-1}$ under different organic amendments. For the Aus season, the NECB was found to be 733–951 kg C ha$^{-1}$.

**Table 4** Input and output carbons and net ecosystem carbon balance as influenced by fertilizer packages under Boro–T. Aman–Aus season rice cultivation

| | | | Carbon gain and loss (kg C ha$^{-1}$) | | |
|---|---|---|---|---|---|
| Seasons (C) | Year (B) | Treatment (A) | Input carbon | Output carbon | Total NECB (kg C ha$^{-1}$) |
| Boro | 2018 | RCF | 5476 | 4526 | 950 |
| | | CD | 7623 | 6512 | 1111 |
| | | VC | 8005 | 6712 | 1293 |
| | 2019 | RCF | 5560 | 4588 | 972 |
| | | CD | 7757 | 6636 | 1121 |
| | | VC | 8050 | 6760 | 1290 |
| T. Aman | 2018 | RCF | 4246 | 3488 | 758 |
| | | CD | 5504 | 4575 | 928 |
| | | VC | 5833 | 4766 | 1067 |
| | 2019 | RCF | 4546 | 3726 | 820 |
| | | CD | 5634 | 4689 | 945 |
| | | VC | 5846 | 4778 | 1077 |
| Aus | 2018 | RCF | 4006 | 3273 | 733 |
| | | CD | 4551 | 3777 | 774 |
| | | VC | 4989 | 4065 | 924 |
| | 2019 | RCF | 4101 | 3353 | 749 |
| | | CD | 4630 | 3838 | 792 |
| | | VC | 5109 | 4158 | 951 |
| Statistical analysis | | | | | |
| A | | | *** | *** | |
| B | | | ns | ns | |
| C | | | *** | *** | |
| A*B | | | ns | ns | |
| A*C | | | *** | *** | |
| A*B*C | | | ns | ns | |

ns non-significant; *** significant at $P < 0.001\%$ levels

The estimation of NECB took into account the organic carbon inputs from different treatments as well as the carbon losses through $CH_4$ and $CO_2$ emissions and the removal of rice biomass. The positive NECB values across all three rice growing seasons indicate net carbon sequestration under the organic amendment treatments.

## Relationship Between Grain Yield and Net Carbon Balance

In Boro season, the net C balance was significantly higher, and grain yields were higher than other crop growing seasons. In Boro season, net C balances were increased by about 20–25% and 30–36% and grain yields by 29–44% and 39–60% compared to T. Aman and Aus seasons, respectively (Fig. 3).

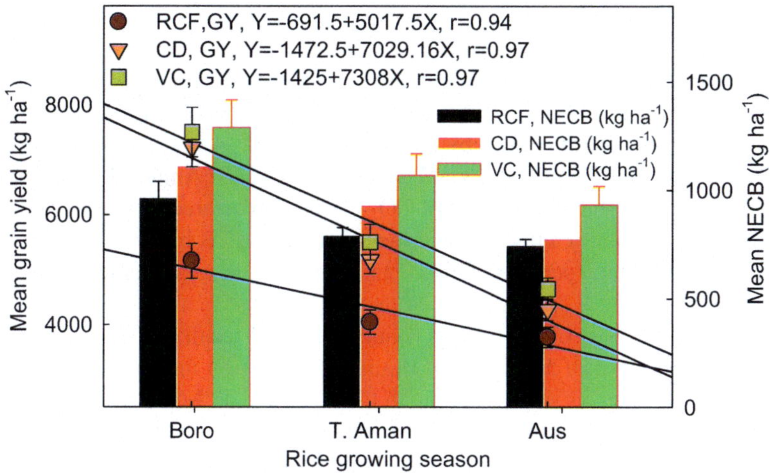


**Fig. 3** Relationship between grain yield and net carbon budget under Boro–T. Aman–Aus season rice cultivation

## Discussion

The continuous use of only chemical fertilizers is one of the causes of declining soil organic carbon or reductions in net carbon balances under different cropping systems such as rice–fallow–rice and rice–wheat–fallow (Sihi et al. 2017; Haque et al. 2019a, b). However, different crop fields or rice fields act as sources of greenhouse gas (GHG) emissions like methane ($CH_4$), carbon dioxide ($CO_2$), and nitrous oxide ($N_2O$). These gases can be produced under aerobic and anaerobic conditions. Any kind of organic amendment provides easily available carbon substrates for soil bacteria, leading to the production of $CO_2$ and $CH_4$ (Haque et al. 2015b). We found significantly higher total emissions of $CO_2$ and $CH_4$ due to organic amendments compared to the recommended sole chemical fertilization (RCF) during the Boro–T. Aman–Aus seasons (Fig. 1).

The effects of organic manure incorporation on increased $CO_2$ and $CH_4$ emissions have also been reported in many field studies (Kim et al. 2013; Haque et al. 2013, 2020, 2021). However, we observed lower $CH_4$ emissions under the Boro–T. Aman–Aus cropping system with vermicompost (VC) than with cow dung (CD) treatments (Fig. 1). The CD-amended plots showed significantly higher $CO_2$-C and $CH_4$-C flux compared to VC and RCF (Table 3), which might be related to enhanced microbial activities (Haque et al. 2017b).

We also observed significantly reduced $CH_4$-C flux from VC-amended soils due to a slower decomposition rate under anaerobic conditions (Haque et al. 2015b). On a seasonal scale, there was a significant increase in the net ecosystem carbon balance (NECB) due to VC and CD application under year-round rice cultivation compared to the use of sole RCF (Table 4). Such higher NECB was related to greater carbon inputs. The removal of carbon by harvesting accounted for around 93–94%

of total carbon outputs, followed by $CH_4$-C and $CO_2$-C (7–8%) under sole RCF conditions.

Although the Boro season significantly increased respiratory carbon loss by 39–62% compared to the T. Aman and Aus seasons, the net primary productivity (NPP) of carbon was influenced by the Boro–T. Aman–Aus seasons (Table 2). The NECB ranged from 733 to 1293 kg C $ha^{-1}$ in the Boro–T. Aman–Aus rice growing seasons, indicating that this amount of $CO_2$-C was sequestered from the atmosphere. The Aus season significantly reduced the seasonal NECB by 6–46% compared to the Boro and T. Aman seasons. This means that the Aus season is less effective for carbon sequestration.

There will be a loss of soil carbon if only chemical fertilizers are used for growing rice, irrespective of VC and CD (Table 3), although the loss would be more significant with the Aus season than with the Boro and T. Aman seasons.

This implies that soil quality will deteriorate in the long run under such conditions. If VC is added, the soil carbon balance becomes more positive, indicating an improvement in soil quality. Therefore, the application of VC is suitable for better yield and soil quality, along with a reduction in carbon loss. The global warming potential can be reduced through the incorporation of VC or other organic substrates in paddy fields (Maris et al. 2016; Win et al. 2015; Sharma et al. 2016; Haque et al. 2021), but no NECB has been defined in combination with CD and VC and in consideration of different rice growing seasons. The results indicate that we cannot only reduce carbon loss from paddy soils but also maintain soil health for better crop production. This finding might be useful for different rice growing environments worldwide.

## Conclusion

The use of vermicompost in paddy soil in Bangladesh was very effective in increasing the net ecosystem carbon balance by 11–26% and improving rice yields compared to cow dung incorporation under the Boro–T. Aman–Aus season and could be a good technology in East Asian regions. Moreover, we found about a 20–49% reduction in carbon loss during triple rice culture with VC. Therefore, it is concluded that the incorporation of VC during Boro–T. Aman–Aus rice cultivation could be an effective soil management practice to increase NECB as well as increase rice productivity.

## References

Allison LE (1965) Organic carbon. In: Methods of soil analysis. Part 2. Chemical and microbiological properties. Wiley, pp 1367–1378

BBS (Bangladesh Bureau of Statistics) (2017) Statistical year book of Bangladesh. Stat. Div., Ministry of Planning, Government of the People's Republic of Bangladesh, Dhaka

Bremner JM, Mulvaney CS (1982) Total nitrogen. In: Page AL, Miller RH, Keeney DR (eds) Methods of soil analysis, Part 2, Chemical and microbiological properties. American Society of Agronomy and Soil Science Society of America, Inc, Madison, pp 595–624

Dong D, Yang M, Wang C, Wang H, Li Y, Luo J, Wu W (2013) Responses of methane emissions and rice yield to applications of biochar and straw in a paddy field. J Soil Sediment 13:1450–1460

Fox RL, Olson RA, Rhoades HF (1964) Evaluating the sulfur status of soil by plant and soil tests. Soil Sci Soc Am Proc 28:243–246

Haque MM, Kim SY, Pramanik P, Kim GY, Kim PJ (2013) Optimum application level of winter cover crop biomass as green manure under considering methane emission and rice productivity in paddy soil. Biol Fertil Soils 49:487–493

Haque MM, Kim SY, Ali MA, Kim PJ (2015a) Contribution of greenhouse gas emissions during cropping and fallow seasons on total global warming potential in mono-rice paddy soils. Plant Soil 387:251–264

Haque MM, Kim SY, Kim GW, Kim PJ (2015b) Optimization of removal and recycling ratio of cover crop biomass using carbon balance to sustain soil organic carbon stocks in a mono-rice paddy system. Agric Ecosyst Environ 207:119–125

Haque MM, Kim GW, Kim PJ, Kim SY (2016a) Comparison of net global warming potential between continuous flooding and midseason drainage in monsoon region paddy during rice cropping. Field Crop Res 193:133–142

Haque MM, Biswas JC, Kim SY, Kim PJ (2016b) Suppressing methane emission and global warming potential from rice fields through intermittent drainage and green biomass amendment. Soil Use Manag 32:72–79

Haque MM, Biswas JC, Waghmode TR, Kim PJ (2016c) Global warming as affected by incorporation of variably aged biomass of hairy vetch for rice cultivation. Soil Res 54:346–353

Haque MM, Biswas JC, Kim SY, Kim PJ (2017a) Intermittent drainage in paddy soil: ecosystem carbon budget and global warming potential. Paddy Water Environ 15:403–411

Haque MM, Biswas JC, Akter M, Kim PJ (2017b) Soil respiration from paddy field in relation to incorporated cover crop biomass composition. J Soil Sci Environ Manage 8:122–129

Haque MM, Biswas JC, Maniruzzaman M, Choudhury AK, Naher UA, Hossain MB, Akhter S, Ahmed F, Kalra N (2017c) Greenhouse gas emissions from selected cropping patterns and adaptation strategies in Bangladesh. Int J Dev Res 07:6832–16838

Haque MM, Biswas JC, Islam MR, Islam A, Kabir MS (2019a) Effect of long-term chemical and organic fertilization on rice productivity, nutrient use-efficiency, and balance under a rice-fallow-rice system. J Plant Nutr 42:2901–2914

Haque MM, Biswas JC, Akter M, Maniruzaman M, Kabir MS (2019b) Carbon budget and aggregate stability of paddy soil under continuous organic amendment. Commun Soil Sci Plant Anal 50:1829–1837

Haque MM, Biswas JC, Hwang HY, Kim PJ (2020) Annual net carbon budget in rice soil. Nutr Cycl Agroecosyst 116:31–40

Haque MM, Biswas JC, Maniruzzaman M, Hossain MB, Islam MR (2021) Water management and soil amendment for reducing emission factor and global warming potential but improving rice yield. Paddy Water Environ. https://doi.org/10.1007/s10333-021-00851-w

Hou H, Peng S, Xu J, Yang S, Mao Z (2012) Seasonal variations of $CH_4$ and $N_2O$ emissions in response to water management of paddy fields located in Southeast China. Chemosphere 89:884–892

Ibrahim M, Cao CG, Zhan M, Li CF, Iqbal J (2015) Changes of $CO_2$ emission and labile organic carbon as influenced by rice straw and different water regimes. J Environ Sci Tech 12:263–274

Jia JX, Ma YC, Xiong ZQ (2012) Net ecosystem carbon budget, net global warming potential and greenhouse gas intensity in intensive vegetable ecosystems in China. Agric Ecosyst Environ 150:27–37

Kabir MS, Salam MU, Chowdhury A, Rahman NMF, Iftekharuddaula KM, Rahman MS, Rashid MH, Dipti SS, Islam A, Latif MA, Islam AKMS, Hossain MM, Nessa B, Ansari TH, Ali MA, Biswas JK (2015) Rice vision for Bangladesh: 2050 and beyond. Ban Rice J 19:1–18

Kim SY, Lee CH, Gutierrez J, Kim PJ (2013) Contribution of winter cover crop amendments on global warming potential in rice paddy soil during cultivation. Plant Soil 366:273–286

Lou Y, Li Z, Zhang T, Liang Y (2004) $CO_2$ emissions from subtropical arable soils of China. Soil Biol Biochem 36:1835–1842

Lu X, Lu X (2017) Tillage and crop residue effects on the energy consumption, input–output costs and greenhouse gas emissions of maize crops. Nutr Cycl Agroecosyst 108:323–337

Ma YC, Kong XW, Yang B, Zhang XL, Yan XY, Yang JC, Xiong ZQ (2013) Net global warming potential and greenhouse gas intensity of annual rice-wheat rotations with integrated soil crop system management. Agric Ecosyst Environ 164:209–219

Maris SC, Teira-Esmatges MR, Catala MM (2016) Influence of irrigation frequency on greenhouse gases emission from a paddy soil. Paddy Water Environ 14:199–210

Olsen RV, Ellis R Jr (1982) Iron. In: Page A et al (eds) Methods of soil analysis, part 2: chemical and microbiological properties. ASA, Madison

SAS Institute (2003) System for windows release 9.1. SAS Institute, Cary

Setyanto P, Makarim AK, Fagi AM, Wassmann R, Buendia LV (2000) Crop management affecting methane emissions from irrigated and rainfed rice in Central Java (Indonesia). Nutr Cycl Agroecosyst 58:85–93

Sharma SK, Singh YV, Tyagi S, Bhatia A (2016) Influence of rice varieties, nitrogen management and planting methods on methane emission and water productivity. Paddy Water Environ 14:325–333

Sihi D, Dari B, Sharma DK, Pathak H, Nain L, Sharma OP (2017) Evaluation of soil health in organic vs. conventional farming of basmati rice in North India. J Plant Nutr Soil Sci 180:389–406

Singh S, Singh J, Kashyap A (1999) Methane flux from irrigated rice fields in relation to crop growth and N-fertilization. Soil Biol Biochem 31:1219–1228

Timsina J, Panaullah GM, Saleque MA, Ishaque M, Pathan ABMBU, Quayyum MA, Connor DJ, Saha PK, Humphreys E, Meisner CA (2006) Nutrient uptake and apparent balances for rice-wheat sequences. I. Nitrogen. J Plant Nutr 29:137–155

Walkley A, Black IA (1934) An examination of digestion method for determining soil organic matter and a proposed modification of the chromic acid titration. Soil Sci 37:29–38

Win KT, Nonaka R, Win AT, Sasada Y, Toyota K, Motobayashi T (2015) Effects of water saving irrigation and rice variety on greenhouse gas emissions and water use efficiency in a paddy field fertilized with anaerobically digested pig slurry. Paddy Water Environ 13:51–60

Xie L, Jing XU, Guo L, Yuxiu XU, Sun X, Zhao H (2017) Impact of water/fertilizer management on methane emission in paddy fields and on global warming potential. Chinese JEco-Agric 25:958–967

Zhao XM, He L, Zhang ZD, Wang HB, Zhao LP (2016) Simulation of accumulation and mineralization ($CO_2$ release) of organic carbon in chernozem under different straw return ways after corn harvesting. Soil Till Res 156:148–154

# Greenhouse Gas Emission from Different Fertilizers (Organic, Inorganic, and Integrated) in Management Agroecosystems

**Mona Nagargade, Preeti Singh, Vishal Tyagi, and Santosh Kumar**

## Introduction

Agriculture is a major contributor to greenhouse gas (GHG) emissions, second only to the energy sector. Farming activities release GHGs from various sources like machinery, diesel fuel, fertilizers, pesticides, and electricity use. Increasing the energy inputs in agricultural practices tends to boost GHG emissions (Nabavi-Pelesaraei et al. 2016). Atmospheric levels of GHGs have been rising steadily, reaching alarming annual averages of 410 ppm for carbon dioxide ($CO_2$), 1866 ppb for methane ($CH_4$), and 332 ppb for nitrous oxide ($N_2O$) in 2019 (IPCC 2021). If $CO_2$ concentrations were to double from current levels, it could lead to a staggering 1.5–4.0 °C increase in average global surface temperatures, along with disruptions in rainfall patterns by the end of this century (Pathak and Wassmann 2007). A study by Al-Amin et al. (2015) predicts that energy-related $CO_2$ emissions will skyrocket from 188.0 million tons in 2010 to 720.5 million tons by 2050. The climate emergency is already unfolding before our eyes through extreme weather events worldwide. From 2001 to 2020, the global surface temperature was a worrying 0.99 (0.84–1.10)°C higher than the 1850–1900 average. The global mean sea level rose by 0.20 (0.15–0.25) m between 1901 and 2018. The rate of sea level rise accelerated from 1.3 (0.6–2.1) mm/year between 1901 and 1971 to 1.9 (0.8–2.9) mm/year from 1971 to 2006 and further increased to an alarming 3.7 (3.2–4.2) mm/year between 2006 and

M. Nagargade (✉)
ICAR-IARI, New Delhi, New Delhi, India

P. Singh · S. Kumar
ICAR-IARI, Gauria Karma, Jharkhand, India

V. Tyagi
ICAR-IARI, New Delhi, New Delhi, India



S. R. Mohanty, B. Kollah (eds.), *Greenhouse Gas Regulating Microorganisms in Soil Ecosystems*, https://doi.org/10.1007/978-3-031-70569-4_4

2018. The observed average rate of heating of the climate system intensified from 0.50 (0.32–0.69) W/$m^2$ for 1971–2006 to 0.79 (0.52–1.06) W/$m^2$ for 2006–2018 (IPCC 2021).

This warming has driven the global mean sea level rise due to ice melt on land and thermal expansion of oceans. India, home to some of the world's most disaster-prone regions, tragically recorded the highest number of deaths from climate crisis-induced disasters in 2018, ranking fifth in the Global Climate Risk Index 2020 (CDP 2020). In India and Bangladesh alone, millions of impoverished people's lives were devastated by floods, cyclone Amphan, and the COVID-19 pandemic. The 2019 monsoon floods from June to October caused a staggering US$10 billion in economic losses and claimed 1750 lives in India. Rising temperatures and changing weather patterns are straining water availability. The water crisis is more severe than anticipated—600 million Indians face high-to-extreme water stress. India's water demand is projected to double the available supply by 2030, putting hundreds of millions at risk of water scarcity and potentially costing the country 6% of its GDP. In 2019, India witnessed its worst droughts on record, affecting over 500 million people across the nation, and major cities like Chennai experienced "zero days" with no water (CDP 2020). The pandemic-induced global lockdown led to a sharp temporary drop in emissions, and India is expected to record 6–10% lower GHG emissions in 2020 compared to 2019 (CDP 2020). However, India's high social and environmental diversity means there is no one-size-fits-all solution for its climate concerns. Setting ambitious climate goals and committing to a low-carbon economy are significant challenges for researchers and policymakers. Adopting climate-resilient nutrient management practices can help reduce emissions at a pace consistent with keeping warming below 1.5 °C/well below 2 °C. Incentive's emission reduction activities through tools like domestic policies and carbon pricing could be effective. Carbon pricing applies a cost to carbon pollution to discourage polluters.

## GHG Dynamics

Agricultural GHG emissions arise mainly from three gases: nitrous oxide ($N_2O$), methane ($CH_4$), and carbon dioxide ($CO_2$). $N_2O$ emissions are primarily caused by applying fertilizers and manure (both spread on fields and excreted by grazing animals) to soils, contributing 12% and 26% to global agricultural GHG emissions, respectively. The main sources of $CH_4$ emissions are enteric fermentation in dairy, beef, and sheep herds, rice cultivation and manure. $CO_2$ emissions are also attributable to manure management and fuel combustion for farming operations. The different atmospheric lifetimes of these GHGs (shown in Table 1) cause variations in their climate change impacts.

**Table 1** Residence timeline of greenhouse gases in the atmosphere

| S. no. | Gases | Mean residence time in the atmosphere |
|---|---|---|
| 1 | $H_2O$ vapor and ozone | Within a day or week |
| 2 | Methane | 12 years |
| 3 | $CO_2$ | 100 years |
| 4 | Nitrous oxide | 120 years |
| 5 | CFCs | Tens of thousands of years |

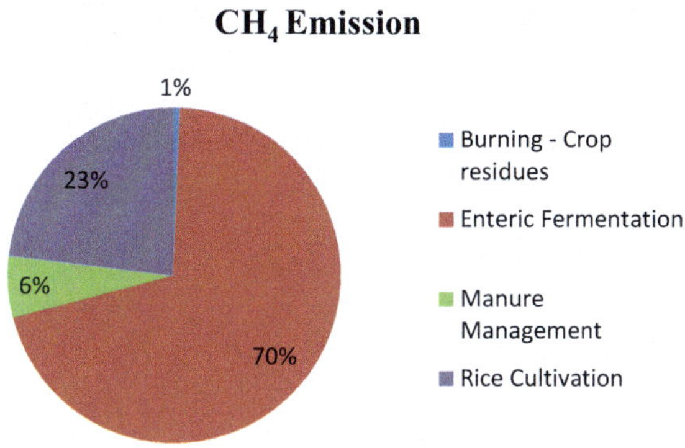


**Fig. 1** Average contribution of different agricultural activities in $CH_4$ emission during 2019 in India. (Source: FAOSTAT 2021)

## Methane Emissions

Methane is a potent greenhouse gas formed in various environments like oceans, wetlands, landfills, and rice paddy fields. Rice paddies are a significant source, accounting for ~12% (24.4 Tg/year) of global methane emissions. Figure 1 depicts the contribution of different agricultural activities to methane emissions.

## Nitrous Oxide Emissions

Nitrous oxide ($N_2O$) ranks as the third most significant long-lived greenhouse gas, following carbon dioxide and methane (UNEP 2013). Its global warming potential is a staggering 298 times higher than that of carbon dioxide (Ravishankara et al. 2009). Emissions of ammonia (NH3) and nitrogen oxides (NOx) impact ozone and particulate matter formation in the atmosphere, adversely affecting human health and terrestrial biodiversity. Furthermore, nitrogen losses to groundwater and surface water degrade drinking water quality and aquatic ecosystem biodiversity (Wim and de Vries 2021). By 2017, the atmospheric concentration of $N_2O$ had reached 330 μg $L^{-1}$, increasing at an annual rate of 0.80 μg $L^{-1}$ (NOAA 2018). Agriculture is responsible for nearly 80% of global anthropogenic $N_2O$ emissions, making it the largest human-induced source of this gas (UNEP 2019). Figure 2 illustrates the various agricultural activities contributing to $N_2O$ emissions.

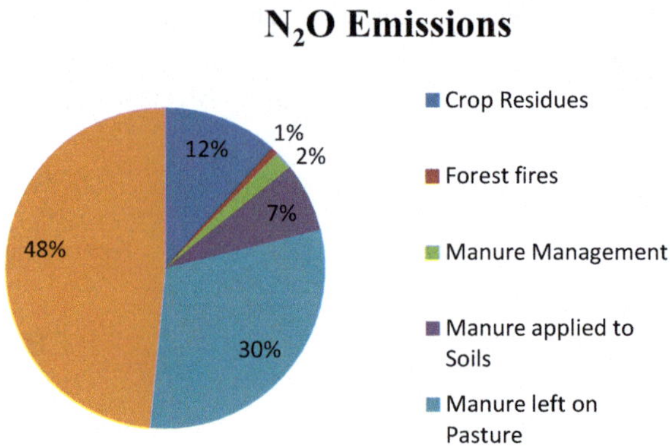


**Fig. 2** Average contribution of different agricultural activities in $N_2O$ emission during 2019 in India. (Source: FAOSTAT 2021)

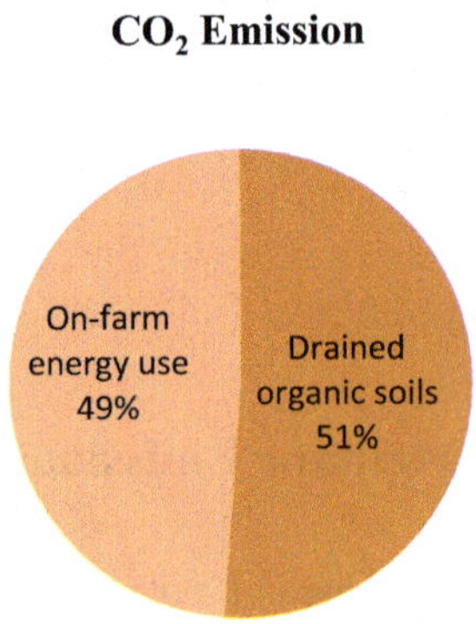


**Fig. 3** Average contribution of different agricultural activities in $CO_2$ emission during 2019 in India. (Source: FAOSTAT 2021)

## Carbon Dioxide

Concerning cumulative anthropogenic (human-caused) carbon dioxide ($CO_2$) emissions, the observed land and ocean sinks absorbed 1430 $GtCO_2$ (59% of total emissions) during the historical period from 1850 to 2019 (IPCC 2021).

Soil organic carbon (SOC) acts as both a source and sink for atmospheric $CO_2$, playing a crucial role in the global carbon cycle. Soil $CO_2$ is produced through the bacterial decomposition of complex organic compounds found in manure under aerobic conditions (Fig. 3).

## Primary Greenhouse Gas Source and Sink Aspects of Soils

Agricultural practices and other land uses have been identified as sources and sinks for the three major greenhouse gases (Mosier et al. 2002). Carbon can be sequestered in soil as inorganic carbon through the formation of secondary carbonates or exist in organic forms like humus, charcoal, or dissolved organic carbon (DOC). Total soil organic carbon content can range from less than 1% in sandy soils to over

20% in wetland or bog soils. The weathering of silicate and carbonate rocks also contributes to soil carbon sinks. Agricultural soils cover approximately 37% of the Earth's surface and store significant amounts of carbon. It is estimated that the top 1 meter of soil contains between 2000 and 2500 Gt (1 billion metric tons) of carbon, with around 60% as organic carbon and 40% as inorganic carbon

## Factors Affecting the Process of Denitrification

### *Temperature, Moisture, and Aeration*

These three factors play a crucial role in determining microbial activity, nitrous oxide ($N_2O$) production, and its eventual diffusion into the atmosphere (Davidson and Swank 1986).

As temperatures rise, the rate of $N_2O$ conversion follows an exponential increase (Hao et al. 2001), leading to seasonal fluctuations in $N_2O$ flux based on soil air temperature. Higher soil temperatures stimulate microbial respiration, resulting in more anaerobic microsites where denitrification can take place. An increase in soil moisture content also tends to enhance $N_2O$ emissions (Giacomini et al. 2006), as both nitrification and denitrification processes rely on adequate moisture levels. However, beyond a certain threshold, excessive moisture can inhibit microbial activity and consequently reduce $N_2O$ production.

Liu et al. (2006) documented significant $N_2O$ emission and diffusion into the atmosphere following rainfall events. According to Sharma and Kumar (2023), when soil aeration is poor, molecular nitrogen ($N_2$) becomes the predominant nitrogen gas released into the atmosphere. The greater the water-filled pore spaces in the soil, the fewer air-filled pores, which promotes $N_2O$ emission via denitrification. The denitrification process will lead to $N_2O$ formation if water-filled pore spaces exceed 70%, while nitrification becomes the dominant process when water-filled pore spaces decrease to 60% (Ruser et al. 2006).

## Soil Properties

### *Soil Type*

Clay soils exhibit higher $N_2O$ emission potential compared to sandy soils (Brentrup et al. 2000), and nitrogen management practices, such as fertilizer application, in clay soils may increase $N_2O$ emissions (Tan et al. 2009). This is because clay soils have fewer macropores, leading to an increase in anaerobic microsites and, consequently, higher $N_2O$ emissions. Sandy soils require higher moisture levels compared to clay soils to generate similar $N_2O$ emissions.

## *Nutrient Status*

The enzymes involved in the denitrification process require various metal cofactors, such as molybdenum (Mo), iron (Fe), copper (Cu), and zinc (Zn). Nitrous oxide reductase enzymes, specifically, require copper for reducing $N_2O$ to molecular nitrogen (N2). Therefore, higher nutrient contents in the soil can potentially lead to higher $N_2O$ emissions.

## *Soil pH*

Soil pH is an important factor because lower pH levels inhibit the activity of nitrous oxide reductase enzymes. Generally, higher pH values tend to decrease soil $N_2O$ emissions during denitrification. At a low pH of around 4.0, Knowles (1982) found that $N_2O$ is the main product of the denitrification process. However, as pH increases, $N_2O$ emissions tend to decrease. The application of acidic fertilizers can enhance $N_2O$ emissions, while liming the soil can help reduce them.

## *Organic Carbon Content*

An increase in soil organic carbon (SOC) can lead to increased $N_2O$ production, as it can stimulate microbial growth and activity. When microbial activity increases, oxygen consumption also rises, generating anaerobic conditions favorable for denitrification. Studies by Passianoto et al. (2003) observed peak $N_2O$ emissions after adding plant residues and labile carbon (dextrose) to the soil. Residues with a lower carbon-to-nitrogen ratio (C/N ratio) of less than 30:1 enhance mineralization, making available nitrogen more accessible for plant uptake.

## *Soil Tillage Practices*

Soil tillage practices affect various soil properties, including structure, aeration, microbial activity, litter decomposition rate, temperature, and moisture content. No-till cultivation practices tend to enhance $N_2O$ emissions compared to conventional tillage (Carvalho et al. 2009), due to an increased population of denitrifying microorganisms (Baggs et al. 2006).

## *Soil as a Sink for $N_2O$*

In addition to contributing to $N_2O$ emissions, soil can also absorb and act as a sink for $N_2O$. Both denitrifying and nitrifying microorganisms possess the ability to consume $N_2O$. Various factors, such as soil oxygen levels, water content, temperature, pH, redox conditions, carbon, and nitrogen availability, are responsible for net negative $N_2O$ flux in soil. Generally, factors that oppose the diffusion of $N_2O$ from soil to the atmosphere seem to increase its consumption.

Net negative $N_2O$ flux can also occur if gross consumption exceeds gross production.

However, further research is needed to fully understand the processes involved in $N_2O$ consumption by soils.

## *Factors Responsible for Carbon Emissions from Soil Include the Following*

### Deforestation

Clearing forest cover will decrease litterfall, litter decomposition, and sequestration. It will expose soil to various disturbances, and there will be less accumulation and more carbon emission. The amount of carbon release will depend upon climatic and soil conditions.

### Cultivation Practices

Carbon emissions are higher in continuous cultivation systems compared to shifting or forest cultivation. Additionally, when wetlands are converted to agricultural or pasture land, the chances of carbon emissions increase due to increased aerobic respiration. Shorter crop rotations deplete soil carbon more rapidly than longer rotations (IPCC 2000).

### Tillage Practices

Tillage makes soil more susceptible to erosion, allowing soil organic carbon (SOC) to decompose quickly due to aeration and increased soil temperature (Beare et al. 1994).

### Soil Erosion

Soil erosion removes aggregated soil particles and organic matter (IPCC 2000). However, erosion does not necessarily enhance emissions, as carbon can accumulate in another landscape.

## Potential Effects of Different Fertilizers on Soil Health and Crop Yield

Climate change has brought carbon farming into focus. Carbon farming is considered successful when carbon gains resulting from enhanced land management or conservation practices exceed carbon losses, especially to the atmosphere. Efficient land management practices and proper fertilization improve the rate at which $CO_2$ is removed from the atmosphere and converted to plant material and soil organic matter. Fertilizer application can affect soil physicochemical properties and functional microbial genes involved in $N_2O$ production through nitrification and denitrification processes (Sun et al. 2014). The amounts of fertilizer applied are driving factors in greenhouse gas emissions. Organic fertilizers have been shown to have positive impacts on soil functional microbial growth (Pereg et al. 2018).

## Effects of Different Types of Organic and Inorganic Fertilizers on Soil GHG Emissions

The use of organic and synthetic fertilizers is a primary source of greenhouse gas emissions from agricultural lands (Li et al. 2013). To alleviate GHG emissions from soil, it is crucial to apply fertilizers judiciously and conserve water resources effectively. Improving nitrogen fertilizer use efficiency through better field management practices is key to protecting crop production, minimizing GHG emissions, sustaining soil resources, and maintaining a healthy agricultural economy (Snyder et al. 2009). Mitigation strategies to combat GHG ($CO_2$ and $N_2O$) emissions while ensuring food production involve restricting nitrogen fertilizer inputs. The application of organic residues in conjunction with synthetic fertilizers has been found to contribute to higher greenhouse gas emissions (Carmo et al. 2013). The potential of *Kappaphycus* seaweed extract (KSWE) in lowering GHGs is manifested by saving at least 260 kg $CO_2$ equivalent per hectare when applied at a 5% concentration (Singh et al. 2018). The application of bio-organic fertilizers enriched with plant growth-promoting microbes (PGPMs: *Trichoderma guizhouense*) tends to decrease $N_2O$ emissions and enhance vegetable yield under protected cultivation (Geng et al. 2021).

The combined application of farmyard manure (FYM) and NPK is considered a viable option for managing soil fertility, moderating GHG emissions, and sustaining rice yield in tropical flooded soils (Bhattacharyya et al. 2013). Nitrous oxide emissions from soils have been observed to be higher in organically fertilized plots (Kaiser and Ruser 2000), most likely due to increased microbial biomass and the availability of a suitable carbon pool for mineralization.

## Relative Contributions to Global Warming Potential (GWP) of $N_2O$ and $CH_4$ from Different Inorganic Fertilizers and Organic Amendments

Methane ($CH_4$) and nitrous oxide ($N_2O$), in addition to $CO_2$, are important greenhouse gases contributing to global climate change. Each greenhouse gas has a specific global warming potential (GWP), defined as the ratio of radiative forcing from 1 kg of a gas to 1 kg of $CO_2$ over a specific time interval. The GWP of $CO_2$ is 1, while the GWP of $CH_4$ is 25, and that of $N_2O$ is 298 (Ritchie et al. 2020) (Table 2).

In the face of intertwined food security issues and global climate change, greenhouse gas (GHG) emissions are increasingly being assessed in relation to crop productivity.

The concept of yield-scaled global warming potential (GWPY) (Van Groenigen et al. 2010), also referred to as GHG intensity (Mosier et al. 2006), serves as a metric to identify efficient cropping systems that can deliver high grain yields while maintaining low global warming potential values. Recent meta-analysis findings suggest that intensive cropping systems managed near their yield potential with high nitrogen use efficiency exhibited low yield-scaled $N_2O$ emissions, and significant increases in $N_2O$ emissions were observed only when fertilizer nitrogen inputs exceeded crop demand (Van Groenigen et al. 2010). When evaluating mitigation strategies and extrapolating emission estimates to regional and national scales, it is essential to consider the interplay between greenhouse gases and productivity levels. The application of cow manure in rice paddy fields can stimulate methane production (Kongchum 2005) due to the persistent reducing conditions in the soils and the presence of organic substrates. Lower methane fluxes are typically observed in fields where less organic manure is applied, with fewer rice residues, multiple aeration periods, poor soil fertility, and low fertilization rates, resulting in poor rice yields (Delwiche and Cicerone 1993). While

**Table 2** Greenhouse gases and their respective GWP values

| S. no. | Gases | Relative GWP/$CO_2$ |
|---|---|---|
| 1 | $CO_2$ | 1 |
| 2 | $CH_4$ | 25 |
| 3 | $N_2O$ | 298 |
| 4 | CFCs | 120–14,800 |

some studies have reported that the addition of nitrogen fertilizers can lead to an increase (Shang et al. 2011) or decrease (Yao et al. 2012) in methane emissions, others have found no effect (Cai et al. 2007).

However, Linquist et al. (2012) discovered that the effect of nitrogen is primarily rate-dependent, with low to moderate rates increasing methane emissions and excessive rates reducing them. Applying organic manure increases soil moisture content, which can enhance nitrogen mineralization, nitrification, and denitrification processes, ultimately influencing $N_2O$ production in the soil (Linn and Doran 1984).

Nitrous oxide emissions from applied fertilizers and manures can vary depending on different environmental factors (e.g., climate and soil conditions), crop factors (e.g., crop type and crop residues), and management practices (e.g., type of manure and fertilizer, application rate, timing of application) (Jan et al. 2011). The most critical factors governing $N_2O$ and methane production are soil redox potential (Eh) and microbial activity. Nitrous oxide can be produced from nitrification under aerobic conditions (high Eh) and denitrification (reduction of nitrate to $N_2O$ and $N_2$) under moderately reducing conditions (lower Eh) where the reduction intensity is not strong enough to reduce nitrate to nitrogen gas ultimately. Significant methane formation generally occurs under strictly reducing conditions (low Eh) where obligate anaerobes thrive and function.

Reduced flooding duration increases $N_2O$ production, whereas continuously flooded soils maintain anaerobic conditions that promote methane production (Neue 1993). Consequently, the cumulative global warming potential in $CO_2$ equivalent (IPCC 2001) from methane and $N_2O$ emissions may not be significantly reduced or may even increase.

High $N_2O$ emissions have been reported in fields subjected to aeration events (e.g., intermittent irrigation or midseason drainage periods) (Zou et al. 2005) or excessive amounts of nitrogen fertilizer (Cai 1997). This is because ammonium ($NH_4^+$) and nitrate ($NO_3^-$) are made available from fertilization and the temporary oxidizing conditions enable the nitrification process (conversion of $NH_4^+$ to $NO_3^-$) to occur (Zhang et al. 2019). The application of mineral nitrogen increases the substrate for nitrification and denitrification processes, which are the primary drivers of $N_2O$ production (Dobbie et al. 1999). Reducing anthropogenic emissions of $N_2O$ and methane while maintaining rice yields through efficient nitrogen fertilizer management remains a challenge for rice production. Balanced fertilizer application can increase crop yield and biomass, leading to enhanced soil carbon sequestration due to the input of biomass into the soil from crop residues and roots, and it can reduce $N_2O$ emissions compared to the sole application of fertilizer (Six et al. 2004). Mukumbuta et al. (2017) studied the effect of organic manure on methane and $N_2O$ emissions and found that while $CO_2$ emissions increased due to manure application, there was no difference in methane and $N_2O$ emissions, and the carbon input from manure application actually reduced the global warming potential. A study conducted by Shimizu et al. (2013) on grassland observed higher $N_2O$ emissions from the combined application of manure and chemical fertilizers compared to the application of chemical fertilizers alone. The application of two types of organic manure, i.e., solid manure and slurry, in grasslands showed no difference in methane and $N_2O$ emissions from the soil (Mori and Hojito 2015).

Some studies have found that soil typically acts as a sink for methane, but methane uptake decreases with the application of nitrogen (Chen et al. 2021). Different organic manures exhibit varying physicochemical characteristics (Mori and Hojto 2015). In the case of methane fermentation, digestive fluid tends to have a higher pH and ammonium nitrogen concentration than slurries. Hence, applying digestive fluid as manure can enhance soil nutrient content and decrease $N_2$ emissions by increasing soil pH (Mukumberta et al. 2017a). However, compared to slurry, digestive fluid had a lower carbon-to-nitrogen (C/N) ratio (Holly et al. 2017), resulting in an increase in $N_2O$ emissions.

## Methods to Study GHGs Emissions and Underlying Processes in Soils

### *Closed Chamber Technique for $N_2O$ Measurement*

- To quantify $N_2O$ exchanges at the boundary between soil–water and the atmosphere, the closed chamber technique, alongside a photoacoustic infrared gas analyzer, is deployed. PVC tubes, measuring 20 cm in length and 19.5 cm in diameter, are embedded 6 cm into the soil a day before conducting measurements.
- On the day of measurement, these tubes are linked to a chamber that is connected via tubing to the gas analyzer, which then assesses the $N_2O$ levels within the chamber after predetermined intervals.
- The accuracy of a single $N_2O$ measurement is noted to be within ±32 ppb.
- $N_2O$ fluxes are derived from the observed linear changes in concentration within the chamber, factoring in the entire volume of the gas analysis setup, including the chamber, the length of the Teflon tubes, and the $CO_2$ and $H_2O$ traps.
- Cumulative $N_2O$ emissions are determined by applying Tilsner et al.'s (2003) method, which involves calculating the $N_2O$ emission rates for consecutive days and accumulating these over the period of the study.
- Similarly, $CH_4$ flux measurements are conducted using the closed chamber method, but $CH_4$ concentrations are ascertained in the lab through FID gas chromatography after sample collection in the field.

## *DNDC Model for GHG Emission Simulation*

- The denitrification–decomposition (DNDC) model is an intricate process-based framework constructed on biogeochemical principles for simulating greenhouse gas emissions.
- At its heart, the DNDC incorporates a soil biogeochemistry model that utilizes the Nernst and Michaelis–Menten equations to monitor microbial behavior.
- Utilizing the Nernst equation, the DNDC model calculates the soil's redox potential (Eh) by considering the concentrations of oxygen, nitrate, and other key electron acceptors. Integrating this with a comprehensive algorithm that includes soil climate, plant growth, and various agricultural management practices—such as tillage, fertilization, manure application, irrigation, and grazing—DNDC can project GHG emissions across a variety of climatic conditions, soil types, and management practices.
- The effectiveness and reliability of the DNDC model have been corroborated by global research, showcasing its utility in understanding and predicting GHG emissions under diverse conditions.

## *Micro-meteorological Measurements*

- Micro-meteorological methods for gauging greenhouse gas (GHG) exchanges are more apt for larger agricultural operations due to the complexity and diversity of smaller farms.
- These smaller plots often feature varied land use and management practices, subtle differences in soil fertility across short distances, and complex cropping patterns, including intercropping, which complicates measurement (Chikowo et al. 2014).
- The toolkit for micro-meteorological GHG flux assessments includes sensors for three-dimensional wind fields and rapid-response gas analyzers, along with additional equipment such as meteorological stations and masts.
- These instruments are essential for accurately measuring fluxes of carbon dioxide ($CO_2$) and energy, methane ($CH_4$)—for which open-path sensors are utilized—and nitrous oxide ($N_2O$), which requires laser spectroscopy for measurement. The financial outlay for such a comprehensive setup is significant, rendering it impractical for smaller farming operations.
- Moreover, the requirement for a reliable energy source to power the instruments presents another layer of complexity.

### *Effects of Different Fertilizers on Nitrifying and Denitrifying Soil Microorganisms*

- The impact of nitrogen (N) or potassium (K) fertilizers on $N_2O$ emissions is significantly influenced by their effect on the population of denitrifying microbes in the soil–plant ecosystem. Notably, the presence of specific bacterial genera such as *Paracoccus*, *Rubrivivax*, *Geobacter*, *Streptomyces*, and *Hyphomicrobium* is pivotal. Their crucial role in $N_2O$ emissions becomes particularly evident during denitrification processes when N and K fertilizers are used together (Li et al. 2021). Additionally, the strategic use of micronutrient fertilizers can alter soil greenhouse gas (GHG) fluxes. The employment of certain synthetic chelates emerges as a beneficial approach, simultaneously reducing $N_2O$ emissions and enhancing zinc (Zn) biofortification. This dual benefit is attributed to the chelation of metal cofactors, mainly copper (Cu) and Zn, which are essential for the enzymatic activities involved in nitrification and denitrification, thereby mitigating $N_2O$ release. Moreover, the application of Zn in conjunction with organic substances like humic and fulvic acids has been observed to stimulate the activity of nitrifying and denitrifying microbial communities, consequently increasing $N_2O$ fluxes (Montoya et al. 2018).
- Transitioning land use from rice paddies to upland fields also introduces a shift in microbial community dynamics due to changes in environmental conditions such as oxygen availability, ammonia concentration, and pH levels. These factors are critical in shaping the community structure of nitrifiers and denitrifiers, affecting the overall GHG emissions from soil (Yang et al. 2016).

## Management Strategies to Reduce $N_2O$ Emissions from Soils Amended with Manure-Based Organic Fertilizers

- Organic fertilizers are praised for their ability to enhance soil nutrient levels, microbial activity, and diversity. They supply a broader array of compounds, including humic substances and carbon, which are not typically found in chemical fertilizers (Yuan et al. 2017).
- The production of nitrous oxide ($N_2O$) after applying manure can occur directly through the processes of nitrification and denitrification or indirectly as nitrogen is lost via ammonia volatilization or nitrate leaching and then transformed into $N_2O$ (Sun et al. 2014). It has been observed that soils treated with manure tend to emit higher levels of $N_2O$ compared to those that have not received manure applications (Graham et al. 2013).

## Critical Options for Improving the Carbon Footprint of Organic and Conventional Cropping Systems and Crop Products

- The concept of a carbon footprint refers to the quantification of the total $CO_2$ emissions produced by the actions of an individual. This metric allows for the assessment of carbon emissions from specific sources. A key strategy within conservation agriculture (CA) is the combination of zero-tillage (ZT) and crop residue retention (CRR) or residue mulching. This approach reduces the need for fuel and energy in tillage operations, thereby decreasing $CO_2$ emissions and lowering the overall carbon footprint (Zhanga et al. 2016).

## Role of Integrated Nutrient Management (Mixtures of Organic and Inorganic Fertilizers) Practices in Mitigating GHG Emissions from Soils

- Certain land management methods that increase soil carbon storage also lead to higher $N_2O$ emissions, ultimately making these practices net contributors to global warming rather than mitigating greenhouse gas emissions (Li et al. 2005). However, adopting integrated organic farming can significantly cut down on greenhouse gas (GHG) emissions and other adverse environmental effects associated with agriculture. Strategies such as reducing tillage and enhancing mineral fertilizers with leguminous cover crops are widely recognized for their potential to elevate the environmental sustainability of agricultural systems (Hoffman et al. 2018). Moreover, employing biofertilizers presents a novel approach to encourage methane-oxidizing bacteria within soil microbial communities.
- While the use of organic residues can boost rice yield, it may also lead to increased $CH_4$ emissions. To both enhance soil and rice productivity and reduce $CH_4$ emissions, a combination of green manure and 50% rice straw is preferred over the use of 100% rice straw (Zhou et al. 2020). Additionally, biochar, a sustainable material produced from green waste, shows promise in lowering $CH_4$ emissions in rice paddy soils (Fu et al. 2021) by promoting the growth of both methanogens and methanotrophs, yet simultaneously inhibiting methanogens (Wang et al. 2019).

## Mitigation and Management Practices

- Promoting energy efficiency and reducing greenhouse gas (GHG) emissions are critical objectives for the agricultural sector (USDA 2015). Various strategies exist to mitigate GHG emissions in agriculture, including crop rotations, efficient manure management for nutrient recycling, improving livestock productivity, enhancing pasture and fodder quality, and maintaining soil fertility (Gerber et al. 2013). Notably, the production and application of fertilizers contribute significantly to energy consumption and GHG emissions. Implementing efficient nutrient management practices such as precision nutrient application, integrated nutrient management, and the use of nano-fertilizers can help reduce emissions associated with fertilizer usage.
- Additionally, managing nitrogen sources through inhibitors or controlled-release fertilizers, along with optimizing application timing and rates, can mitigate GHG emissions per unit area. Tillage practices in agriculture are significant contributors to GHG emissions, particularly conventional tillage methods, which result in elevated $CO_2$ and $N_2O$ losses (Smith et al. 2012). Transitioning to no-tillage systems proves to be an effective strategy for reducing GHG emissions from rice fields and contributing to global warming mitigation (Ahmad et al. 2009). Organic farming and conservation tillage also play roles in lowering GHG emissions.
- In rice paddy fields, various soil amendments such as rice straw, slag, industrial or agricultural wastes, and biochar are applied to regulate $CH_4$ emissions (Fu et al. 2021). Furthermore, adopting urea application based on leaf color chart (LCC) assessments can significantly reduce the global warming potential (GWP) of rice–wheat systems compared to blanket application methods (Bhatiya et al. 2012). These diverse strategies underscore the importance of sustainable agricultural practices in addressing the challenges posed by climate change.

## Conclusion

Excessive application of nitrogen (N) leading to potential $NO_3$–N leaching poses a dual threat: groundwater pollution and indirect contributions to greenhouse gas (GHG) emissions. The increased global warming associated with GHG emissions can disrupt the carbon and nitrogen cycles globally, altering the structure and functions of ecosystems. Implementing measures such as carbon sequestration and efficient nutrient management not only reduces GHG emissions but also helps maintain the sustainability of agroecosystems. These strategies play a crucial role in mitigating the impacts of excessive nitrogen application on both environmental and agricultural systems.

## References


Ahmad S, Li C, Dai G, Zhan M, Wang J, Pan S, Cao C (2009) Greenhouse gas emission from direct seeding paddy field under different rice tillage systems in central China. Soil Tillage Res 106(1):54–61

Al-Amin AQ, Rasiah R, Chenayah S (2015) Prioritizing climate change mitigation: an assessment using Malaysia to reduce carbon emissions in future. Environ Sci Pol 50:24–33. https://doi.org/10.1016/j.envsci.2015.02.002

Baggs EM, Chebii J, Ndufa JK (2006) A short-term investigation of trace gas emissions following tillage and no-tillage of agroforestry residues in western Kenya. Soil Tillage Res 90(1–2):69–76

Beare MH, Cabrera ML, Hendrix PF, Coleman DC (1994) Aggregate-protected and unprotected organic matter pools in conventional and no-tillage soils. Soil Sci Soc Am J 58:787–795

Bhattacharyya P, Nayak AK, Mohanty S, Tripathi R, Shahid M, Kumar A, Raja R, Panda BB, Roy KS, Neogi S, Dash PK, Shukla AK, Rao KS (2013) Greenhouse gas emission in relation to labile soil C, N pools and functional microbial diversity as influenced by 39 years long-term fertilizer management in tropical rice. Soil Tillage Res 129:93–105

Brentrup F, Küsters J, Lammel J, Kuhlmann H (2000) Methods to estimate on-field nitrogen emissions from crop production as an input to LCA studies in the agricultural sector. Int J Life Cycle Assess 5(6):349–357

Cai ZC (1997) A category for estimation of $CH_4$ emission from rice paddy fields in China. Nutr Cycl Agroecosyst 49:171–179

Cai Z, Shan Y, Xu H (2007) Effects of nitrogen fertilization on CH4 emissions from rice fields. Soil Sci and Plant Nut. 53:353–361

Carmo JB et al (2013) Infield greenhouse gas emissions from sugarcane soils in Brazil: effects from synthetic and organic fertilizer application and crop trash accumulation. GCB-Bioenergy 5(3):267–280. https://doi.org/10.1111/j.1757-1707.2012.01199.x

Carvalho JLN et al (2009) Conversion of Cerrado into agricultural land in the southwestern Amazon: carbon stocks and soil fertility. Sci Agric (Piracicaba, Braz) 66(2):233–241

CDP (2020) Building Back Greener, India Inc. Demonstrates Climate Resilience, India Annual Report

Chen J, Feng M, Cui Y, Liu G (2021) The impacts of nitrogen addition on upland soil methane uptake: A global metaanalysis. Science of The Total Environment 795:148863. https://doi.org/10.1016/j.scitotenv.2021.148863

Chikowo R, Zingore S, Nyamangara J, Bekunda M, Messina J, Snapp S (2014) Approaches to reinforce crop productivity under rain-fed conditions in sub-humid environments in Sub Saharan Africa. In: Rattan L, Singh BR, Mwaseba DL, Kraybill D, Hansen DO, Eik LO (eds) Sustainable intensification to advance food security and enhance climate resilience in Africa. Springer, pp 235–253. https://doi.org/10.1007/978-3-319-09360-4_12

Davidson EA, Swank WT (1986) Environmental parameters regulating gaseous nitrogen losses from two forested ecosystems via nitrification and denitrification. Appl Environ Microbiol 52(6):1287–1292

de Vries W (2021) Impacts of nitrogen emissions on ecosystems and human health: a mini-review. Curr Opin Environ Sci Health 21:100249. https://doi.org/10.1016/j.coesh.2021.100249

Delwiche CC, Cicerone RJ (1993) Factors affecting methane production under rice. Global Biogeochem Cycles 7(1):143–155

Dobbie KE, McTaggart IP, Smith KA (1999) Nitrous oxide emissions from intensive agricultural systems: variations between crops and seasons, key driving variables, and mean emission factors. J Geophys Res 104:26891–26899

FAOSTAT (2021) Emissions –agriculture. Rome. http://faostat3.fao.org/browse/G1/*/E

Fu L, Lu Y, Tang L, Hu Y, Xie Q, Zhong L, Fan C, Liu Q, Zhang S (2021) Dynamics of methane emission and archaeal microbial community in paddy soil amended with different types of biochar. Appl Soil Ecol 162:103892

Geng YJ, Yuan YM, Miao YC, Zhi JZ, Huang MY, Zhang YH, Wang H, Shen QR, Zou JW, Li SQ (2021) Decreased nitrous oxide emissions associated with functional microbial genes under bio-organic fertilizer application in vegetable fields. Pedosphere 31(2):279–288
Gerber PJ, Steinfeld H, Henderson B, Mottet A, Opio C, Dijkman J, Falcucci A, Tempio G (2013) Tackling climate change through livestock – A global assessment of emissions and mitigation opportunities. FAO, Rome. https://www.fao.org/3/i3437e/i3437e.pdf
Giacomini SJ, Jantalia CP, Aita C, Urquiaga SS, Alves BJR (2006) Nitrous oxide emissions following pig slurry application in soil under no-tillage system. Pesquisa Agropecuaria Brasileira 41(11):1653–1661
Graham CJ, Van Es HM, Melkonian JJ (2013) Nitrous oxide emissions are greater in silt loam soils with a legacy of manure application than without. Biol Fertil Soils 49(8):1123–1129
Hao X, Chang C, Carefoot JM, Janzen HH, Ellert BH (2001) Nitrous oxide emissions from an irrigated soil as affected by fertilizer and straw management. Nutr Cycl Agroecosyst 60(1):1–8
Hoffman E, Cavigelli MA, Camargo G, Ryan M, Ackroyd VJ, Richard TL, Mirsky S (2018) Energy use and greenhouse gas emissions in organic and conventional grain crop production: accounting for nutrient inflows. Agric Syst 162:89–96
Holly MA, Larson RA, Mark Powell J, Ruark MD, Aguirre-Villegas H (2017) Greenhouse gas and ammonia emissions from digested and separated dairy manure during storage and after land application. Agric Ecosyst Environ 239:410–419
IPCC (2000) Land use, land-use change, and forestry special report. Cambridge University Press, New York
IPCC (2001) The Third Assessment Report, Climate Change 2001. Cambridge Univ. Press, New York
IPCC (2013) Climate change 2013: the physical science basis. Cambridge University Press
IPCC (2021) Summary for policymakers. In: Climate change 2021: the physical science basis. Contribution of working group I to the sixth assessment report of the Intergovernmental Panel on Climate Change [Masson-Delmotte V P et al (eds)]. Cambridge University Press, Cambridge, United Kingdom and New York, NY, USA. p 3–32. https://doi.org/10.1017/9781009157896.001
Jan PL, Gerard LV, Wim de V, Johannes K (2011) Differentiation of nitrous oxide emission factors for agricultural soils. Environ Pollut 159:3215–3222. https://doi.org/10.1016/j.envpol.2011.04.001
Kaiser EA, Ruser R (2000) N2O emissions from arable soils in Germany—an evaluation of six long-term experiments. J Plant Nutr Soil Sci 163:249–260. https://doi.org/10.1002/1522-2624(200006)163:33.0.CO;2-Z
Knowles R (1982) Denitrification. Microbiol Rev 46(1):43–70
Kongchum M (2005) Effect of plant residue and water management practices on soil redox chemistry, methane emission, and rice productivity. A Dissertation, he Graduate Faculty of the Louisiana State University and Agricultural and Mechanical College
Li C, Frolking S, Butterbach-Bahl K (2005) Carbon sequestration in arable soils is likely to increase nitrous oxide emissions, offsetting reductions in climate radiative forcing. Clim Change 72:321–338
Li D, Watson CJ, Yan MJ, Lalor S, Rafique R, Hyde B, Lanigan G, Richards KG, Holden NM, Humphreys J (2013) A review of nitrous oxide mitigation by farm nitrogen management in temperate grassland-based agriculture. J Environ Manag 128:893–903
Li Z, Li L, Xia S, Zhang R, Zhang R, Chen P, Pan J, Yi L (2021) K fertilizer alleviates $N_2O$ emissions by regulating the abundance of nitrifying and denitrifying microbial communities in the soil-plant system. J Environ Manag 291:112579
Linn DM, Doran JW (1984) Effect of water-filled pore space on carbon dioxide and nitrous oxide production in tilled and non-tilled soils. Soil Sci Soc Am J 48:1267–1272
Linquist BA, Adviento-Borbe MA, Pittelkow CM, van Kessel C, van Groenigen KJ (2012) Fertilizer management practices and greenhouse gas emissions from rice systems: A quantitative review and analysis. Field Crops Res 135:10–21

Liu XJ, Mosier AR, Halvorson AD, Zhang FS (2006) The impact of nitrogen placement and tillage on NO, N 2 O, CH 4 and CO 2 fluxes from a clay loam soil. Plant Soil 280(1):177–188

Montoya M, Castellano-Hinojosa A, Vallejo A, Álvarez JM, Bedmar EJ, Recio J, Guardia G (2018) Zinc fertilizers influence greenhouse gas emissions and nitrifying and denitrifying communities in a non-irrigated arable cropland. Geoderma 325:208–217

Mori A, Hojito M (2015) Effect of dairy manure type on the carbon balance of mowed grassland in Nasu, Japan: comparison between manure slurry plus synthetic fertilizer plots and farmyard manure plus synthetic fertilizer plots. Soil Sci Plant Nutr 61:736–746

Mosier AR, Morgan JA, King JY, Le Cain D, Milchunas DG (2002) Soil-atmosphere exchange of CH4, CO2, NOx and N2O in the Colorado shortgrass steppe under elevated CO2. Plant Soil 240:201–211

Mosier AR, Halvorson AD, Reule CA, Liu XJ (2006) Net Global Warming Potential and Greenhouse Gas Intensity in Irrigated Cropping Systems in Northeastern Colorado. J Env Quality 35(4):1584. https://doi.org/10.2134/jeq2005.0232

Mukumbuta I, Shimizu M, Hatano R (2017) Mitigating global warming potential and greenhouse gas intensities by applying composted manure in cornfield: a 3-year field study in an Andosol soil. Agriculture 7:1–20. https://doi.org/10.3390/agriculture7020013

Nabavi-Pelesaraei A, Abdi R, Rafiee S (2016) Neural network modeling of energy use and greenhouse gas emissions of watermelon production systems. J Saudi Soc Agric Sci 15(1):38–47. https://doi.org/10.1016/j.jssas.2014.05.001

National Oceanic and Atmospheric Administration (NOAA) (2018) Nitrous oxide ($N_2O$)—combined data set. Available online at https://www.esrl.noaa.gov/gmd/hats/combined/N$_2$O.html (verified on May10, 2018)

Neue H (1993) Methane emission from rice fields: wetland rice fields may make a major contribution to global warming. Bioscience 43(7):466–473

Passianoto CC, Ahrens T, Feigl BJ, Steudler PA, Do Carmo JB, Melillo JM (2003) Emissions of CO 2, N 2 O, and NO in conventional and no-till management practices in Rondônia, Brazil. Biol Fertil Soils 38(4):200–208

Pathak H, Wassmann R (2007) Introducing greenhouse gas mitigation as a development objective in rice-based agriculture: I. Generation of technical coefficients. Agric Syst 94(3):807e25. https://doi.org/10.1016/j.agsy.2006.11.015

Pereg L, Morugán-Coronado A, McMillan M, García-Orenes F (2018) Restoration of nitrogen cycling community in grapevine soil by a decade of organic fertilization. Soil Till Res 179:11–19

Ravishankara AR, Daniel JS, Portmann RW (2009) Nitrous oxide ($N_2O$): the dominant ozone-depleting substance emitted in the 21st century. Science 326:123–125

Ruser R, Flessa H, Russow R, Schmidt G, Buegger F, Munch JC (2006) Emission of $N_2O$, $N_2$ and $CO_2$ from soil fertilized with nitrate: effect of compaction, soil moisture and rewetting. Soil Biol Biochem 38(2):263–274

Ritchie H, Rosado P, Roser M (2020) Greenhouse gas emissions. Our World in Data. https://ourworldindata.org/greenhouse-gas-emissions

Shang QY, Yang XX, Gao CM, Wu PP, Liu JJ, Xu YC, Shen QR, Zou JW, Guo SW (2011) Net annual global warming potential and greenhouse gas intensity in Chinese double rice-cropping systems: A 3-year field measurement in long-term fertilizer experiment. Global Change Bio 17:2196–2210

Sharma PK, Kumar S (2023) Soil Air and Plant Growth. In: Soil Physical Environment and Plant Growth. Springer, Cham. https://doi.org/10.1007/978-3-031-28057-3_6 Download citation

Shimizu M, Hatano R, Arita T, Kouda Y, Mori A, Matsuura S, Niimi M, Jin T, Desyatkin AR, Kawamura O (2013) The effect of fertilizer and manure application on $CH_4$ and $N_2O$ emissions from managed grasslands in Japan. Soil Sci Plant Nutr 59:69–86

Singh I, Gopalakrishnan VAK, Solomon S, Shukla SK, Rai R, Zodape ST, Ghosh A (2018) Can we not mitigate climate change using seaweed based biostimulant: a case study with sugarcane cultivation in India. J Clean Prod. https://doi.org/10.1016/j.jclepro.2018.09.070

Six J, Ogle ST, Breidt FJ, Contant RT, Mosier AR, Paustian K (2004) The potential to mitigate global warming with no-tillage management is only realized when practised in the long term. Glob Chang Biol 10:155–160

Smith K, Watts D, Way T, Torbert H, Prior S (2012) Impact of tillage and fertilizer application method on gas emissions in a corn cropping system. Pedosphere 22(5):604–615

Snyder CS, Bruulsema TW, Jensen TL, Fixen PE (2009) Review of greenhouse gas emissions from crop production systems and fertilizer management effects. Agric Ecosyst Environ 133:247–266. https://doi.org/10.1016/j.agee.2009.04.021

Sun F, Harrison JH, Ndegwa PM, Johnson K (2014) Effect of manure treatment on ammonia and greenhouse gases emissions following surface application. Water Air Soil Pollut 225(4):1923–1946

Tan IY, van Es HM, Duxbury JM, Melkonian JJ, Schindelbeck RR, Geohring LD, Hively WD, Moebius BN (2009) Single-event nitrous oxide losses under maize production as affected by soil type, tillage, rotation, and fertilization. Soil Tillage Res 102(1):19–26

Tilsner J, Wrage N, Lauf J, Gebauer G (2003) Emission of gaseous nitrogen oxides from an extensively managed grassland in NE Bavaria, Germany I. Annual budgets of $N_2O$ and NOx emissions. Biogeochemistry 63:229–247

UNEP (2013) Drawing down $N_2O$ to protect climate and the ozone layer. A UNEP synthesis. United Nations Environment Programme (UNEP), Nairobi

United Nations Environment Programme (UNEP) (2019) UN Environment releases frontiers 2018/19: emerging issues of environmental concern. UNEP, Nairobi

USDA (2015) COMET-Farm. U.S. Department of Agriculture, Natural Resources Conservation Service. Prepared by Colorado State University, Fort Collins, CO. [Available at http://comet-farm.nrel.colostate.edu]

Van Groenigen JW, Velthof GL, Oenema O, Van Groenigen KJ, Van Kessel C (2010) Towards an agronomic assessment of $N_2O$ emissions: a case study for arable crops. Eur J Soil Sci 61:903–913

Wang C, Shen J, Liu J, Qin H, Yuan Q, Fan F, Hu Y, Wang J, Wei W, Li Y, Wu J (2019) Microbial mechanisms in the reduction of $CH_4$ emission from double rice cropping system amended by biochar: a four-year study. Soil Biol Biochem 135:251–263

Yang H, Sheng R, Zhang Z, Wang L, Wang Q, Wei W (2016) Responses of nitrifying and denitrifying bacteria to flooding-drying cycles in flooded rice soil. Appl Soil Ecol 103:101–109

Yuan J, Sha Z-M, Hassani D, Zhao Z, Cao L-K (2017) Assessing environmental impacts of organic and inorganic fertilizer on daily and seasonal greenhouse gases effluxes in rice field. Atmos Environ 155:119–128. https://doi.org/10.1016/j.atmosenv.2017.02.007

Yao Z, Zheng X, Dong H, Wang R, Mei B, Zhu J (2012) A 3-year record of N2O and CH4 emissions from a sandy loam paddy during rice seasons as affected by different nitrogen application rates. Agri Ecosystems Env 152:1–9

Zhanga XQB, Pua C, Zhaoa X, Xuea JF, Zhanga R, Niec ZJ (2016) Tillage effects on carbon footprint and ecosystem services of climate regulation in a winter wheat-summer maize cropping system of the North China plain. Ecol Ind 67:821–829

Zhou G, Gao S, Xu C, Dou F, Shimizu K, Cao W (2020) Rational utilization of leguminous green manure to mitigate methane emissions by influencing methanogenic and methanotrophic communities. Geoderma 361:114071

Zou JW, Huang Y, Jiang JY, Zheng XH, Sass RL (2005) A 3 year field measurement of methane and nitrous oxide emissions from rice paddies in China: effects of water regime, crop residue, and fertilizer application. Global Biogeochem Cycles 19:GB2021

Zhang K, Wu Y, Hang H (2019) Differential contributions of NO3−/NH4+ to nitrogen use in response to a variable inorganic nitrogen supply in plantlets of two Brassicaceae species in vitro. Plant Methods 15:86. https://doi.org/10.1186/s13007-019-0473-1

# Greenhouse Gas Emissions from Agroecosystems Under Conservation Agriculture

**M. Homeshwari Devi, Immanuel C. Haokip, Rakesh Parmar, S. Kalidas-Singh, and M. Neshwari Devi**

## Introduction

The Industrial Revolution of the eighteenth century led to a continual increase in the atmospheric concentration of greenhouse gases (GHG). The Intergovernmental Panel on Climate Change (IPCC) (2023) estimates that the present atmospheric levels of carbon dioxide ($CO_2$), methane ($CH_4$), and nitrous oxide ($N_2O$) are, respectively, 50%, 173%, and 24.4% greater than preindustrial levels. The global net anthropogenic GHG emissions include $CO_2$ from fossil fuel combustion and industrial processes (64%); net $CO_2$ from land use, land use change, and forestry (11%); $CH_4$ (18%); $N_2O$ (4%); and fluorinated gases (2%) (IPCC 2023). Although $CH_4$ and $N_2O$ are emitted in lower quantities than $CO_2$, their threat lies in that their global warming potential (GWP) is 21 and 310 times greater than $CO_2$, respectively (Thangarajan et al. 2013). In 2019, the energy sector accounted for over 34% (20 Gt $CO_2$e) of all net anthropogenic GHG emissions, followed by industry at 24% (14 Gt $CO_2$e); agriculture, forestry, and other land use at 22% (13 Gt $CO_2$e); transportation at 15% (8.7 Gt $CO_2$e); and buildings at 6% (3.3 Gt $CO_2$e). Approximately 14%, 41%, and 69% of the total net anthropogenic emissions of $CO_2$, $CH_4$, and $N_2O$, respectively, are attributed to agriculture, forestry, and other land uses (Nabuurs et al. 2023).

The agriculture-related activities that contributed to GHG emissions include changes in land use patterns and forest cover, biomass burning, animal manure and

M. H. Devi (✉) · I. C. Haokip · R. Parmar
ICAR-Indian Institute of Soil Science, Bhopal, MP, India

S. Kalidas-Singh
Department of SSAC, Rajiv Gandhi University,
Rono Hills, Doimukh, Arunachal Pradesh, India

M. N. Devi
Department of Chemistry, Waikhom Mani Girls' College, Thoubal, Manipur, India

 
S. R. Mohanty, B. Kollah (eds.), *Greenhouse Gas Regulating Microorganisms in Soil Ecosystems*, https://doi.org/10.1007/978-3-031-70569-4_5

enteric fermentation in ruminants, lowland rice farming, chemical fertilizers, and different soil management techniques (Beach et al. 2015). In recent years, the growing public awareness of the agricultural sector's role in climate change has sparked interest in maximizing agricultural operations' ability to reduce GHG emissions. Potential agricultural methods that could reduce GHG emissions comprise restoring degraded areas, managing cropland and grazing land better, enhancing soil carbon (C) storage, managing animals well, and increasing the effectiveness of fertilizers and manure, among other things. Apart from this, proper management of tillage and crop residues could be a great mitigation strategy. Accordingly, conservation agriculture (CA), a production system based on minimum mechanical soil disturbance such as the absence of tillage and reduced tillage, maintaining continuous covering of soil with living (cover crop) or dead plant materials (mulch), and crop species diversification through rotation or intercropping can help mitigate GHG emission from soil and has been adopted worldwide. In 2016, about 12.5% (180 M ha) of the total global cropland was under CA, which expanded to 14.7% (205.4 M ha) at present (Kassam et al. 2018, 2022). The increase in area under CA is a testament to its potential to improve crop productivity and soil health, reduce soil degradation, and improve soil C sequestration (Pittelkow et al. 2015). Also, CA has gained significant attention from policymakers, researchers, and farmers due to its potential to mitigate GHG emissions from agroecosystems in recent years.

The CA systems also encourage lowering the consumption of inputs derived from fossil fuels and improve fertilizer efficiency. Nonetheless, little is still known about how well CA works to lower GHG emissions from agroecosystems. This is caused in part by the intricate relationships that exist between the many elements of the agroecosystem as well as the absence of reliable monitoring and measurement methods. The purpose of this chapter is to give a summary of GHG emissions from agroecosystems vis-à-vis CA systems. It will look at how well CA works to cut GHG emissions from a variety of agricultural systems and investigate emissions under various CA management situations.

## $CO_2$ Emissions from Agroecosystems

Agroecosystems, which combine crops and/or livestock in a particular area, are a net emitter as well as a sink of atmospheric $CO_2$. Currently, agriculture-related emissions of $CO_2$ account for around 14% of global GHG emissions (Nabuurs et al. 2023). Agroecosystems release $CO_2$ into the atmosphere through a variety of processes, such as soil respiration, crop residue burning and decomposition, the use of chemical fertilizers and organic amendments, and the operation of farm machinery powered by fossil fuels. The primary process in soil that releases $CO_2$ is root and microbial respiration which occurs as a result of crop residue and soil organic matter (SOM) decomposition (Davidson et al. 2000a, b; Schlesinger and Andrews 2000). Approximately 99% of the $CO_2$ released during the breakdown of SOM is contributed by soil microorganisms (Rahman 2013). Burning agricultural residue releases enormous amounts of $CO_2$, $CH_4$, and $N_2O$ into the atmosphere in addition to aerosols, trace gases, and particulate matter, even though it is a rapid and economical way for farmers to get rid of the massive amounts of crop residue they generate (Sahai et al. 2011). Between 2011 and 2020, the burning of crop residues increased India's overall $CO_2$ emissions by 75% (Deshpande et al. 2023). According to Venkatramanan et al. (2021), the quantity of crop residues generated in India in 2017–2018 is around 516 million tons, of which 23% is burnt releasing approximately 176.1 Tg of $CO_2$.

Soil microbes also released $CO_2$ as a consequence of crop residue decomposition. There is a significant correlation between the evolution of $CO_2$ and the loss of litter C during litter decomposition (Ntonta et al. 2022). Both the manufacturing process and the on-field post-application of synthetic and organic fertilizers generate a sizable quantity of $CO_2$ (Walling and Vaneeckhaute 2020). Additionally, the emission factors for the production of synthetic nitrogen (N) range from 1 to 10 kg $CO_2e$ per kg of N, depending on the type of fertilizer and the hydrogen source used in the steam reforming process. On the other hand, the emission factors for the production of phosphorus (P) fertilizer are typically between 1 and 1.5 kg $CO_2e$ per kg of $P_2O_5$, and the emissions from the production of potassium (K) fertilizer are largely unknown (Walling and Vaneeckhaute 2020). Zhai et al. (2011) reported the highest average and cumulative $CO_2$ emissions in the plots treated with NPK, followed by NP and NK treatment, and the control plot had the lowest emissions. In addition to improving crop production, soil C content, microbial biomass, and plant growth, organic amendments added to soils through composts, manures, and wastes also considerably increase atmospheric $CO_2$ emissions (Shrestha et al. 2013). Global meta-analysis revealed that while organic N fertilizers greatly reduced $N_2O$ emissions, they increased $CH_4$ and $CO_2$ emissions (Sosulski et al. 2020). Chemical N fertilizers were replaced with organic N fertilizers. Greater rates of application resulted in increased $CO_2$ emissions from the soil; cumulative $CO_2$ emissions from soil modified with chicken manure were higher than those from soil treated with dairy manure (Rahman 2013).

The use of fossil fuels in agricultural processes including tillage, planting, harvesting, transporting, pumping water and irrigation, drying grain, using forestry equipment, and operating fishing vessels produces fossil fuel-based $CO_2$ emissions (Lal 2004; Houshyar 2017). According to Guru et al. (2022), the associated C equivalents for tillage, sowing and transplanting, harvesting, straw management, and other machinery were around 15.38–71.60, 11.39–54.94, 2.81–10.72, 6.70–57.08, and 25.13–86.83 kg $CO_2e$ $ha^{-1}$, respectively. Approximately 1.5% of GHG emissions are attributable to energy use in the agriculture sector (Herzog 2009). According to Davidson et al. (2000b), the yearly emissions from the main forest, secondary forest, active pasture, and degraded pasture were 2.0, 1.8, 1.5, and 1.0 kg C $m^{-2}$ $yr^{-1}$, respectively. The degradation of tropical forests releases 2.1 billion tons of $CO_2$, which comes from harvesting timber (53%), harvesting of firewood (30%), and forest fires (17%) (Pearson et al. 2017). Over the past three decades, the loss of 138.4 $km^2$ of Sundarbans mangrove ecosystem in India has resulted in the emission of about 2028.4 ± 713.7 Gg of $CO_2$ (Das et al. 2023). Grazing land is also a significant contributor to $CO_2$ emissions. The mean $CO_2$ emission rates across all grazing systems were 199.7 mg C $m^{-2}$ $h^{-1}$, and $CO_2$ emission was much greater in grazing-dominated enclosure and contractual grazing enclosure systems than in open grazing rangeland (Oduor et al. 2018).

## Factors Affecting $CO_2$ Emission in Different Agroecosystems

Factors that govern $CO_2$ emission from agroecosystems include variations in land use and changes in forest cover, soil management techniques, fossil fuel combustion from farm machinery, and climate change.

### *Land Use Change*

The release of C contained in vegetation and SOM occurs when forests, grasslands, or wetlands are converted to farmland or grazing lands, which increases GHG emissions. Deforestation is the primary cause of the net $CO_2$ flux, which is also caused by logging, forest degradation, jhum cultivation, timber and firewood harvest, expansion of agriculture, and urbanization (Friedlingstein et al. 2022). According to Friedlingstein et al. (2022), changes in global land use and forestry could emit about $4.5 \pm 2.6$ Gt $CO_2$ during 2012–2021 alone. The conversion of agricultural land, forests, and grasslands into urban development also has a favorable and significant impact on the overall $CO_2$ emissions (Tang and Hu 2021).

### *Soil Management Practices*

Soil management techniques such as tillage, crop rotation, nutrient management, irrigation, and precipitation are examples of agricultural techniques that have an impact on the soil's $CO_2$ flow and SOM decomposition. The effect of tillage on $CO_2$ emissions is significant. Due to the acceleration of the SOM oxidation and mineralization processes, conventional tillage may result in a loss of around 50% of the soil organic carbon (SOC) content (Rutkowska et al. 2018). By directly influencing $CO_2$ emission rates, management techniques can have an impact on soil temperature and water content and are the most significant environmental variables influencing soil $CO_2$ emissions (Bauchmann 2000; Davidson et al. 2000a). Soil disturbance in conventional tillage through its influence in boosting the soil temperature increased the $CO_2$ flux by 58% compared with no-tillage; the availability of water consequently upon irrigation could increase the $CO_2$ flux by 27% (Sainju et al. 2021). Increased $CO_2$ emission through enhanced soil respiration also results from fertilization (Song and Zhang 2009; Zhai et al. 2011). However, soil amendments had a greater effect on the total $CO_2$ emissions compared to soil temperature and moisture (Li et al. 2013). Elevated $CO_2$ soil emissions may arise from high N availability in the soil and are closely linked to enhanced microbial C metabolism in the soil (Dai et al. 2020). The application of organic amendments affects $CO_2$ emissions significantly as well since it gives soil microorganisms abundant substrate (Li et al. 2013). Ren et al. (2017) also reported a significant increase in $CO_2$ emission with manure amendment in comparison to chemical fertilizers. The emission of $CO_2$ is also influenced by cropping system, which impact the quality and quantity of residue reintroduced to the soil (Curtin et al. 2000; Amos et al. 2005; Sainju et al. 2005), as well as the management practices employed (Kong et al. 2020).

## *Fossil Fuel*

The usage of fossil fuels for energy in agriculture results in large GHG emissions during crop production stages (Pratibha et al. 2015). According to Flammini et al. (2021), the yearly emissions from energy consumption in agriculture in 2019 alone were around 523 Mt $CO_2e\ yr^{-1}$, but this increased to 1029 Mt $CO_2e\ yr^{-1}$ when electricity was included. In the EU alone, open-field agriculture consumed 1431 PJ, which is 3.7% of their total yearly energy consumption (Paris et al. 2022). Energy consumption in agriculture is predicted to soar in tandem with the anticipated 50% increase in global agricultural output by 2050 to fulfil anticipated increases in food demand (Calicioglu et al. 2019). This will necessitate a swift shift from conventional, vast agri-food systems to contemporary, intense production methods, which are marked by increased energy use inside the farm (Smil 2008; Sims et al. 2015).

## *Climate Change Impacts*

Global temperature and precipitation are affected by climate change. With rise in temperature, soil respiration rate is expected to accelerate (Davidson and Janssens 2006; Bond-Lamberty and Thomson 2010; Frey et al. 2013). Elevated $CO_2$ levels can augment soil respiration through enhanced C inputs in soil (Adair et al. 2011; Hungate et al. 1997; van Groenigen et al. 2014). Not only this, the productivity and composition of plant communities, the timing and intensity of soil respiration, and extreme weather events like floods, drought, heat waves, and cold waves influence $CO_2$ emissions (Semeraro et al. 2023). Climate change-induced increases in rainfall and droughts will have an impact on cycles of soil drying and rewetting as well as related processes including soil GHG fluxes in temperate climate zones (Gelfand et al. 2015). Elevated temperatures increase soil respiration in all types of ecosystems, although the extent of this effect is dependent on the characteristics of the soil (Dacal et al. 2022). Global heterotrophic respiration is greatly impacted by climate change, and from 1987 to 2016, microbial decomposition of soil C increased, indicating accumulating losses of soil C (Lei et al. 2021). Simulation model showed that heterotrophic respiration has been rising globally since the 1980s at a rate of roughly 2% per decade and is expected to increase by 40% at the end of the century (Nissan et al. 2023). Runion et al. (2009) reported that soil $CO_2$ efflux in elevated $CO_2$ was significantly higher on average throughout the measurement period. Analysis of soil $CO_2$ gradients and fluxes throughout the soil profile indicated that elevated $CO_2$ primarily enhanced soil $CO_2$ production within the primary rooting zone, whereas warming influenced soil $CO_2$ production in deeper soil layers as well (Reinthaler et al. 2021).

## *$CO_2$ Emission Under Conservation Agriculture*

Conventional agriculture uses the intensive tillage method, which changes the soil environment, breaks down soil aggregates, and exposes SOC to microbial activities, releases more $CO_2$, requires more energy, modifies microbial activity, and ultimately deteriorates soil health (Naorem et al. 2023). As a promising sustainable crop management alternative for reducing global warming, CA-based tillage methods like minimum, reduced, and no-tillage have been widely adopted as a $CO_2$ emissions mitigation strategy (Yue et al. 2023). Reducing or no tillage minimizes soil disturbance, shields SOC from microbial exposure, and lowers SOC decomposition, enhances soil aggregation, stores soil C and reduces C loss from soil through $CO_2$ fluxes or leaching, restricts soil temperature increases, etc. (Six et al. 2002, 2004; Nyambo et al. 2020; Wang et al. 2020; Jayaraman et al. 2022). However, the effect of tillage on soil $CO_2$ emissions is largely inconsistent and is largely controlled by soil and environmental variables (Abdalla et al. 2016; Ogle et al. 2019). Furthermore, Ogle et al. (2019) found that soil features and climatic conditions play a major role in the impact of no-till management on soil C emission and storage. Meta-analysis showed that tilled soils released 21% more $CO_2$ than no-till soils, and emission is dependent on the background SOC content (Abdalla et al. 2016). In sandy soils from arid climates with low SOC (<1%) and poor soil moisture, the emission increased to 29%; however, in clayey soils with high background SOC (>3%), tillage did not affect $CO_2$ fluxes. Bregaglio et al. (2022) also showed the largest decrease in $CO_2$ emissions as a result of no-tillage in long-term studies carried out in arid settings with clayey soils containing low SOC. Tilled soil has a larger potential $CO_2$ flow than no-till soil (Mangalassery et al. 2014). Reduced tillage does not influence $CO_2$ emissions globally, according to Yue et al. (2023), while no-tillage reduces $CO_2$ and GWP by an average of 15.1% and 22.6%, respectively, with no impact on crop production. Switching from conventional to minimum tillage could lower overall $CO_2$ emissions by 15–29% (Koga et al. 2003). Silva-Olaya et al. (2013) reported that the average $CO_2$ emissions from conventional till, reduced tillage, minimum tillage, and no-tillage were 0.82, 0.56, 0.51, and 0.49 g $m^{-2}$ $h^{-1}$, respectively. In the cultivation of vetch, no-tillage and reduced-tillage methods produced, on average, 68.7% and 42.2% lesser emissions than conventional tillage

(151.1 kg $CO_2$ $ha^{-1}$), respectively (Gozubuyuk et al. 2020). However, increased microbial activity brought on by increased soil water availability and conservation tillage may also result in higher $CO_2$ emissions (Plaza-Bonilla et al. 2014).

Maintaining permanent soil cover on the soil surface through crop residues, CA system supports microbial growth, and conflicting effects on $CO_2$ emission have been reported (Lenka and Lal 2013). Quantitative research using data from China revealed that, in comparison to no mulch, crop residue mulching increased the GWP by 11.4% (Guo and Liu 2022). Lenka and Lal (2013) also reported higher diurnal fluxes of $CO_2$ under high mulch rates in no-till soils, and the mulch rate affected the flux more than did the fertilizer rates. However, the impact of residue on enhancing $CO_2$ fluxes is less pronounced when residue is thoroughly mixed with soil (Curtin et al. 1998). Due to slower decomposition of crop residues present on the surface, no-tillage reduced $CO_2$ fluxes more than conventional tillage (Curtin et al. 2000). In addition to reducing $CO_2$ emissions, straw mulching enhances crop yields and improves water use efficiency in dry regions (Hu et al. 2016).

Conservational tillage practices and controlled traffic systems in CA have been reported to reduce GHG emissions through reduced fossil fuel consumption in field preparation compared with conventional till systems (Kern and Johnson 1993; Petersen et al. 2008, 2011; Jayaraman et al. 2022). Thus, the CA system can reduce the energy problem and GHG emissions from farm production systems since it demands less frequent and intensive machinery use (Ince and Guler 2020). According to FAO, CA can cut fossil fuel emissions by as much as 60% with a 20% long-term reduction in the use of agrochemicals and fertilizers when compared to traditional agriculture. According to Filipovic et al. (2006), no-till cereal cultivation results in significantly lower diesel consumption (7.35 l $ha^{-1}$) and $CO_2$ emissions (20.21 kg $ha^{-1}$) compared to conventional tillage system (60.99 l $ha^{-1}$ and 167.72 kg $ha^{-1}$). When compared to puddled transplanted rice, direct seeding reduced energy consumption (48.50%), specific energy (49.63%), C footprint (16.48%), and cultivation costs (35%) (Yadav et al. 2020). No-tillage in maize curtailed energy consumption by 38.4% and 20.1% when compared to deep tillage and conventional tillage, respectively, and saved 1044 MJ of energy per ton of maize grain produced (Nisar et al. 2021). Adoption of zero tillage in pigeon pea–castor systems lowers energy input by 30% compared to conventional tillage (Pratibha et al. 2015). Thus, CA results in lower fuel consumption, thereby lowering GHG emissions (Jayaraman et al. 2022).

## *$CH_4$ Emissions from Agroecosystems*

The two main sources of $CH_4$ in agroecosystems are lowland rice cultivation and livestock (FAO 2023). Flooded rice paddy fields account for 12% of global anthropogenic $CH_4$ emissions, whereas livestock, primarily via enteric fermentation in ruminants and manure management systems, contribute 30–32% of the emissions

(Gerber et al. 2013; ADB 2023). In addition, incomplete combustion during crop residue burning also releases $CH_4$ although in minute quantities (Romasanta et al. 2017).

## *Methane Emission from Livestock*

Enteric $CH_4$ emissions produced by methanogens present in rumens of livestock such as goats, sheep, cattle, and buffalo are subsequently belched out. Over 90% of the $CH_4$ that ruminants create worldwide originates from intestinal fermentation, while manures make up the remaining portion (Gerber et al. 2013; Opio et al. 2013). The vast majority of manure-derived $CH_4$ emissions emanated from pigs (42%) and cattle (41%) (Gerber et al. 2013). In general, cattle create around seven and nine times as much $CH_4$ as sheep and goats, respectively, making them the top ruminant emitters per animal daily (Shibata et al. 1993). Eggleston et al. (2006) estimate that a single cow generates 154–264 pounds of $CH_4$ gas annually. Level of intake, kind and quality of feeds, age of the animal, properties of manure, and management system are the main factors influencing the production of $CH_4$ from livestock (IPCC 2006; Shibata and Terada 2010). In general, $CH_4$ generation rises with an increase in daily feed intake, and the yield of $CH_4$ was higher in cows fed a wheat diet than in cows fed a maize diet. Increasing the fat concentration from 2% to 6% dry matter decreased $CH_4$ yield, regardless of the type of grain consumed (Alvarez-Hess et al. 2019). Compared to dry cows and heifers, lactating cows produced greater $CH_4$. Age was substantially correlated with body weight, milk output, and $CH_4$ emissions per unit of food. When the cows became older, their development was marked by a rise that peaked at 2000 days of age and then a fall. According to Swainson et al. (2007), sheep younger than a year old consistently produce 20% less $CH_4$ than older adult sheep. Heifers grazing on fertilized grassland released more $CH_4$ into the atmosphere (Broucek 2014). The functional group determined the $CH_4$ yield; other forbs produced the lowest output while grasses and sedges/rushes produced the greatest. According to Melts et al. (2014), grasses and sedges/rushes had a larger $CH_4$ yield because of their lower K, Mg, and lignin content, but legumes and other grassy fodder produced $CH_4$ more quickly at first. Organic substances present in livestock manure, such as proteins, carbohydrates, and lipids, serve as substrate and energy for the development of anaerobic bacteria. $CH_4$ is produced during manure storage through a process akin to enteric fermentation. According to Broucek (2014), $CH_4$ from ruminant manure and nonruminant livestock produces 0.4% and 2% of global GHG emissions, respectively. The physical form of the feces, the amount of digestible material, climate, and the length of time they remain intact all affect how much $CH_4$ is emitted from manure. Where manure from the dairy industry is managed in liquid systems, which results in higher amounts of $CH_4$ emissions, manure emissions are rather significant (Moeletsi and Tongwane 2004; Gerber et al. 2013). Excessive levels of readily digestible organic matter in the manure will cause acidogenic microorganisms to grow more quickly than methanogens, which will lead to the buildup of volatile fatty acids and a lower pH (Clemens et al. 2002; Bastami et al. 2016). This may inhibit methanogens and almost completely stop the production of $CH_4$ (Bastami et al. 2016). Like other biological processes, methanogenesis and hydrolysis are temperature-sensitive. The average daily $CH_4$ emissions from manure were in the following order: summer > spring > autumn > winter. The temperature and $CH_4$ emissions showed a positive correlation and increased ninefold when the temperature rose from 15 °C to 35 °C (Dalby et al. 2021).

## *Methane Emission from Paddy Soils*

In lowland rice paddies, methanogenic archaea bacteria break down SOM and rice root exudates under an anaerobic environment and at low redox potentials (< −200 mV) (Philippot et al. 2009). According to Yuan et al. (2018), there are two primary pathways for the creation of $CH_4$: acetoclastic, which uses acetate as a substrate and accounts for 80% of $CH_4$ synthesis, and hydrogenotrophic, which uses $H_2/CO_2$ as a substrate and accounts for 10–30% of $CH_4$ formation. The $CH_4$ thus formed is then released into the atmosphere by diffusion through standing water, ebullition in the form of bubbles, and aerenchyma in rice plants (Tokida et al. 2013; Green 2013). About 90% of emissions come via the aerenchyma pathway, with the remaining 8% and 2% coming from diffusion and ebullition through the floodwater, respectively (Smartt et al. 2016). $CH_4$ emissions from farmland are affected by several factors such as soil redox potential (Eh) and pH, temperature and water content, SOM, fertilizers, and rice crop cultivars (Xu et al. 2020). Methanogens are obligative anaerobes and require highly reduced conditions for growth with critical (−150 to −160 mV) and showed a negative exponential association between $CH_4$ production and soil Eh when soil Eh is in the range of −230 and −150 mV (Wang et al. 1993, 2018). Acid sensitivity and neutrophilia are other characteristics of methanogenic bacteria. The ideal pH for producing $CH_4$ was when close to neutrality, and a modest increase in soil pH led to an augmentation of $CH_4$ production, while a tiny fall in pH caused by the introduction of acidic materials dramatically inhibited $CH_4$ production (Wang et al. 1993). Temperature affects soil $CH_4$ production through its effect on the methanogenic community and microbial oxidation of SOM (Peng et al. 2018). The temperature optima of most methanogenic bacteria are around 30–35 °C, although their growth is also reported from 4 °C to 100 °C (Parashar et al. 1993; Lu et al. 2015). The $Q_{10}$ within this range of temperature was about 1.5–28 and averaged around 4.1 (Ding and Cai 2003). The water regime especially flooding duration is one of the most important elements regulating $CH_4$ emissions from paddy fields since methanogenesis is a purely anaerobic process. Following rice paddies flooding, a fall in Eh values is observed which leads to an increase in soil $CH_4$ concentration and fluxes to the atmosphere (Wang et al. 1993). In comparison to mid-season drainage and intermittent wet and dry spells, long-term flooded rice fields resulted in the highest emission of $CH_4$ (Zhou et al. 2015). Methanogen abundance decreases and methanotroph abundance increases as a result of alternate wetting and drying cycles that hinder $CH_4$ synthesis and encourage $CH_4$ oxidation (Ma and Lu 2011).

As a primary substrate for methanogenesis, SOM content influences the production of $CH_4$ in paddy fields, and a positive correlation between $CH_4$ emission and SOM, total C content, and dissolved SOC was observed (Qin et al. 2015; Humphreys et al. 2019). The type and quantity of organic matter addition affects $CH_4$ emission and methanogenic community in paddy rhizosphere soil (Yuan et al. 2018). Substrates with a high C:N ratio are rich in labile C and are optimum for anaerobic degradation by methanogenic microbes for $CH_4$ production (Le Mer and Roger 2001; Anshori et al. 2021). Although N fertilizers have a direct and indirect impact on every step that results in $CH_4$ emission from paddy soils, the effects are erratic and contradicting (Cai et al. 2003; Schimel 2000; Li et al. 2021c). Application of N fertilizers increases crop growth as well as alters $CH_4$ methanogens and methanotrophs in soil, thereby producing complex effects on $CH_4$ emissions (Banger et al. 2012). Emissions were stimulated at low N application rates but inhibited at high N rates (Sun et al. 2016). A meta-analysis revealed that N fertilizers increased $CH_4$ emissions in 98 out of 155 data pairs in rice soils, with the emissions per kg N fertilizer being higher at lower application rates compared to higher rates (Banger et al. 2012).

Varietal characteristics of the rice plant also affect $CH_4$ emission due to root exudations and secretions in the rhizosphere, root traits and biomass, photosynthetic C allocation, and number of leaf blades, leaf sheaths, and culms which provide an effective channel for gas exchange (Qin et al. 2015; Li et al. 2022). In China, $CH_4$ emissions from the rice–rice system were significantly and positively correlated with rice straw biomass, tiller number, and chlorophyll content (Qin et al. 2015). Between different cultivars, pure-line cultivars emit more $CH_4$ than hybrid rice, and total $CH_4$ emissions were larger in rice–rice rotation than in rice–soybean rotation (Rogers et al. 2014; Brye et al. 2017).

## *$CH_4$ Emission Under Conservation Agriculture*

The intensity of tillage, handling of residues, and use of cover crops affect $CH_4$ emissions from soil. The CA systems may be a crucial tactic for reducing $CH_4$ emissions to the environment by strengthening agricultural soils as $CH_4$ sinks (Ussiri et al. 2009; Bayer et al. 2012). Because tillage creates an aerobic environment, methanogens do not thrive there. Undisturbed anaerobic soils typically function as a net sink of $CH_4$, but when soils are used for agriculture, $CH_4$ oxidation rates drastically drop (Baggs and Blum 2004; Suwanwaree and Robertson 2005; Mojeremane et al. 2012). Tillage practices affect soil Eh, aeration, structure, moisture status, microbial diversity and richness, and crop productivity which can all have an impact on $CH_4$ fluxes (Hutsch 2001; Mangalassery et al. 2014; Maucieri et al. 2021). According to Wihardjaka et al. (2023), no-tillage might cut $CH_4$ emissions by 15.58% when compared to intensive tillage. Global metadata showed that no-tillage reduces $CH_4$ emissions without affecting crop production, while reduced tillage raises $CH_4$ emissions (Yue et al. 2023). Compared to moldboard plough and chisel till, no-tillage could abate $N_2O$ and $CH_4$ emissions by more than 50% (Ussiri et al. 2009).

The various rice establishment techniques also had an impact on the total seasonal $CH_4$ emissions. Seasonal $CH_4$ emissions from rice cultivation in Bangladesh showed that direct seeding (91.52 kg $CH_4$ $ha^{-1}$), system rice intensification (110.35 kg $CH_4$ $ha^{-1}$), and alternate wetting and drying (128.11 kg $CH_4$ $ha^{-1}$) significantly reduce $CH_4$ emissions as compared to continuous flooding (147.85 kg $CH_4$ $ha^{-1}$) (Borna et al. 2022). When compared to traditional puddled transplanted rice, the system rice intensification techniques decreased the $CH_4$ emission by 64% (Jain et al. 2014). Converting from traditional rice crop establishment techniques to CA practices also results in a reduction in diesel and power usage (Tirol-Padre et al. 2016).

Residue management in agricultural systems is thought to be an efficient approach to improve soil quality and mitigate climate change. Crop residue can provide easily labile C substrates for methanogens and lower the soil Eh to optimum condition, thereby significantly increasing the emission of $CH_4$ from paddy systems (Ma et al. 2009). The amount of residue input, timing, and technique of application determine the cumulative $CH_4$ produced from soil. Crop residue incorporation treatments enhanced the cumulative $CH_4$ fluxes by at least 59.7% compared to the synthetic N application alone (Wang and Luo 2018). Applying straw in the autumn greatly reduced the $CH_4$ flux by an average of 24–65% compared to applying straw in the spring, and the seasonal $CH_4$ flux was also roughly 24% lower when combining straw with soil following autumn harvesting than when spreading straw over the surface (Song et al. 2019). Cover crops are regarded as sustainable management strategies that enhance soil quality and have a major impact on GHG emissions (Kim et al. 2012; Paustian et al. 2016; Wang et al. 2022). The C:N ratio, residue assimilation, and fertilizer management of cover crops are the main factors influencing $CH_4$ (Kaye and Quemada 2017). Low C:N ratio cover crops, like legume-based cover crops, reduce the impact of $CH_4$ emissions and boost rice yield more than high C:N ratio cover crops, like cereal-based cover crops (Kim et al. 2012). This is explained by the fact that, in anaerobic environments, their increased labile C content enhanced $CH_4$ emissions (Lu et al. 2000; Le Mer and Roger 2001). In the mono-rice farming system, fallow season cover crops raise the amount of $CH_4$ released from paddy soils; more $CH_4$ is released from cereal-based cover crops than from legume-based cereal crops (Li et al. 2021a, b). One of the drawbacks of returning straw to the soil is it may lead to an increase in $CH_4$ emissions (Pandey et al. 2012).

## *$N_2O$ Emissions from Agroecosystems*

The $N_2O$ gas is the third most important GHG after $CO_2$ and $CH_4$ and is 273 times more potent than $CO_2$ in a 100-year time scale with a current atmospheric lifetime of 116 ± 9 years (Vallero 2019). From 270 parts per billion in 1750 to 335 parts per billion in 2020, the concentration of atmospheric $N_2O$ has increased by more than 20%, making up 6% of GHG released (Ritchie et al. 2020). It is crucial to the equilibrium of atmospheric radiation, the chemistry of stratospheric ozone, and climate change. Its emission into the atmosphere is largely due to human-induced activities such as agriculture, land use change, transportation, and industry and natural sources such as those released from release of wetlands and marine and tropical forests. Globally, agricultural activities through soil fertility management and emissions from natural soils account two-thirds of the total anthropogenic emissions (Fowler et al. 2015).

In soils, both biotic and abiotic processes are responsible for the production of $N_2O$ gas. The two key biochemical reactions that control $N_2O$ production in both terrestrial and aquatic ecosystems are nitrification and denitrification (Zhu et al. 2013). Denitrification serves as the primary mechanism for $N_2O$ production, contributing to 60% of total agricultural $N_2O$ losses globally (Saggar et al. 2009). The remaining 40% is ascribed to nitrification and nitrifier denitrification (Kroeze et al. 1999). The abiotic production of $N_2O$, and particularly nitric oxide (NO), occurs primarily through chemodenitrification (Venterea et al. 2005). Nitrification is the microbial oxidation of ammonium ($NH_4^+$) to nitrate ($NO_3^-$) under aerobic conditions, with $N_2O$ emitted as a by-product (Guo et al. 2018). Another process responsible for the emission of $N_2O$ during nitrification is nitrifier denitrification within the same autotrophic nitrifiers. Evidence indicates that nitrifier denitrification occurs at relatively high $NH_4^+$ concentrations, low $O_2$ and low C contents, and potentially low pH (Velthof and Rietra 2018). Denitrification is the reduction of $NO_3^-$ to $N_2$ gas under anoxic and suboxic conditions where facultative anaerobes (fungi, archaea, and bacteria) use $NO_3^-$ and $NO_2^-$ as an alternative electron acceptor and reduce it to $N_2O$ as the intermediate and finally to $N_2$ gas. The abiotic production of $N_2O$ via chemodenitrification occurs when $NO_2^-$ and $NH_2OH$ are chemically reduced by $N_2O$ and NO. Studies suggested that this process might account for 6.8–75% of the total $N_2O$ production in agricultural soils (Wang et al. 2019).

## Factors Influencing $N_2O$ Emission from Agroecosystems

Several factors controlled $N_2O$ emissions from the agroecosystem during the nitrification and denitrification processes. These variables include N availability, soil temperature, moisture content, C:N ratio, amount of fertilizer applied, and land use management.

## *N Availability*

In agroecosystems, a huge quantity of synthetic N fertilizer is applied to obtain higher production and productivity to feed the ever-increasing global population. This increases the availability and concentration of $NO_3^-$ in the soil, which serves as the fundamental substrate for nitrification and denitrification, and is the most significant factor influencing $N_2O$ emission (Conrad 1996). Examples of both organic and inorganic N sources in agriculture include atmospheric N fixation, manures, slurry, legumes or postharvest crop residues, and synthetic fertilizers (Butterbach-Bahl et al. 2011). Numerous field investigations have documented increased $N_2O$ emissions due to N application (Sapkota et al. 2020; Thies et al. 2020). Hoben et al. (2011) observed that $N_2O$ emission factors per unit of applied N varied from 0.6% to 1.5% across different locations and years. Meta-analysis using 78 published studies from 233 site years concluded that $N_2O$ emission displayed a significantly faster than linear growth pattern for N synthetic fertilizers (Shcherbak et al. 2014). Also, application times, cropping systems, and measurement are some factors induced by N fertilizer that affect $N_2O$ emissions (Sun et al. 2016).

## *Soil Moisture and Aeration*

Soil moisture is a major driver of $N_2O$ emissions as it controls the amount of $O_2$ available to soil microbes, transports $NO_3^-$ and $NH_4^+$, and facilitates the diffusion of NO, $N_2O$, and $N_2$ out of the soil (Pilegaard 2013). Water-filled pore space (WFPS) is an important indicator of soil water content that can be used to connect changes in soil water content to $N_2O$ emissions (Rochette et al. 2004). When WFPS >60%, denitrification is the dominant source of $N_2O$ emission, and <60% WFPS means chemoautotrophic nitrification is predominant (Davidson et al. 2000a, b). Li et al. (2005) suggested a high level of $N_2O$ emission could be expected beyond 50% of soil water content. Because the intermediate in denitrification, $N_2O$, is deoxidized to $N_2$ under stringent anaerobic conditions, a drop in $N_2O$ emissions may be seen at nearly full pore volume filling. Ciarlo et al. (2007) found that in laboratory settings, the maximum $N_2O$ emissions were measured at 80% WFPS. These circumstances were not reductive enough to allow for the full reduction to $N_2$. The highest $N_2O$ fluxes from grazing soils were observed by Dobbie and Smith (2003) when WFPS levels exceeded 60% and $NO_3^-$ concentration was not limiting. Addition of labile sources of N and C at 70% WFPS had the largest $N_2O$ emission fluxes (Wei et al. 2022). Under field conditions, variations in soil water content after irrigation and precipitation can have an impact on $N_2O$ emissions (Simojoki and Jaakkola 2000). Higher $N_2O$ emissions were only a result of the irrigation events when there was sufficient nitrogen available. According to Scheer et al. (2008), due to decreased soil moisture, less water is applied during fewer irrigation events, which reduces $N_2O$ emissions. Due to the extreme drying of the soil profile in between irrigation events,

Mumford et al. (2019) detected increased $N_2O$ emission in low-frequency irrigation compared to high-frequency and medium-frequency irrigation.

## *Soil Temperature*

Soil temperatures have an impact on $N_2O$ emissions through kinetic reaction, microbial community population, and respiration by forming anaerobic zones, even if they are not the primary driver of $N_2O$ emissions (Butterbach-Bahl et al. 2013). In general, nitrification rates rise with temperature and peak between 20 °C and 35 °C (Parton et al. 2001); however, peaks between 35 °C and 40 °C have also been documented (Lai et al. 2019). According to Ngoc and Maeda (2018), $N_2O$ emissions peaked at 25 °C and fell to their lowest point at 35 °C. This is probably because nitrification and denitrification contribute differently to $N_2O$ emissions. An exponential association was shown by Benoit et al. (2015) between $N_2O$ emissions and temperature at 5–45 °C and 12–32 °C, respectively.

## *Organic Matter*

The presence of organic substrates is critical for regulating nitrification and denitrification in soil. Nitrifiers require C as a source of energy which serves as an electron donor for denitrifiers. Also, the decomposition of SOM by soil microbes may enhance microbial respiration, leading to anaerobic conditions that favor denitrification reaction and $N_2O$ emissions (Chen et al. 2013). The rate of denitrification also increases with increase in the amount of C available, provided that there are sufficient levels of $NO_3^-$ (Senbayram et al. 2012; Friedl et al. 2018).

## *Soil pH*

Soil pH influences $N_2O$ emissions by affecting the activity of nitrifying and denitrifying bacteria in the soil. It also determines whether $NO_2^-$ and $NO_3^-$ chemically decompose into $N_2O$ or $N_2$ and subsequently influences the emission ratio of $N_2/N_2O$. Under acidic conditions (pH < 6.0), a greater proportion of $N_2O$ relative to $N_2$ is emitted, but at higher pH values, $N_2$ becomes the principal product (Wang et al. 2021). At pH lower than 5, abiotic $N_2O$ emissions is the dominant form, but they may not be as significant as biotic systems as it contributes <3% of total $N_2O$ emissions (Su et al. 2019). This showed that soil pH exhibited a negative correlation with the $N_2O$ emission ratio, whereas a positive correlation was observed with the abundance of total denitrification genes (Samad et al. 2016). An increase of one unit pH may result in a decrease of 0.2 units in the molar fraction of emitted $N_2O$ (Stevens and Laughlin 1998). Thus, liming acidic soils to neutrality reduces $N_2O$ emissions. Imbalance and excessive use of nitrogenous fertilizers in agriculture that cause soil acidity would enhance $N_2O$ emissions, while soil amendments that can neutralize soil acidity would mitigate the emission (Thomson et al. 2012). The nitrification process, which occurs under aerobic conditions, typically leads to a decrease in soil pH, as observed by Signor and Cerri (2013). On the other hand, denitrification operates independently of soil pH. Liming acid soil decreased $N_2O$ emission in acidic soil by 37% in upland soil and 44% in paddy soil, respectively (Khaliq et al. 2019).

### *Soil Texture*

The predominant influence of soil texture on $N_2O$ emissions lies in its ability to determine the prevalence of either aerobic or anaerobic conditions within the soil (Meurer et al. 2016). Compared to sandy soils, fine-textured soils have a higher microspore and may retain more water, which causes anaerobic conditions to last longer and significantly increase $N_2O$ emission (Parton et al. 1996). Consequently, $N_2O$ emissions are often higher in clayey soils than in sandy soils (Pelster et al. 2012), owing to the fact that they have more capillary pores within soil aggregates compared to sandy soils (Lesschen et al. 2011). However, reduced gas diffusivity in fine-textured soils allows greater time for $N_2O$ to be completely reduced to $N_2$ gas (Weitz et al. 2001). In contrast, Skiba and Ball (2002) reported that maximum emissions were always from the sandy freely drained soil.

### *$N_2O$ Emission Under Conservation Agriculture*

Although several studies concluded the beneficial effects of CA on soil C sequestration, mitigating $CH_4$ and $CO_2$ emission, there is much disagreement on the effects of CA on $N_2O$ emissions (Rochette 2008). There is an impression that CA-associated $N_2O$ emission may offset a reduction in $CH_4$ and $CO_2$ if unabated (Guenet et al. 2021). Reduced soil disturbance in CA influenced the soil structure, aeration, moisture, temperature, C and N availability, microbial activity, and $O_2$ consumption, thereby affecting the nitrification and denitrification processes (Monti et al. 2021), which likely affected the $N_2O$ emissions. In no-till, extremely high $N_2O$ fluxes were observed and were linked to decreased gas diffusivity and air-filled porosity (Ball et al. 1999). Likewise, Rochette et al. (2008) also reported that $N_2O$ emissions in no-tillage are more than doubled of fall moldboard ploughing in the heavy clay soil, whereas the average emissions in loam soil in the 3 years were similar. These findings suggested that the effectiveness of no-till in reducing net $N_2O$ emissions could be constrained in fine-textured soils with high water content and limited aeration. Enhanced $N_2O$ emissions in the no-till system were due to greater denitrification, the presence of anaerobic conditions, and the availability of labile C for denitrifiers

(Mirzaei et al. 2022). Drury et al. (2006) reported comparable $N_2O$ emissions under varying tillage when N was placed at a shallow depth, but under deep placement N, zone-tillage produced average $N_2O$ emissions that were 20% lower than no-tillage and 38% lower than moldboard plough tillage. After 7 years of treatment, Lognoul et al. (2017) found that the average $N_2O$ emissions in the reduced tillage were considerably higher than those in the conventional tillage. Mei et al. (2018) reported from metadata that conservation tillage significantly increased soil $N_2O$ emissions by 17.8% compared to conventional tillage, with the greatest increase observed in tropical climates where short-term application of conservation tillage was practiced. In contrast, Mutegi et al. (2010) reported that no-tillage and reduced tillage yielded lower $N_2O$ emissions compared to conventional tillage, respectively. Yoo et al. (2016) showed that $N_2O$ emission was significantly reduced in no-tillage soils, compared to those in conventional tillage soils. A meta-analysis by Van Kessel et al. (2013) revealed that no-tillage and reduced tillage could offset $N_2O$ emissions when it is sustained for a prolonged period (>10 years), particularly in dry climates, whereas in humid climates, deep placement of fertilizer N in no-tillage and reduced tillage is recommended to mitigate $N_2O$ emission. Conservation agriculture requires permanent retention of crop residue on at least 30% of the surface soil. Crop residues are a source of C, organic N, and energy, thereby influencing aeration and microbial respiration and affecting soil $N_2O$ emission (Chen et al. 2013; Surey et al. 2020). Contradictory results on the effect of residue on $N_2O$ emission have been reported in the literature (Xia et al. 2018). Crop residue retention could enhance or counter $N_2O$ emission depending on residue level, type of residue, and climate conditions (Ranaivoson et al. 2017). Crop residue with a low C:N ratio will promote mineralization and increase substrate availability, whereas residue with a high C:N ratio on the soil surface may increase the immobilization of mineral N (Nicolardot et al. 2001). Schmatz et al. (2020) reported enhanced $N_2O$ emissions with increasing amounts of low C:N residues, but not when the amount of high C:N residues is increased. Meta-analysis showed that crop residue mulch and retention on no-till had positive effects on soil $N_2O$ emission compared to residue incorporation with plough tillage (Hu et al. 2019). Enhanced $N_2O$ emissions with the increasing rate of residue under both conventional till and no-till systems have been reported (Mirzaei et al. 2022). In their study, Hu et al. (2019) demonstrated that higher amounts of crop residues led to increased $N_2O$ emissions in tropical and subtropical regions. A meta-analysis also revealed that higher concentrations of total N and water-soluble C are characteristics of immature residues. Reduced relative proportions of cellulose, hemicelluloses, lignin, and a low C:N ratio greatly increased $N_2O$ emissions, whereas mature residues with the reverse properties exhibited little effects (Abalos et al. 2022). The increased denitrification, presence of anaerobic conditions, and availability of easily degradable C in the no-till systems were identified as contributors to the elevated $N_2O$ emissions, as indicated by Mirzaei et al. (2022). It is reported that $N_2O$ emission from soil is enhanced when crop residues are applied to soils with pH 5.5–6.5 or above 7.5 with low clay content (Li et al. 2021a, b).

Crop diversification involving a balanced mix of legume and nonlegume crops has been adopted as an important practice in CA. These types of crop rotations have an intricate effect on $N_2O$ emissions through organic matter and N fertilizer inputs (Machado et al. 2021). The inclusion of legume cover crops in a no-tillage CA system has the possibility of enhanced $N_2O$ emission in low-input farming systems (Vangeli et al. 2022). Long-term simulations showed lower $N_2O$ in diversified maize rotation compared to monoculture (Jiang et al. 2021). Emission also depends upon the type of legume (Rochette et al. (2004) reported higher emissions in alfalfa systems compared to soybean systems) and is attributed to frequent cutting and harvesting of aboveground plant material and litter in alfalfa. Although legumes have been reported as an $N_2O$ mitigation strategy in many studies since no synthetic N is added to the soil (Jensen et al. 2012), sizeable losses of $N_2O$ have been reported from legume-based pastures and crops (Jensen et al. 2010). Enhanced $N_2O$ emissions from legumes are most probably a result of N release from root exudates during the growing season and crop residue decomposition (Rochette and Janzen 2005; Wichern et al. 2008).

## Conclusion

A significant amount of GHG is emitted from different agroecosystems, including croplands, grasslands, and agroforestry systems. The use of cover crops, reduced tillage, and diverse crop rotations in CA have been found to decrease emissions of $CO_2$, $N_2O$, and $CH_4$ from agricultural soils while also enhancing soil health and productivity. However, the effectiveness of CA in reducing GHG emissions may vary depending on local environmental conditions and management practices. Nevertheless, CA is an important step towards sustainable land use and climate change mitigation, and its adoption should be encouraged through policy incentives and education programs. More research is needed to understand the long-term impacts of CA on GHG emissions from different ecosystems. To inform agricultural policymakers and facilitate the use of CA, an efficient tool is needed to evaluate its cost–benefit under different circumstances.

## References

Abalos D, Rittl TF, Recous S, Thiébeau P, Topp CF, van Groenigen KJ, Butterbach-Bahl K, Thorman RE, Smith KE, Ahuja I, Olesen JE (2022) Predicting field $N_2O$ emissions from crop residues based on their biochemical composition: a meta-analytical approach. Sci Total Environ 812:152532

Abdalla K, Chivenge P, Ciais P, Chaplot V (2016) No-tillage lessens soil $CO_2$ emissions the most under arid and sandy soil conditions: results from a meta-analysis. Biogeosciences 13(12):3619–3633

Adair CE, Reich PB, Trost JJ, Hobbie SE (2011) Elevated $CO_2$ stimulates grassland soil respiration by increasing carbon inputs rather than by enhancing soil moisture. Glob Chang Biol 17(12):3546–3563

ADB (2023) How can we incentivize reducing methane emission in rice farming in Asia? ADB Agriculture, Food, Nature and Rural Development Webinar Series. Asian Development Bank. Manila

Alvarez-Hess PS, Williams SR, Jacobs JL, Hannah MC, Beauchemin KA, Eckard RJ, Wales WJ, Morris GL, Moate PJ (2019) Effect of dietary fat supplementation on methane emissions from dairy cows fed wheat or corn. J Dairy Sci 102(3):2714–2723

Amos B, Arkebauer TJ, Doran JW (2005) Soil surface fluxes of greenhouse gases in an irrigated maize-based agroecosystem. Soil Sci Soc Am J 69(2):387–395

Anshori A, Sunarminto BH, Haryono E, Pramono A (2021) Effect of organic fertilizers on $CH_4$ and $N_2O$ production from organic paddy field. IOP Conference Series: Environ Earth Sci 724(1):012056

Baggs EM, Blum H (2004) $CH_4$ oxidation and emissions of $CH_4$ and $N_2O$ from *Lolium perenne* swards under elevated atmospheric $CO_2$. Soil Biol Biochem 36(4):713–723

Ball BC, Scott A, Parker JP (1999) Field $N_2O$, $CO_2$ and $CH_4$ fluxes in relation to tillage, compaction and soil quality in Scotland. Soil Tillage Res 53(1):29–39

Banger K, Tian H, Lu C (2012) Do nitrogen fertilizers stimulate or inhibit methane emissions from rice fields? Glob Chang Biol 18(10):3259–3267

Bastami MS, Jones DL, Chadwick DR (2016) Reduction of methane emission during slurry storage by the addition of effective microorganisms and excessive carbon source from brewing sugar. J Environ Qual 45(6):2016–2022

Bayer C, Gomes J, Vieira FC, Zanatta JA, de Cassia PM, Dieckow J (2012) Methane emission from soil under long-term no-till cropping systems. Soil Tillage Res 124:1–7

Beach RH, Creason J, Ohrel SB, Ragnauth S, Ogle S, Li C, Ingraham P, Salas W (2015) Global mitigation potential and costs of reducing agricultural non-$CO_2$ greenhouse gas emissions through 2030. J Integr Environ Sci 12(sup1):87–105

Benoit M, Garnier J, Billen G (2015) Temperature dependence of nitrous oxide production of a luvisolic soil in batch experiments. Process Biochem 50(1):79–85

Bond-Lamberty B, Thomson A (2010) Temperature-associated increases in the global soil respiration record. Nature 464(7288):579–582

Borna SN, Siddique IA, Al Mahmud A, Khatun R, Hossain M, Islam S, Meharg AA, Islam MR (2022) Influence of rice establishment methods on water productivity, methane emissions and rice grain heavy metals content from irrigated rice paddies in Bangladesh. Eur J Agric Food Sci? 4(5):112–119

Bregaglio S, Mongiano G, Ferrara RM, Ginaldi F, Lagomarsino A, Rana G (2022) Which are the most favourable conditions for reducing soil $CO_2$ emissions with no-tillage? Results from a meta-analysis. Int Soil Water Conserv Res 10(3):497–506

Broucek J (2014) Production of methane emissions from ruminant husbandry: a review. J Environ Prot 5(15):1482

Brye KR, Rogers CW, Smartt AD, Norman RJ, Hardke JT, Gbur EE (2017) Methane emissions as affected by crop rotation and rice cultivar in the Lower Mississippi River Valley, USA. Geoderma Reg 11:8–17

Buchmann N (2000) Biotic and abiotic factors controlling soil respiration rates in *Picea abies* stands. Soil Biol Biochem 32(11–12):1625–1635

Butterbach-Bahl K, Gundersen P, Ambus P, Augustin J, Beier C, Boeckx P, Dannenmann M, Sanchez Gimeno B, Ibrom A, Kiese R, Kitzler B (2011) Nitrogen processes in terrestrial ecosystems. In: The European nitrogen assessment: sources, effects and policy perspectives. Cambridge University Press, pp 99–125

Butterbach-Bahl K, Baggs EM, Dannenmann M, Kiese R, Zechmeister-Boltenstern S (2013) Nitrous oxide emissions from soils: how well do we understand the processes and their controls? Philos Trans R Soc B: Biol 368(1621):20130122

Cai Z, Tsuruta H, Gao M, Xu H, Wei C (2003) Options for mitigating methane emission from a permanently flooded rice field. Glob Chang Biol 9(1):37–45

Calicioglu O, Flammini A, Bracco S, Bellù L, Sims R (2019) The future challenges of food and agriculture: an integrated analysis of trends and solutions. Sustain For 11(1):222

Chen H, Li X, Hu F, Shi W (2013) Soil nitrous oxide emissions following crop residue addition: a meta-analysis. Glob Chang Biol 19(10):2956–2964

Ciarlo E, Conti M, Bartoloni N, Rubio G (2007) The effect of moisture on nitrous oxide emissions from soil and the $N_2O/(N_2O + N_2)$ ratio under laboratory conditions. Biol Fertil Soils 43:675–681

Clemens J, Bergmann S, Vandré R (2002) Reduced ammonia emissions from slurry after self-acidification with organic supplements. Environ Technol 23(4):429–435

Conrad R (1996) Soil microorganisms as controllers of atmospheric trace gases ($H_2$, CO, $CH_4$, OCS, $N_2O$, and NO). Microbiol Rev 60(4):609–640

Curtin D, Selles F, Wang H, Biederbeck VO, Campbell CA (1998) Carbon dioxide emissions and transformation of soil carbon and nitrogen during wheat straw decomposition. Soil Sci Soc Am J 62(4):1035–1041

Curtin D, Wang H, Selles F, McConkey BG, Campbell CA (2000) Tillage effects on carbon fluxes in continuous wheat and fallow–wheat rotations. Soil Sci Soc Am J 64(6):2080–2086

Dacal M, Delgado-Baquerizo M, Barquero J, Berhe AA, Gallardo A, Maestre FT, García-Palacios P (2022) Temperature increases soil respiration across ecosystem types and soil development, but soil properties determine the magnitude of this effect. Ecosystems 25(1):184–198

Dai Z, Yu M, Chen H, Zhao H, Huang Y, Su W, Xia F, Chang SX, Brookes PC, Dahlgren RA, Xu J (2020) Elevated temperature shifts soil N cycling from microbial immobilization to enhanced mineralization, nitrification and denitrification across global terrestrial ecosystems. Glob Chang Biol 26(9):5267–5276

Dalby FR, Hafner SD, Petersen SO, VanderZaag AC, Habtewold J, Dunfield K, Chantigny MH, Sommer SG (2021) Understanding methane emission from stored animal manure: a review to guide model development. J Environ Qual 50(4):817–835

Das N, Mondal A, Saha NC, Ghosh S, Mandal S (2023) Decadal loss of above-ground biomass and subsequent potential $CO_2$ emission from the Sundarbans mangrove ecosystem, India. Acta Ecol Sin 43(3):452–458

Davidson EA, Janssens IA (2006) Temperature sensitivity of soil carbon decomposition and feedbacks to climate change. Nature 440(7081):165–173

Davidson EA, Keller M, Erickson HE, Verchot LV, Veldkamp E (2000a) Testing a conceptual model of soil emissions of nitrous and nitric oxides: using two functions based on soil nitrogen availability and soil water content, the hole-in-the-pipe model characterizes a large fraction of the observed variation of nitric oxide and nitrous oxide emissions from soils. Bioscience 50(8):667–680

Davidson EA, Verchot LV, Cattânio JH, Ackerman IL, Carvalho JE (2000b) Effects of soil water content on soil respiration in forests and cattle pastures of eastern Amazonia. Biogeochemistry 48:53–69

Deshpande MV, Kumar N, Pillai D, Krishna VV, Jain M (2023) Greenhouse gas emissions from agricultural residue burning have increased by 75% since 2011 across India. Sci Total Environ 904:166944

Ding W, Cai Z (2003) Effect of temperature on methane production and oxidation in soils. Effect of temperature on methane production and oxidation in soils. Ying yong sheng tai xue bao: J Appl Ecol 14(4):604–608

Dobbie KE, Smith KA (2003) Nitrous oxide emission factors for agricultural soils in Great Britain: the impact of soil water-filled pore space and other controlling variables. Glob Change Biol 9(2):204–218

Drury CF, Reynolds WD, Tan CS, Welacky TW, Calder W, McLaughlin NB (2006) Emissions of nitrous oxide and carbon dioxide: influence of tillage type and nitrogen placement depth. Soil Sci Soc Am J 70(2):570–581

Eggleston HS, Buendia L, Miwa K, Ngara T, Tanabe K (2006) IPCC guidelines for national greenhouse gas inventories. Institute for Global Environmental Strategies (IGES), Hayama, Kanagawa

FAO (2023) Methane emissions in livestock and rice systems—Sources, quantification, mitigation and metrics. Food and Agriculture Organization of the United Nations, Rome. https://doi.org/10.4060/cc7607en

Filipovic D, Kosutic S, Gospodaric Z, Zimmer R, Banaj D (2006) The possibilities of fuel savings and the reduction of CO2 emissions in the soil tillage in Croatia. Agric Ecosyst Environ 115(1–4):290–294

Flammini A, Pan X, Tubiello FN, Qiu SY, Rocha Souza L, Quadrelli R, Bracco S, Benoit P, Sims R (2021) Emissions of greenhouse gases from energy use in agriculture, forestry and fisheries: 1970–2019. Earth Syst Sci Data Discuss 14:811–821

Fowler D, Steadman CE, Stevenson D, Coyle M, Rees RM, Skiba UM, Sutton MA, Cape JN, Dore AJ, Vieno M, Simpson D (2015) Effects of global change during the 21st century on the nitrogen cycle. Atmos Chem Phys 15(24):13849–13893

Frey SD, Lee J, Melillo JM, Six J (2013) The temperature response of soil microbial efficiency and its feedback to climate. Nat Clim Chang 3(4):395–398

Friedl J, De Rosa D, Rowlings DW, Grace PR, Müller C, Scheer C (2018) Dissimilatory nitrate reduction to ammonium (DNRA), not denitrification dominates nitrate reduction in subtropical pasture soils upon rewetting. Soil Biol Biochem 125:340–349

Friedlingstein P, O'sullivan M, Jones MW, Andrew RM, Gregor L, Hauck J, Le Quéré C, Luijkx IT, Olsen A, Peters GP, Peters W (2022) Global carbon budget 2022. Earth Syst Sci Data Discuss 14:4811–4900

Gelfand I, Cui M, Tang J, Robertson GP (2015) Short-term drought response of N2O and CO2 emissions from mesic agricultural soils in the US Midwest. Agric Ecosyst Environ 212:127–133

Gerber PJ, Steinfeld H, Henderson B, Mottet A, Opio C, Dijkman J, Falcucci A, Tempio G (2013) Tackling climate change through livestock: a global assessment of emissions and mitigation opportunities. Food and Agriculture Organization of the United Nations (FAO), Rome

Gozubuyuk Z, Sahin U, Celik A (2020) Tillage and irrigation impacts on the efficiency of fossil fuel utilization for Hungarian vetch production and fuel-related $CO_2$ emissions. Environ Eng Sci 37(3):201–213

Green SM (2013) Ebullition of methane from rice paddies: the importance of furthering understanding. Plant Soil 370(1):31–34

Guenet B, Gabrielle B, Chenu C, Arrouays D, Balesdent J, Bernoux M, Bruni E, Caliman JP, Cardinael R, Chen S, Ciais P (2021) Can $N_2O$ emissions offset the benefits from soil organic carbon storage? Glob Chang Biol 27(2):237–256

Guo C, Liu X (2022) Effect of soil mulching on agricultural greenhouse gas emissions in China: a meta-analysis. PLoS One 17(1):e0262120

Guo G, Wang Y, Hao T, Wu D, Chen GH (2018) Enzymatic nitrous oxide emissions from wastewater treatment. Front Environ Sci Eng 12:1–2

Guru PK, Shrivastava AK, Tiwari P, Khandai S, Chandel NS (2022) Estimation of carbon emissions of agricultural machinery use in India. ORYZA- Int J Rice 59(3):260–268

Herzog T (2009) World greenhouse gas emissions in 2005. World Resources Institute

Hoben JP, Gehl RJ, Millar N, Grace PR, Robertson GP (2011) Nonlinear nitrous oxide ($N_2O$) response to nitrogen fertilizer in on-farm corn crops of the US Midwest. Glob Change Biol 17(2):1140–1152

Houshyar E (2017) Environmental impacts of irrigated and rain-fed barley production in Iran using life cycle assessment (LCA). Span J Agric Res 15(2):6

Hu F, Gan Y, Cui H, Zhao C, Feng F, Yin W, Chai Q (2016) Intercropping maize and wheat with conservation agriculture principles improves water harvesting and reduces carbon emissions in dry areas. Eur J Agron 74:9–17

Hu N, Chen Q, Zhu L (2019) The responses of soil $N_2O$ emissions to residue returning systems: a meta-analysis. Sustain For 11(3):748

Humphreys J, Brye KR, Rector C, Gbur EE (2019) Methane emissions from rice across a soil organic matter gradient in Alfisols of Arkansas, USA. Geoderma Reg 16:e00200
Hungate BA, Holland EA, Jackson RB, Chapin FS III, Mooney HA, Field CB (1997) The fate of carbon in grasslands under carbon dioxide enrichment. Nature 388(6642):576–579
Hütsch BW (2001) Methane oxidation in non-flooded soils as affected by crop production. Eur J Agron 14(4):237–260
Ince E, Güler MA (2020) On the advantages of the new power-split infinitely variable transmission over conventional mechanical transmissions based on fuel consumption analysis. J Clean Prod 244:118795
IPCC (2023) Sixth assessment report of the Intergovernmental Panel on Climate Change. IPCC, Geneva
IPCC (2006) 2006 IPCC Guidelines for National Greenhouse Gas Inventories (eds. Eggleston, H.S., Buendia, L., Miwa, K., Ngara, T. & Tanabe, K.). Prepared by the National Greenhouse Gas Inventories Programme, IGES, Japan
Jain N, Dubey R, Dubey DS, Singh J, Khanna M, Pathak H, Bhatia A (2014) Mitigation of greenhouse gas emission with system of rice intensification in the Indo-Gangetic Plains. Paddy Water Environ 12:355–363
Jayaraman S, Sahu M, Sinha NK, Mohanty M, Chaudhary RS, Yadav B, Srivastava LK, Hati KM, Patra AK, Dalal RC (2022) Conservation agricultural practices impact on soil organic carbon, soil aggregation and greenhouse gas emission in a vertisol. Agriculture 12(7):1004
Jensen ES, Peoples MB, Hauggaard-Nielsen H (2010) Faba bean in cropping systems. Field Crop Res 115(3):203–216
Jensen ES, Peoples MB, Boddey RM, Gresshoff PM, Hauggaard-Nielsen H, Alves JR, Morrison MJ (2012) Legumes for mitigation of climate change and the provision of feedstock for biofuels and biorefineries. A review. Agron Sustain Dev 32:329–364
Jiang R, Yang JY, Drury CF, He W, Smith WN, Grant BB, He P, Zhou W (2021) Assessing the impacts of diversified crop rotation systems on yields and nitrous oxide emissions in Canada using the DNDC model. Sci Total Environ 759:143433
Kassam A, Freidrich T, Derpsch R (2018) Global spread of conservation agriculture. Int J Environ Sci 76(1):29–51
Kassam A, Friedrich T, Derpsch R (2022) Successful experiences and lessons from conservation agriculture worldwide. Agronomy 12(4):769
Kaye JP, Quemada M (2017) Using cover crops to mitigate and adapt to climate change. A review. Agron Sustain Dev 37:1–7
Kern JS, Johnson MG (1993) Conservation tillage impacts on national soil and atmospheric carbon levels. Soil Sci Soc Am J 57(1):200–210
Khaliq MA, Tarin MW, Jingxia G, Yanhui C, Guo W (2019) Soil liming effects on $CH_4$, $N_2O$ emission and Cd, Pb accumulation in upland and paddy rice. Environ Pollut 248:408–420
Kim SY, Gutierrez J, Kim PJ (2012) Considering winter cover crop selection as green manure to control methane emission during rice cultivation in paddy soil. Agric Ecosyst Environ 161:130–136
Koga N, Tsuruta H, Tsuji H, Nakano H (2003) Fuel consumption-derived $CO_2$ emissions under conventional and reduced tillage cropping systems in northern Japan. Agric Ecosyst Environ 99(1–3):213–219
Kong D, Liu N, Ren C, Li H, Wang W, Li N, Ren G, Feng Y, Yang G (2020) Effect of nitrogen fertilizer on soil $CO_2$ emission depends on crop rotation strategy. Sustain For 12(13):5271
Kroeze C, Mosier A, Bouwman L (1999) Closing the global $N_2O$ budget: a retrospective analysis 1500–1994. Global Biogeochem Cycles 13(1):1–8
Lai TV, Farquharson R, Denton MD (2019) High soil temperatures alter the rates of nitrification, denitrification and associated $N_2O$ emissions. J Soils Sediments 19:2176–2189
Lal R (2004) Carbon emission from farm operations. Environ Int 30(7):981–990
Le Mer J, Roger P (2001) Production, oxidation, emission and consumption of methane by soils: a review. Eur J Soil Biol 37(1):25–50

Lei J, Guo X, Zeng Y, Zhou J, Gao Q, Yang Y (2021) Temporal changes in global soil respiration since 1987. Nat Commun 12(1):403

Lenka NK, Lal R (2013) Soil aggregation and greenhouse gas flux after 15 years of wheat straw and fertilizer management in a no-till system. Soil Tillage Res 126:78–89

Lesschen JP, Velthof GL, de Vries W, Kros J (2011) Differentiation of nitrous oxide emission factors for agricultural soils. Environ Pollut 159(11):3215–3222

Li ZA, Zou B, Xia HP, Ding YZ, Tan WN, Ma ZR (2005) Effect of fertilizer and water content on $N_2O$ emission from three plantation soils in south China. J Environ Sci 17(6):970–976

Li LJ, You MY, Shi HA, Ding XL, Qiao YF, Han XZ (2013) Soil CO2 emissions from a cultivated Mollisol: effects of organic amendments, soil temperature, and moisture. Eur J Soil Biol 55:83–90

Li Z, Reichel R, Xu Z, Vereecken H, Brüggemann N (2021a) Return of crop residues to arable land stimulates $N_2O$ emission but mitigates $NO_3^-$ leaching: a meta-analysis. Agron Sustain Dev 41:1–7

Li J, Wang S, Shi Y, Zhang L, Wu Z (2021b) Do fallow season cover crops increase $N_2O$ or $CH_4$ emission from paddy soils in the mono-rice cropping system? Agronomy 11(2):199

Li Y, Clough TJ, Moinet GY, Whitehead D (2021c) Emissions of nitrous oxide, dinitrogen and carbon dioxide from three soils amended with carbon substrates under varying soil matric potentials. Eur J Soil Sci 72(5):2261–2275

Li S, Chen L, Han X, Yang K, Liu K, Wang J, Chen Y, Liu L (2022) Rice cultivar renewal reduces methane emissions by improving root traits and optimizing photosynthetic carbon allocation. Agriculture 12(12):2134

Lognoul M, Theodorakopoulos N, Hiel MP, Regaert D, Broux F, Heinesch B, Bodson B, Vandenbol M, Aubinet M (2017) Impact of tillage on greenhouse gas emissions by an agricultural crop and dynamics of $N_2O$ fluxes: insights from automated closed chamber measurements. Soil Tillage Res 167:80–89

Lu Y, Wassmann R, Neue HU, Huang C (2000) Dissolved organic carbon and methane emissions from a rice paddy fertilized with ammonium and nitrate. American Society of Agronomy, Crop Science Society of America, and Soil Science Society of America

Lu Y, Fu L, Lu Y, Hugenholtz F, Ma K (2015) Effect of temperature on the structure and activity of a methanogenic archaeal community during rice straw decomposition. Soil Biol Biochem 81:17–27

Ma K, Lu Y (2011) Regulation of microbial methane production and oxidation by intermittent drainage in rice field soil. FEMS Microbiol Ecol 75(3):446–456

Ma J, Ma E, Xu H, Yagi K, Cai Z (2009) Wheat straw management affects $CH_4$ and $N_2O$ emissions from rice fields. Soil Biol Biochem 41(5):1022–1028

Machado PV, Farrell RE, Deen W, Voroney RP, Congreves KA, Wagner-Riddle C (2021) Contribution of crop residue, soil, and fertilizer nitrogen to nitrous oxide emissions varies with long-term crop rotation and tillage. Sci Total Environ 767:145107

Mangalassery S, Sjögersten S, Sparkes DL, Sturrock CJ, Craigon J, Mooney SJ (2014) To what extent can zero tillage lead to a reduction in greenhouse gas emissions from temperate soils? Sci Rep 4(1):4586

Maucieri C, Tolomio M, McDaniel MD, Zhang Y, Robatjazi J, Borin M (2021) No-tillage effects on soil $CH_4$ fluxes: a meta-analysis. Soil Tillage Res 212:105042

Mei K, Wang Z, Huang H, Zhang C, Shang X, Dahlgren RA, Zhang M, Xia F (2018) Stimulation of $N_2O$ emission by conservation tillage management in agricultural lands: a meta-analysis. Soil Tillage Res 182:86–93

Melts I, Normak A, Nurk L, Heinsoo K (2014) Chemical characteristics of biomass from nature conservation management for methane production. Bioresour Technol 167:226–231

Meurer KH, Franko U, Stange CF, Dalla Rosa J, Madari BE, Jungkunst HF (2016) Direct nitrous oxide ($N_2O$) fluxes from soils under different land use in Brazil—a critical review. Environ Res Lett 11(2):023001

Mirzaei M, Gorji Anari M, Taghizadeh-Toosi A, Zaman M, Saronjic N, Mohammed S, Szabo S, Caballero-Calvo A (2022) Soil nitrous oxide emissions following crop residues management in corn-wheat rotation under conventional and no-tillage systems. Air Soil Water Res 15:11786221221128789

Moeletsi ME, Tongwane MI (2004) methane and nitrous oxide emissions from manure management in South Africa. Animals 5(2):193–205

Mojeremane W, Rees RM, Mencuccini M (2012) The effects of site preparation practices on carbon dioxide, methane and nitrous oxide fluxes from a peaty gley soil. Forestry 85(1):1–5

Monti M, Badagliacca G, Romeo M, Gelsomino A (2021) No-Till and solid digestate amendment selectively affect the potential denitrification activity in two Mediterranean orchard soils. Soil Syst 5(2):31

Mumford MT, Rowlings DW, Scheer C, De Rosa D, Grace PR (2019) Effect of irrigation scheduling on nitrous oxide emissions in intensively managed pastures. Agric Ecosyst Environ 272:126–134

Mutegi JK, Munkholm LJ, Petersen BM, Hansen EM, Petersen SO (2010) Nitrous oxide emissions and controls as influenced by tillage and crop residue management strategy. Soil Biol Biochem 42(10):1701–1711

Nabuurs GJ, Mrabet R, Hatab AA, Bustamante M, Clark H, Havlík P, House JI, Mbow C, Ninan KN, Popp A, Roe S (2023) Agriculture, forestry and other land uses (AFOLU). In: Climate change 2022: mitigation of climate change. Cambridge University Press, pp 747–860

Naorem A, Jayaraman S, Sinha NK, Mohanty M, Chaudhary RS, Hati KM, Mandal A, Thakur JK, Patra AK, Srinivasarao C, Chaudhari SK (2023) Eight-year impacts of conservation agriculture on soil quality, carbon storage, and carbon emission footprint. Soil Tillage Res 232:105748

Ngoc THV, Maeda M (2018) Interactive effects of ammonium application rates and temperature on nitrous oxide emission from tropical agricultural soil. J Soil Sci Plant Nut 64(6):767–773

Nicolardot B, Recous S, Mary B (2001) Simulation of C and N mineralisation during crop residue decomposition: a simple dynamic model based on the C: N ratio of the residues. Plant Soil 228(1):83–103

Nisar S, Benbi DK, Toor AS (2021) Energy budgeting and carbon footprints of three tillage systems in maize-wheat sequence of north-western Indo-Gangetic Plains. Energy 229:120661

Nissan A, Alcolombri U, Peleg N, Galili N, Jimenez-Martinez J, Molnar P, Holzner M (2023) Global warming accelerates soil heterotrophic respiration. Nat Commun 14(1):3452

Ntonta S, Mathew I, Zengeni R, Muchaonyerwa P, Chaplot V (2022) Crop residues differ in their decomposition dynamics: review of available data from world literature. Geoderma 419:115855

Nyambo P, Cornelius C, Araya T (2020) Carbon dioxide fluxes and carbon stocks under conservation agricultural practices in South Africa. Agriculture 10(9):374

Oduor CO, Karanja N, Onwong'a R, Mureithi S, Pelster D, Nyberg G (2018) Pasture enclosures increase soil carbon dioxide flux rate in Semiarid Rangeland, Kenya. Carbon Balance Manage 13:24

Ogle SM, Alsaker C, Baldock J, Bernoux M, Breidt FJ, McConkey B, Regina K, Vazquez-Amabile GG (2019) Climate and soil characteristics determine where no-till management can store carbon in soils and mitigate greenhouse gas emissions. Sci Rep 9(1):11665

Opio C, Gerber P, Mottet A, Falcucci A, Tempio G, MacLeod M, Vellinga T, Henderson B, Steinfeld H (2013) Greenhouse gas emissions from ruminant supply chains–A global life cycle assessment. Food and Agriculture Organization of the United Nations

Pandey D, Agrawal M, Bohra JS (2012) Greenhouse gas emissions from rice crop with different tillage permutations in rice–wheat system. Agric Ecosyst Environ 159:133–144

Parashar DC, Gupta PK, Rai J, Sharma RC, Singh N (1993) Effect of soil temperature on methane emission from paddy fields. Chemosphere 26(1–4):247–250

Paris B, Vandorou F, Balafoutis AT, Vaiopoulos K, Kyriakarakos G, Manolakos D, Papadakis G (2022) Energy use in open-field agriculture in the EU: a critical review recommending energy efficiency measures and renewable energy sources adoption. Renew Sust Energ Rev 158:112098

Parton WJ, Mosier AR, Ojima DS, Valentine DW, Schimel DS, Weier K, Kulmala AE (1996) Generalized model for $N_2$ and $N_2O$ production from nitrification and denitrification. Global Biogeochem Cycles 10(3):401–412

Parton WJ, Holland EA, Del Grosso SJ, Hartman MD, Martin RE, Mosier AR, Ojima DS, Schimel DS (2001) Generalized model for NOx and $N_2O$ emissions from soils. J Geophys Res Atmos 106(D15):17403–17419

Paustian K, Lehmann J, Ogle S, Reay D, Robertson GP, Smith P (2016) Climate-smart soils. Nature 532(7597):49–57

Pearson TR, Brown S, Murray L, Sidman G (2017) Greenhouse gas emissions from tropical forest degradation: an underestimated source. Carbon Balance Manag 12:1–1

Pelster DE, Chantigny MH, Rochette P, Angers DA, Rieux C, Vanasse A (2012) Nitrous oxide emissions respond differently to mineral and organic nitrogen sources in contrasting soil types. J Environ Qual 41(2):427–435

Peng J, Wegner CE, Bei Q, Liu P, Liesack W (2018) Metatranscriptomics reveals a differential temperature effect on the structural and functional organization of the anaerobic food web in rice field soil. Microbiome 6:169

Petersen SO, Schjønning P, Thomsen IK, Christensen BT (2008) Nitrous oxide evolution from structurally intact soil as influenced by tillage and soil water content. Soil Biol Biochem 40(4):967–977

Petersen SO, Mutegi JK, Hansen EM, Munkholm LJ (2011) Tillage effects on $N_2O$ emissions as influenced by a winter cover crop. Soil Biol Biochem 43(7):1509–1517

Philippot L, Hallin S, Börjesson G, Baggs EM (2009) Biochemical cycling in the rhizosphere having an impact on global change. Plant Soil 321:61–81

Pilegaard K (2013) Processes regulating nitric oxide emissions from soils. Philos Trans Biol Sci 368:20130126

Pittelkow CM, Liang X, Linquist BA, Van Groenigen KJ, Lee J, Lundy ME, Van Gestel N, Six J, Venterea RT, Van Kessel C (2015) Productivity limits and potentials of the principles of conservation agriculture. Nature 517(7534):365–368

Plaza-Bonilla D, Álvaro-Fuentes J, Arrúe JL, Cantero-Martínez C (2014) Tillage and nitrogen fertilization effects on nitrous oxide yield-scaled emissions in a rainfed Mediterranean area. Agric Ecosyst Environ 189:43–52

Pratibha G, Srinivas I, Rao KA, Raju BM, Thyagaraj CR, Korwar GR, Venkateswarlu B, Shanker AK, Choudhary DK, Rao KS, Srinivasarao C (2015) Impact of conservation agriculture practices on energy use efficiency and global warming potential in rainfed pigeonpea–castor systems. Eur J Agron 66:30–40

Qin X, Li YE, Wang H, Li J, Wan Y, Gao Q, Liao Y, Fan M (2015) Effect of rice cultivars on yield-scaled methane emissions in a double rice field in South China. J Integr Environ Sci 12(sup1):47–66

Rahman MM (2013) Carbon dioxide emission from soil. Agric Res 2:132–139

Ranaivoson L, Naudin K, Ripoche A, Affholder F, Rabeharisoa L, Corbeels M (2017) Agro-ecological functions of crop residues under conservation agriculture. A review. Agron Sustain Dev 37:1–7

Reinthaler D, Harris E, Poetsch EM, Herndl M, Richter A, Wachter H, Bahn M (2021) Responses of grassland soil $CO_2$ production and fluxes to drought are shifted in a warmer climate under elevated $CO_2$. Soil Biol Biochem 163:108436

Ren F, Zhang X, Liu J, Sun N, Wu L, Li Z, Xu M (2017) A synthetic analysis of greenhouse gas emissions from manure amended agricultural soils in China. Sci Rep 7(1):8123

Ritchie H, Roser M, Rosado P (2020) $CO_2$ and greenhouse gas emissions. Published online at OurWorldInData.org. Retrieved from: https://ourworldindata.org/co2-and-other-greenhouse-gas-emissions [Online Resource]

Rochette P (2008) No-till only increases $N_2O$ emissions in poorly-aerated soils. Soil Tillage Res 101(1–2):97–100

Rochette P, Janzen HH (2005) Towards a revised coefficient for estimating $N_2O$ emissions from legumes. Nutr Cycl Agroecosyst 73:171–179

Rochette P, Angers DA, Bélanger G, Chantigny MH, Prévost D, Lévesque G (2004) Emissions of $N_2O$ from alfalfa and soybean crops in eastern Canada. Soil Sci Soc Am J 68(2):493–506

Rochette P, Angers DA, Chantigny MH, Bertrand N (2008) Nitrous oxide emissions respond differently to no-till in a loam and a heavy clay soil. Soil Science Society of America Journal 72(5):1363–1369

Rogers CW, Brye KR, Smartt AD, Norman RJ, Gbur EE, Evans-White MA (2014) Cultivar and previous crop effects on methane emissions from drill-seeded, delayed-flood rice production on a silt-loam soil. Soil Sci 179(1):28–36

Romasanta RR, Sander BO, Gaihre YK, Alberto MC, Gummert M, Quilty J, Castalone AG, Balingbing C, Sandro J, Correa T Jr, Wassmann R (2017) How does burning of rice straw affect $CH_4$ and $N_2O$ emissions? a comparative experiment of different on-field straw management practices. Agric Ecosyst Environ 239:143–153

Runion BG, Allen Torbert H, Prior SA, Rogers HH (2009) Effects of elevated atmospheric carbon dioxide on soil carbon in terrestrial ecosystems of the Southeastern United States. In: Soil carbon sequestration and the greenhouse effect, 2nd edn. SSSA Special Publication 57. ASA–CSSA–SSSA, 677 S, Madison

Rutkowska B, Szulc W, Sosulski T, Skowrońska M, Szczepaniak J (2018) Impact of reduced tillage on $CO_2$ emission from soil under maize cultivation. Soil Tillage Res 180:21–28

Saggar S, Luo J, Giltrap DL, Maddena M (2009) Nitrous oxide emission from temperate grasslands: process, measurements, modelling and mitigation. In: Sheldon AI, Barnhart EP (eds) Nitrous oxide emissions research progress. Nova Science Publishers, New York, pp 1–66

Sahai S, Sharma C, Singh SK, Gupta PK (2011) Assessment of trace gases, carbon and nitrogen emissions from field burning of agricultural residues in India. Nutr Cycl Agroecosyst 89:143–157

Sainju UM, Whitehead WF, Singh BP (2005) Biculture legume-cereal cover crops for enhanced biomass yield and carbon and nitrogen. Agron J 97:1403–1412

Sainju UM, Ghimire R, Dangi S (2021) Soil carbon dioxide and methane emissions and carbon balance with crop rotation and nitrogen fertilization. Sci Total Environ 775:145902

Samad MS, Biswas A, Bakken LR, Clough TJ, de Klein CA, Richards KG, Lanigan GJ, Morales SE (2016) Phylogenetic and functional potential links pH and $N_2O$ emissions in pasture soils. Sci Rep 6(1):35990

Sapkota TB, Singh LK, Yadav AK, Khatri-Chhetri A, Jat HS, Sharma PC, Jat ML, Stirling CM (2020) Identifying optimum rates of fertilizer nitrogen application to maximize economic return and minimize nitrous oxide emission from rice–wheat systems in the Indo-Gangetic Plains of India. Arch Agron Soil Sci 66(14):2039–2054

Scheer C, Wassmann R, Kienzler K, Ibragimov N, Eschanov R (2008) Nitrous oxide emissions from fertilized, irrigated cotton (*Gossypium hirsutum* L.) in the Aral Sea Basin, Uzbekistan: influence of nitrogen applications and irrigation practices. Soil Biol Biochem 40(2):290–301

Schimel J (2000) Global change: rice, microbes and methane. Nature 403:375–377

Schlesinger WH, Andrews JA (2000) Soil respiration and the global carbon cycle. Biogeochemistry 48:7–20

Schmatz R, Recous S, Weiler DA, Pilecco GE, Schu AL, Giovelli RL, Giacomini SJ (2020) How the mass and quality of wheat and vetch mulches affect drivers of soil $N_2O$ emissions. Geoderma 372:114395

Semeraro T, Scarano A, Leggieri A, Calisi A, De Caroli M (2023) Impact of climate change on agroecosystems and potential adaptation strategies. Land 12(6):1117

Senbayram M, Chen R, Budai A, Bakken L, Dittert K (2012) $N_2O$ emission and the $N_2O/(N_2O+N_2)$ product ratio of denitrification as controlled by available carbon substrates and nitrate concentrations. Agric Ecosyst Environ 147:4–12

Shcherbak I, Millar N, Robertson GP (2014) Global metaanalysis of the nonlinear response of soil nitrous oxide ($N_2O$) emissions to fertilizer nitrogen. Proc Natl Acad Sci USA 111(25):9199–9204

Shibata M, Terada F (2010) Factors affecting methane production and mitigation in ruminants. Anim Sci J 81(1):2–10

Shibata M, Terada F, Kurihara M, Nishida T, Iwasaki K (1993) Estimation of methane production in ruminants. J Anim Sci Technol 64:790

Shrestha RK, Lal R, Rimal B (2013) Soil carbon fluxes and balances and soil properties of organically amended no-till corn production systems. Geoderma 197:177–185

Signor D, Cerri CE (2013) Nitrous oxide emissions in agricultural soils: a review. Pesqui Agropecu Trop 43:322–338

Silva-Olaya AM, Cerri CE, La Scala JN, Dias CT, Cerri CC (2013) Carbon dioxide emissions under different soil tillage systems in mechanically harvested sugarcane. Environ Res Lett 8(1):015014

Simojoki A, Jaakkola A (2000) Effect of nitrogen fertilization, cropping and irrigation on soil air composition and nitrous oxide emission in a loamy clay. Eur J Soil Sci 51(3):413–424

Sims R, Flammini A, Puri M, Bracco S (2015) Opportunities for agri-food chains to become energy-smart. FAO USAID

Six J, Conant RT, Paul EA, Paustian K (2002) Stabilization mechanisms of soil organic matter: implications for C-saturation of soils. Plant Soil 241:155–176

Six J, Bossuyt H, Degryze S, Denef K (2004) A history of research on the link between (micro) aggregates, soil biota, and soil organic matter dynamics. Soil Tillage Res 79(1):7–31

Skiba U, Ball B (2002) The effect of soil texture and soil drainage on emissions of nitric oxide and nitrous oxide. Soil Use Manag 18(1):56–60

Smartt AD, Brye KR, Rogers CW, Norman RJ, Gbur EE, Hardke JT, Roberts TL (2016) Previous crop and cultivar effects on methane emissions from drill-seeded, delayed-flood rice grown on a clay soil. Appl Environ Soil Sci 9542361:1–13

Smil V (2008) Energy in nature and society- general energetic of complex systems. MIT Press, Cambridge, MA, p 512

Song C, Zhang J (2009) Effects of soil moisture, temperature, and nitrogen fertilization on soil respiration and nitrous oxide emission during maize growth period in northeast China. Acta Agric Scand Sect B Soil Plant Sci 59(2):97–106

Song HJ, Lee JH, Jeong HC, Choi EJ, Oh TK, Hong CO, Kim PJ (2019) Effect of straw incorporation on methane emission in rice paddy: conversion factor and smart straw management. Appl Biol Chem 62:1–3

Sosulski T, Stępień W, Wąs A, Szymańska M (2020) $N_2O$ and $CO_2$ emissions from bare soil: effect of fertilizer management. Agriculture 10(12):602

Stevens RJ, Laughlin RJ (1998) Measurement of nitrous oxide and di-nitrogen emissions from agricultural soils. Nutr Cycl Agroecosyst 52:131–139

Su Q, Domingo-Félez C, Jensen MM, Smets BF (2019) Abiotic nitrous oxide ($N_2O$) production is strongly pH dependent, but contributes little to overall $N_2O$ emissions in biological nitrogen removal systems. Environ Sci Technol 53(7):3508–3516

Sun BF, Zhao Hong ZH, LüYiZhong L, Lu Fei LF, Wang XiaoKe WX (2016) The effects of nitrogen fertilizer application on methane and nitrous oxide emission/uptake in Chinese croplands. J Integr Agric 15(2):440–450

Surey R, Schimpf CM, Sauheitl L, Mueller CW, Rummel PS, Dittert K, Kaiser K, Böttcher J, Mikutta R (2020) Potential denitrification stimulated by water-soluble organic carbon from plant residues during initial decomposition. Soil Biol Biochem 147:107841

Suwanwaree P, Robertson GP (2005) Methane oxidation in forest, successional, and no-till agricultural ecosystems: effects of nitrogen and soil disturbance. Soil Sci Soc Am J 69(6):1722–1729

Swainson NM, Hoskin SO, Clark H, Lopez-Villalobos N (2007) The effect of age on methane emissions from young, weaned red deer (*Cervus elaphus*) stags grazing perennial-ryegrass (*Lolium perenne*)-based pasture. NZ J Agric Res 50(3):407–416

Tang M, Hu F (2021) How does land urbanization promote $CO_2$ emissions reduction? Evidence from Chinese prefectural-level cities. Front Environ Sci 9:766839

Thangarajan R, Bolan NS, Tian G, Naidu R, Kunhikrishnan A (2013) Role of organic amendment application on greenhouse gas emission from soil. Sci Total Environ 465:72–96

Thies S, Joshi DR, Bruggeman SA, Clay SA, Mishra U, Morile-Miller J, Clay DE (2020) Fertilizer timing affects nitrous oxide, carbon dioxide, and ammonia emissions from soil. Soil Sci Soc Am J 84(1):115–130

Thomson AJ, Giannopoulos G, Pretty J, Baggs EM, Richardson DJ (2012) Biological sources and sinks of nitrous oxide and strategies to mitigate emissions. Philos Trans R Soc Lond Ser B Biol Sci 367(1593):1157–1168

Tirol-Padre A, Rai M, Kumar V, Gathala M, Sharma PC, Sharma S, Nagar RK, Deshwal S, Singh LK, Jat HS, Sharma DK (2016) Quantifying changes to the global warming potential of rice wheat systems with the adoption of conservation agriculture in northwestern India. Agric Ecosyst Environ 219:125–137

Tokida T, Cheng W, Adachi M, Matsunami T, Nakamura H, Okada M, Hasegawa T (2013) The contribution of entrapped gas bubbles to the soil methane pool and their role in methane emission from rice paddy soil in free-air [$CO_2$] enrichment and soil warming experiments. Plant Soil 364:131–143

Ussiri DA, Lal R, Jarecki MK (2009) Nitrous oxide and methane emissions from long-term tillage under a continuous corn cropping system in Ohio. Soil Tillage Res 104(2):247–255

Vallero DA (2019) Air pollution calculations. Elsevier Science

Van Groenigen KJ, Qi X, Osenberg CW, Luo Y, Hungate BA (2014) Faster decomposition under increased atmospheric CO2 limits soil carbon storage. Science 344(6183):508–509

Van Kessel C, Venterea R, Six J, Adviento-Borbe MA, Linquist B, van Groenigen KJ (2013) Climate, duration, and N placement determine $N_2O$ emissions in reduced tillage systems: a meta-analysis. Glob Change Biol 19(1):33–44

Vangeli S, Restovich S, Posse G (2022) Cover crop species affect $N_2O$ emissions at hotspot moments of summer crops. Front Soil Sci 2:903387

Velthof GL, Rietra RP (2018) Nitrous oxide emission from agricultural soils. Wageningen Environmental Research. Report; No. 2921. The Netherlands

Venkatramanan V, Shah S, Rai AK, Prasad R (2021) Nexus between crop residue burning, bioeconomy and sustainable development goals over North-Western India. Front Energy Res 8:614212

Venterea RT, Rolston DE, Cardon ZG (2005) Effects of soil moisture, physical, and chemical characteristics on abiotic nitric oxide production. Nutr Cycl Agroecosyst 72:27–40

Walling E, Vaneeckhaute C (2020) Greenhouse gas emissions from inorganic and organic fertilizer production and use: a review of emission factors and their variability. J Environ Manag 276:111211

Wang XG, Luo Y (2018) Crop residue incorporation and nitrogen fertilizer effects on greenhouse gas emissions from a subtropical rice system in Southwest China. J Mt Sci 15(9):1972–1986

Wang ZP, Delaune RD, Patrick WH Jr, Masscheleyn PH (1993) Soil redox and pH effects on methane production in a flooded rice soil. Soil Sci Soc Am J 57(2):382–385

Wang J, Akiyama H, Yagi K, Yan X (2018) Controlling variables and emission factors of methane from global rice fields. J Atmos Chem Phys 18(14):10419–10431

Wang M, Xiong W, Zou Y, Lin M, Zhou Q, Xie X, Sun Y (2019) Evaluating the net effect of sulfadimidine on nitrogen removal in an aquatic microcosm environment. Environ Pollut 248:1010–1019

Wang H, Wang S, Yu Q, Zhang Y, Wang R, Li J, Wang X (2020) No tillage increases soil organic carbon storage and decreases carbon dioxide emission in the crop residue-returned farming system. J Environ Manag 261:110261

Wang C, Amon B, Schulz K, Mehdi B (2021) Factors that influence nitrous oxide emissions from agricultural soils as well as their representation in simulation models: a review. Agronomy 11(4):770

Wang Y, Saikawa E, Avramov A, Hill NS (2022) Agricultural greenhouse gas fluxes under different cover crop systems. Front Clim 3:742320

Wei L, Liu X, Qin C, Xing W, Gu Y, Wang X, Bai L, Li J (2022) Impacts of soil moisture and fertilizer on $N_2O$ emissions from cornfield soil in a karst watershed, SW China. Atmosphere 13(8):1200

Weitz AM, Linder E, Frolking S, Crill PM, Keller M (2001) $N_2O$ emissions from humid tropical agricultural soils: effects of soil moisture, texture and nitrogen availability. Soil Biol Biochem 33(7–8):1077–1093

Wichern F, Eberhardt E, Mayer J, Joergensen RG, Müller T (2008) Nitrogen rhizodeposition in agricultural crops: methods, estimates and future prospects. Soil Biol Biochem 40(1):30–48

Wihardjaka A, Harsanti ES, Viandari NA, Zu'amah H (2023) Effect of soil tillage and nitrogen fertilizer management on methane emissions from irrigated rice fields in Central Java, Indonesia. Chil J Agric Res 83(3):347–357

Xia L, Lam SK, Wolf B, Kiese R, Chen D, Butterbach-Bahl K (2018) Trade-offs between soil carbon sequestration and reactive nitrogen losses under straw return in global agroecosystems. Glob Chang Biol 24(12):5919–5932

Xu P, Zhou W, Jiang M, Khan I, Shaaban M, Jiang Y, Hu R (2020) Nitrogen fertilizer application in the rice-growing season can stimulate methane emissions during the subsequent flooded fallow period. Sci Total Environ 744:140632

Yadav GS, Babu S, Das A, Mohapatra KP, Singh R, Avasthe RK, Roy S (2020) No-till and mulching enhance energy use efficiency and reduce carbon footprint of a direct-seeded upland rice production system. J Clean Prod 271:122700

Yoo J, Woo SH, Park KD, Chung KY (2016) Effect of no-tillage and conventional tillage practices on the nitrous oxide ($N_2O$) emissions in an upland soil: soil $N_2O$ emission as affected by the fertilizer applications. Appl Biol Chem 59:787–797

Yuan J, Yuan Y, Zhu Y, Cao L (2018) Effects of different fertilizers on methane emissions and methanogenic community structures in paddy rhizosphere soil. Sci Total Environ 627:770–781

Yue K, Fornara DA, Heděnec P, Wu Q, Peng Y, Peng X, Ni X, Wu F, Peñuelas J (2023) No tillage decreases GHG emissions with no crop yield tradeoff at the global scale. Soil Tillage Res 228:105643

Zhai LM, Liu HB, Zhang JZ, Huang J, Wang BR (2011) Long-term application of organic manure and mineral fertilizer on $N_2O$ and $CO_2$ emissions in a red soil from cultivated maize-wheat rotation in China. Agric Sci China 10(11):1748–1757

Zhou M, Zhu B, Brüggemann N, Wang X, Zheng X, Butterbach-Bahl K (2015) Nitrous oxide and methane emissions from a subtropical rice–rapeseed rotation system in China: a 3-year field case study. Agric Ecosyst Environ 212:297–309

Zhu X, Burger M, Doane TA, Horwath WR (2013) Ammonia oxidation pathways and nitrifier denitrification are significant sources of $N_2O$ and NO under low oxygen availability. Proc Natl Acad Sci USA 110:6328–6333

# Factors Affecting Carbon Sequestration and Greenhouse Gas Emission in Agroecosystems

**Sudeshna Bhattacharjya, Asha Sahu, Shubham Singh, and A. B. Singh**

## Introduction

Soil serves a range of ecosystem functions, including supporting agriculture and forestry. Its role will become increasingly important in the next millennium if we are to meet the food, fiber, and fuel needs of a population expected to reach 8–10 billion by 2050. These needs cannot be met by simply expanding the area under cultivation, so intensification of currently managed land is inevitable. However, agricultural intensification can damage soil functions and threaten the sustainability of agroecosystems. It can also alter other parts of the environment through emissions to water or air. The world's cultivated soils have already lost 25–75% of their original carbon stock (Lal et al. 2018; Lorenz and Lal 2018) in the form of $CO_2$, due to unsustainable land management practices that result in land degradation and amplify climate change.

Degradation of the land lowers the ability of soil to store carbon, which causes climate change and costs trillions of dollars annually (Davies 2017). The challenge of offsetting global greenhouse gas (GHG) emissions can be addressed by implementing established soil organic carbon (SOC)-oriented soil management strategies in soils that have the capacity to absorb carbon, such as croplands and degraded soils. Sustainable soil management practices that prioritize soil organic carbon (SOC) not only reduce greenhouse gas emissions but also yield several advantages such as increased farm income, decreased poverty and malnourishment, and the provision of vital ecosystem services such as regulating climate and hydrology, protecting biodiversity, and nutrient cycling. Contribution to the Sustainable Development Goals' (SDGs') accomplishment heightened resistance to adverse weather events.

S. Bhattacharjya (✉) · A. Sahu · S. Singh · A. B. Singh
Division of Soil Biology, ICAR-Indian Institute of Soil Science, Bhopal, MP, India



S. R. Mohanty, B. Kollah (eds.), *Greenhouse Gas Regulating Microorganisms in Soil Ecosystems*, https://doi.org/10.1007/978-3-031-70569-4_6

## Importance of SOC in Climate Change Mitigation

One important measure of soil health is soil organic carbon, or SOC. It forms the greatest terrestrial carbon reservoir, performs a variety of soil activities, and offers a wide range of ecosystem services. As a result, through controlling biogeochemical reactions and the exchange of greenhouse gases (GHGs), SOC is essential to the global carbon balance. After taking into consideration land and ocean sinks, the net global increase in anthropogenic $CO_2$ emissions during the last 10 years (2009–2018) was 4.9 Gt C $yr^{-1}$. Due to land use changes and unsustainable farming methods, it is estimated that humans have released significant amounts of $CO_2$ into the atmosphere, resulting in a historical SOC depletion of 115–154 Gt C (average of 135 Gt C) (Lal 2018).

In addition to significantly lowering emissions from fossil fuels, it is now commonly acknowledged that negative emission technologies will be necessary to combat the ensuing climate change. Because of this, we have to make the most of our extensive knowledge of SOC and the soil's natural ability to store carbon and reduce $CO_2$ emissions. According to estimates, the worldwide technical potential for sequestering SOCs is 1.45–3.44 Gt C (5.3–12.6 Gt $CO_2$) annually (Lal 2018). This amounts to 38–91% of the fossil fuel emissions from the electricity industry worldwide, 67–100% of the fossil fuel emissions from the transportation sector globally (Muntean et al. 2018), and 9–23% of the global total emissions (53 Gt $CO_2$) from all sectors in 2017 (UNEP 2018).

Given the significant capacity of soils to stabilize the climate, sequestering organic carbon in soil may be able to remove between 0.79 and 1.54 Gt C $yr^{-1}$ from the atmosphere in a technically possible manner. But in order to realize this potential and amass SOC in globally important amounts, the international community must create laws and financial incentives.

## SOC in Policy Decision

### *Global Soil Partnership (GSP) Under FAO*

In light of the threats posed by global warming and food insecurity, the Global Soil Partnership (GSP) started proactively attempting to quantify, track, manage, preserve, and/or boost soil organic carbon (SOC) stocks after SOC loss was identified as the second biggest global threat to soil sustainability in the 2015 Status of the World's Soil Resources report. A step in toward this goal is the release of the most extensive worldwide soil organic carbon (SOC) map by the UN Food and Agriculture Organization (FAO), which shows the quantity of carbon stores in the soil. Additionally, the FAO has started a program called RECSOIL to recarbonize soils.

Protecting additional soil organic carbon (SOC) losses from carbon-rich soils (peatlands, black soils, and permafrost) and, when practical, restoring SOC stocks in agriculture and degraded soils are the primary goals of RECSOIL. Other goals include boosting farmer income through increased soil productivity, promoting better food security and nutrition, and mitigating the impact of climate change within the guidelines of the United Nations Framework Convention on Climate Change (UNFCCC).

## Convention on Biological Diversity

Although they are separate entities, soil organic matter (SOM), soil organic carbon (SOC), and soil biodiversity are intimately related. The wide range bacteria, fungus, protozoa, insects, worms, various invertebrates, and vertebrates actually compose soil biodiversity. The top 2–3 cm of the soil, where organic matter and root concentrations are highest, is where the majority of soil biodiversity may be found. For instance, in drylands all over the world, biological soil crusts consisting of communities of mosses, lichens, and microbes near the soil's surface are crucial to maintaining soil fertility. By breaking down plant residues, soil biodiversity helps to create aboveground vegetation by transforming organically stored nutrients into inorganic forms that plants can absorb. Moreover, soil resilience—the capacity of the soil to bounce back from shocks—is boosted by soil biodiversity. It has been estimated that soil biodiversity contributes between USD 1.5 and 13 trillion a year to ecosystem services globally.

## UNCCD

The aim of Sustainable Development Goal 15.3 is to "combat desertification, restore degraded land and soil, and strive toward a world devoid of land degradation by 2030." There is a growing discrepancy between the Nationally Determined Contributions of the Paris Agreement and the emission levels realistically required to maintain global warming within 1.5 °C, because the commitments of the 2030 Agenda for Sustainable Development are still not being met. In order to solve these issues, neutrality on land degradation must be achieved. Due to its sensitivity to land management techniques and the potential for large social, economic, and environmental benefits, soil organic carbon, or SOC, can serve as an important indicator and driver of success in this area. SOC was chosen as one of the three indicators for land degradation neutrality because of its multifaceted roles and responsiveness to land management.

## UNFCCC

Following the recent recognition of soils as a key subject of the global carbon strategy for climate change adaptation as well as mitigation, the UNFCCC launched high-level initiatives like the “4p1000 initiative” at COP21. The aforementioned program highlights the significance of soil organic matter (SOM) in tackling the triple problem of ensuring food and nutrition security, adapting to climate change, and reducing greenhouse gas (GHG) emissions caused by anthropogenic activity. The Paris Agreement, reached COP21 of the United Nations Framework Convention on Climate Change, advocates for limiting global warming well below 2 °C, ideally around 1.5 °C, and for strengthening adaptation to the negative effects of climate change. Additionally, it calls for the development of low-GHG emissions and climate resilience through measures that won't jeopardize food supplies. The Koronivia Joint Work on Agriculture (KJWA), which was established during the COP23 to the UNFCCC in 2018, was a significant development for agricultural negotiations. Under the UNFCCC, system soils have been designated as a major agricultural challenge for the first time with the following action: “Improved soil carbon, soil health and soil fertility under grassland and cropland as well as integrated systems, including water management.”

## SOC to Achieve Sustainable Development Goals

The International Year of Soils was celebrated in 2015, and the first Intergovernmental Technical Panel on Soils (ITPS) report was published, demonstrating a greater understanding of soils among the world's population and policymakers (FAO and ITPS 2015). The contribution of soils to the attainment of several Sustainable Development Goals (SDGs) of the United Nations is now widely acknowledged (Keesstra et al. 2016). There are three SDGs that directly benefit from soil carbon sequestration: Goals 2, 13, and 15 of the FAO (2017) are zero hunger, climate action, and life on land. It can also help achieve SDGs 6 and 12, which are concerned with clean water and sanitation and responsible consumption and production, respectively. In the short to medium term, sustainable soil management techniques can efficiently store more carbon in the soil, enhancing food production and livelihoods and supporting in the 2030 SDG accomplishment.

## 4 Per 1000 Initiatives

With 103 nations establishing adaptation and mitigation goals for agriculture and 129 countries establishing targets for other land uses, including forests and degraded land, COP21 represented an important milestone (Richards et al. 2016). The lack of clarity in the suggested activities, however, makes it necessary to set ambitious and common goals for land-based mitigation and adaptation. Only then can we hope to remain under the 2 °C global warming threshold. In this regard, the "4 per 1000 Initiative: Soils for Food Security and Climate" was introduced by the French Ministry of Agriculture in 2015 in advance of the United Nations Framework Convention on Climate Change (UNFCCC)'s Paris climate summit, or COP21. The global aspirational objective stated by this ambitious program is to raise SOC stocks by 4 per 1000, or 0.4%, per year in all land uses, including forests. By 2030, if this target is fully realized, it will lead to sequestration of C twice the amount needed for achieving the mitigation level covered by COP21 Nationally Determined Contributions as compared to a middle-of-the-road baseline scenario of the fifth assessment report of the IPCC (IPCC, AR5; UNEP 2015). In order to raise SOC stocks at a pace of 4 per 1000 (0.4%) of the initial value per year for 20 years, the effort aspires to encourage more sustainable land management practices. Originally, it was proposed that if this rate of carbon sequestration could be reached for all soils worldwide down to a depth of 40 cm, the amount of carbon extracted from the atmosphere would be equivalent to 8.9 gigatons of annual $CO_2$ emissions from fossil fuels, thereby "halting the annual increase in $CO_2$ in the atmosphere."

## SOC Sequestration Through Integrated Nutrient Management (INM)

Integrated nutrient management (INM) is one of the most important ways to challenge the threat of climate change by sequestering carbon through various agricultural management practices, such as balanced fertilization, manuring, composting, crop residues, mulching, and diverse cropping systems. Carbon enters the soil in agricultural fields through roots, trash, and harvest residues. Most of it is deposited as soil organic matter. According to the IPCC's Fourth Assessment Report, soil carbon sinks can be improved through crop land management, grazing land management, bioenergy, the restoration of organic soils and degraded lands, and water management, accounting for 89% of agriculture's technological mitigation potential. According to Majumder et al. (2008), a continuous rice–wheat cropping system in the subtropical Indo-Gangetic Plains of India is most suitable for balanced nitrogen, phosphorus, and potassium (NPK) fertilizer application combined with a reasonable quantity of farmyard manure (FYM) to sustain crop production and boost the SOC.

Right cropping patterns and balanced fertilization are two effective management options for enhancing SOC storage in agricultural soils (Lal 2005). Theoretically, atmospheric $CO_2$ could be reduced by up to 2% with a 1% increase in mean SOC levels (Azeez 2009). Applying organic fertilizer considerably raised the SOC content in comparison to conventional management techniques by increasing the amount of plant biomass added and the amount of lignin-containing organic input.

These two factors together lead to a greater buildup of SOC per unit of carbon input as FYM (Paustian et al. 1992).

Ghosh et al. (2017) found that integrated nutrient management (INM), which combines NPK fertilizer with farmyard manure (FYM), is the best way to sequester soil carbon compared to plots that receive only NPK fertilizer. INM increased total soil organic carbon (SOC) concentrations by increasing the amount of microaggregates, recalcitrant C pools, and both macroaggregates and labile C pools. INM is especially effective at changing the distribution of SOC among particle size separates and increasing SOC accumulation in a tilled maize and wheat production system. In multiple cropping systems, manuring and fertilizer application can sequester carbon in the soil and improve its nutrient status. Integrated nutrient management, or INM, is a method of combining organic manure with inorganic fertilizers that has been shown to be effective in accumulating SOC and enhancing soil health over time, according to a recent study by Trivedi et al. (2020). In the surface soil of FYM + lime-treated plots, a continuous increase in organic inputs led to a greater overall SOC stock. Improving C encapsulation and aggregation in the soil aggregates of manure-treated plots could be the cause of the elevated SOC stock. However, greater quantities of SOC in the approved fertilizer doses combined with lime may be associated with increased yearly biomass production and recycling in the deeper soil layers.

## Effect of Long-Term Fertilization on Soil C Pools and Sequestration

To gain a deeper knowledge of the dynamics of SOC and nitrogen, long-term research is necessary. The most effective management strategy for improving SOC sequestration and reducing the rise in greenhouse gas emissions is long-term fertilization along with manure application. The greatest potential for enhancing SOC storage in agricultural soils is found in cropping systems, balanced fertilization, and different management tactics. According to Majumder et al. (2008), the addition of organic matter and chemical fertilizers (NPK) enhanced SOC in a sandy loam soil by 24.3% when compared to the control treatment.

Ten-year long-term trials at IARI, New Delhi, showed a considerable increase in SOC sequestration over the time as follows NPK + FYM > 100% NPK > 150% NPK = 100% NP = 100% N = 50% NPK = control. Because of the slower rate of FYM breakdown, this might have contributed to the higher SOC content and C sequestration in FYM-treated plot (Table 1). Because of the substantial increment that occurred when combined application of 100% NPK and FYM has been made, this treatment was the most successful management technique for sequestering SOC. In the semiarid subtropical region of India, switching to a high-fertility treatment (100% NPK + FYM) in the maize, wheat, and cowpea cropping system might sequester 1.83Tg C annually.

**Table 1** Amount of C sequestered in different soil C fractions in 0–45 cm depth in a long-term fertilizer experiment

| | Organic C sequestered (kg ha$^{-1}$ yr$^{-1}$) | | |
|---|---|---|---|
| Treatments | SOC | POC | $KMnO_4$-C |
| 50% NPK | 58 | 10 | 10 |
| 100% NPK | 135 | 21 | 46 |
| 150% NPK | 553 | 37 | 62 |
| 100% NP | 120 | 19 | 27 |
| 100% N | 101 | 14 | 24 |
| 100% NPK + FYM | 997 | 70 | 76 |

As chemical fertilizers (NPK) and FYM are applied consistently, the soil's potential to absorb nitrogen and store SOC is increased. According to a 10-year study focusing on integrated application of organic and inorganic inputs by Dutta et al. (2018), all cropping systems that employed these strategies contributed in sequestering carbon in the soil. Following a decade, the SOC content (0–15 cm) improved by 84% under organic application, 28.8% under inorganic application, and 68.3% under integrated nutrient management in comparison to the starting point (2004). In addition to other management approaches, the study claims that cropping systems which include green manure crop and oilseed crop greatly enhance SOC storage as compared to only cereal-based system (Table 2).

## Soil Organic Status Under Continuous Long-Term Fertilizer Applications

Analysis of various approaches of nutrient management revealed that, with the exception of Ranchi (Alfisols) and Pantnagar (Mollisols), crop cultivation raised SOC at 11 locations. SOC at Ranchi was comparable to the starting value for the majority of nutrition management approaches. At Pantnagar, on the other hand, SOC decreased (Table 3), most likely as a result of the soils' increased oxidation upon agricultural use. Both of these areas were covered in forests in the past; the state of SOC was close to equilibrium. A new state of equilibrium of SOC was established as a result of fertilizer application and land preparation. Under integrated balanced nutrition management, the rise in SOC was maximum. Higher doses of fertilizer resulted in greater plant biomass and higher crop yield specifically due to increased SOC content in soil. Comparison of two long-term fertilizer experiment (LTFE) soils of the Vertisol order (LTFE Akola and LTFE Raipur) showed that SOC status declined in the control treatment but increased under increasing nutrient application from 100% N to 100% NPK + FYM after 12 years of study in LTFE Akola (Table 4). The 100% NPK + FYM (INM) treatment had the highest net positive change in SOC status. In contrast, SOC status declined in the control and imbalanced fertilizer application treatments (100% N and 100% NP)

**Table 2** Comparative effect of long-term application of organic inputs, inorganic inputs, and integrated nutrient management on C sequestration in soil

| | SOC pools ($Mg\ ha^{-1}$) | | | | Carbon sequestration ($Mg\ ha^{-1}$) | | | | Carbon sequestration rate ($Mg\ ha^{-1}$) | | | |
|---|---|---|---|---|---|---|---|---|---|---|---|---|
| Treatments | Basmati rice–wheat–*Sesbania* | Basmati rice–lentil–*Sesbania* | Basmati rice–vegetable pea–*Sesbania* | Basmati rice–wheat–*Sesbania* | Basmati rice–wheat–*Sesbania* | Basmati rice–lentil–*Sesbania* | Basmati rice–vegetable pea–*Sesbania* | Basmati rice–wheat–*Sesbania* | Basmati rice–wheat–*Sesbania* | Basmati rice–lentil–*Sesbania* | Basmati rice–vegetable pea–*Sesbania* | Basmati rice–*Brassica napus*–*Sesbania* |
| Organic | 24.78 | 23.99 | 22.88 | 25.29 | 14.62 | 13.63 | 12.14 | 15.36 | 1.46 | 1.36 | 1.21 | 1.54 |
| Inorganic | 15.98 | 19.08 | 17.93 | 17.71 | 2.97 | 7.01 | 5.56 | 5.27 | 0.30 | 0.70 | 0.56 | 0.53 |
| Integrated | 23.94 | 20.74 | 23.34 | 23.41 | 23.14 | 9.31 | 12.76 | 12.49 | 1.36 | 0.93 | 1.28 | 1.25 |

**Table 3** Sequestration or depletion of SOC stock (kg $ha^{-1}$ $yr^{-1}$) in surface (0–20 cm) soil layer in various nutrient management treatments

| LTFE locations | Control | 100% N | 100% NP | 100% NPK | 150% NPK | 100% NPK + FYM |
|---|---|---|---|---|---|---|
| Pantnagar (after 40 years) | −539.4 | −432.7 | −372.0 | −432.2 | −417.5 | 50.3 |
| Barrackpore (after 40 years) | −51.9 | −4.1 | 26.2 | 93.9 | 107.9 | 160.2 |
| Raipur (after 20 years) | −173.1 | −26.1 | 58.7 | 53.5 | 79.3 | 156.6 |
| Akola (after 20 years) | −45.0 | −5.0 | 45.0 | 125.0 | 175.0 | 245.0 |
| Jabalpur (after 34 years) | 58.8 | 123.5 | 135.3 | 214.7 | 241.2 | 311.8 |
| Ranchi (after 40 years) | −10.0 | −22.5 | −40.0 | 15.0 | 40.0 | 90.0 |
| Ludhiana (after 40 years) | 5.0 | 85.0 | 105.0 | 117.5 | 117.5 | 167.5 |

**Table 4** SOC content (kg $ha^{-1}$) in long-term fertilizer experiments in India

| Locations | Initial | NP | NPK | NPK + FYM |
|---|---|---|---|---|
| Akola (Vertisol) | 4.6 | 4.9 | 5.1 | 6.8 |
| Bangalore (Alfisol) | 4.6 | 4.7 | 4.9 | 5.7 |
| Barrackpore (Inceptisol) | 7.0 | 4.7 | 4.6 | 6.8 |
| Coimbatore (Inceptisol) | 3.0 | 6.7 | 7.0 | 9.8 |
| Jagtial (Inceptisol) | 7.9 | 10.0 | 10.0 | 10.1 |
| Jabalpur (Vertisol) | 5.7 | 6.5 | 7.4 | 9.6 |
| Ludhiana (Inceptisol) | 2.2 | 3.3 | 3.6 | 5.2 |
| New Delhi (Inceptisol) | 4.1 | 3.9 | 4.4 | 5.5 |
| Palampur (Alfisol) | 7.8 | 9.4 | 9.7 | 1.3 |
| Pantnagar (Mollisol) | 15.0 | 8.5 | 8.7 | 16.0 |
| Ranchi (Alfisol) | 4.5 | 3.8 | 3.9 | 4.5 |
| Raipur (Inceptisol) | 6.2 | 6.0 | 6.1 | 7.0 |
| Udaipur (Inceptisol) | 6.9 | 7.1 | 7.8 | 9.9 |

after 9 years of study in LTFE Raipur. However, SOC status increased in both the balanced fertilizer treatment (100% NPK) and the INM treatment (100% NPK + FYM) (Table 5).

**Table 5** Carbon input–output (Mg ha$^{-1}$) balance in two different LTFE soils

| Carbon status | Treatments | | | | |
|---|---|---|---|---|---|
| | Control | 100% N | 100% NP | 100% NPK | 100% NPK + FYM |
| Vertisol of Akola (sorghum–wheat: after 12 years) | | | | | |
| Initial SOC | 10.12 | 10.12 | 10.12 | 10.12 | 10.12 |
| Present SOC | 9.9 | 11.66 | 12.32 | 12.98 | 15.40 |
| Net change | (−0.22) | (+1.54) | (+2.22) | (+2.86) | (+5.28) |
| Vertisol of Raipur (rice–wheat: after 9 years) | | | | | |
| Initial SOC | 14.52 | 14.52 | 14.52 | 14.52 | 14.52 |
| Present SOC | 13.20 | 13.86 | 13.64 | 14.64 | 14.92 |
| Net change | (−1.32) | (−0.66) | (−0.88) | (+0.08) | (+0.40) |

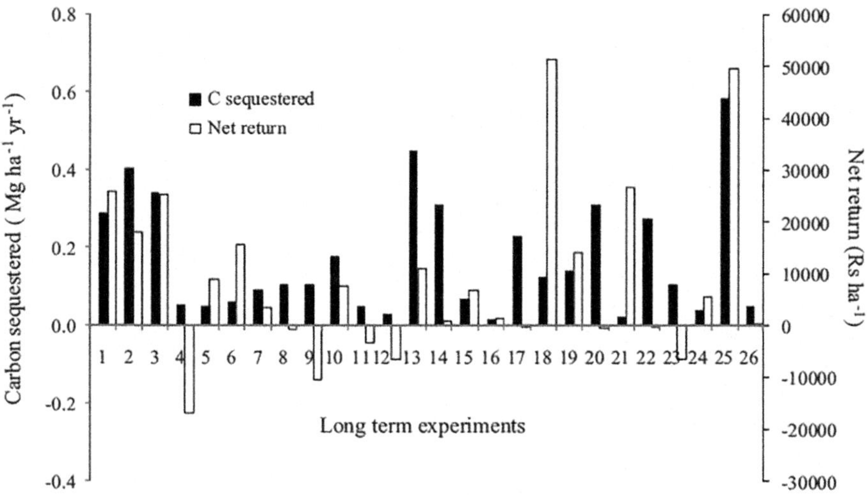


**Fig. 1** C sequestration as compared to the variation in net return under the integrated nutrient management and balanced nutrient management in long-term experiments in India. (Source: Pathak et al. 2011).

## Economic Analysis of Carbon Sequestration Potential: Experience from Long-Term Experiments

In order to estimate long-term shifts in crop yields, SOC, and the possible capacity of farmland to sequester carbon, long-term experiments are necessary (Ladha et al. 2013). Additionally, long-term trials offer an invaluable database for calculating the value of carbon sequestration that can aid in the decision-making process about the viability of the practice (Lal 2005). Thus, to determine the potential and expense of carbon sequestration, an overall assessment of 26 long-term experiments from various agroclimatic zones across India was carried out. The research results, as reported by Pathak et al. (2011), are as follows (Fig. 1): in comparison to the control, the

100% NPK + FYM enhanced SOC concentration by 52% and carbon sequestration potential by 105% on average. In 17 of 26 LTEs, the 100% NPK + FYM outperformed the 100% NPK treatment in terms of net return. SOC sequestration in the 100% NPK + FYM was achieved in the remaining nine long-term experiments, albeit at a lower net return. This implies that in certain regions, producers could have to pay more to accomplish C sequestration using 100% NPK + FYM treatment. Overall, the results of this study point to the 100% NPK + FYM treatment as an efficient approach for raising SOC and soil carbon sequestration capacity. Still, further research is required to create and improve nutrient management plans that are both financially sustainable for farmers and efficient in storing carbon.

Cause-and-effect relationships regarding SOC dynamics under the effect of fertilizer and manure application and experience from different long-term fertilizer experiments of India are presented in Fig. 2.

## *Soil Organic C Dynamics and SOC Stock in Three LTFE Soils: Modeling Approach (Application of RothC Model)*

Numerous simulation models have been developed for analyzing the dynamics and turnover of SOC and to forecast SOC changes under varying climatic and management situations. Although it is a widely employed C turnover simulation model, the Rothamsted carbon model, or RothC 26.3, has not been extensively evaluated in Indian settings. Three long-term fertilizer experiment in Jabalpur (Vertisol), Ludhiana (Inceptisol), and Palampur (Alfisol) were compared by Jha et al. (2021) using the RothC model. Through this model, it has been shown that 100% NPK treatment can increase SOC stock by 12–61% and 100% NPK + FYM can increase SOC stock by 30–107%, at 0–30 cm soil depth. In the 100% NPK and 100% NPK + FYM treatments in Vertisol, Alfisol, and Inceptisol, SOC stocks attained an equilibrium state in 48–95 years and 68–116 years, respectively.

## *Soil Carbon Sequestration Through the Arbuscular Mycorrhizal Fungi*

Soil aggregates are one of the primary factors influencing SOC sequestration. Aggregates in the soil preserve the stability of the soil's structure, foster or obstruct connections, and thereby preserve or release carbon. For aggregate formation and stabilization, arbuscular mycorrhizal fungi (AMF) are the most significant type of fungi. AMF produces globulin-related soil proteins (GRSPs), which are critical for soil aggregation. GRSPs improve soil quality, preserve soil structure, and store carbon in the form of the glycoprotein glomalin. Typically, 3–5% N, 36–59% C, and 0.03–0.1% P are found in glomalin. GRSPs are hydrophobic glycoproteins that are

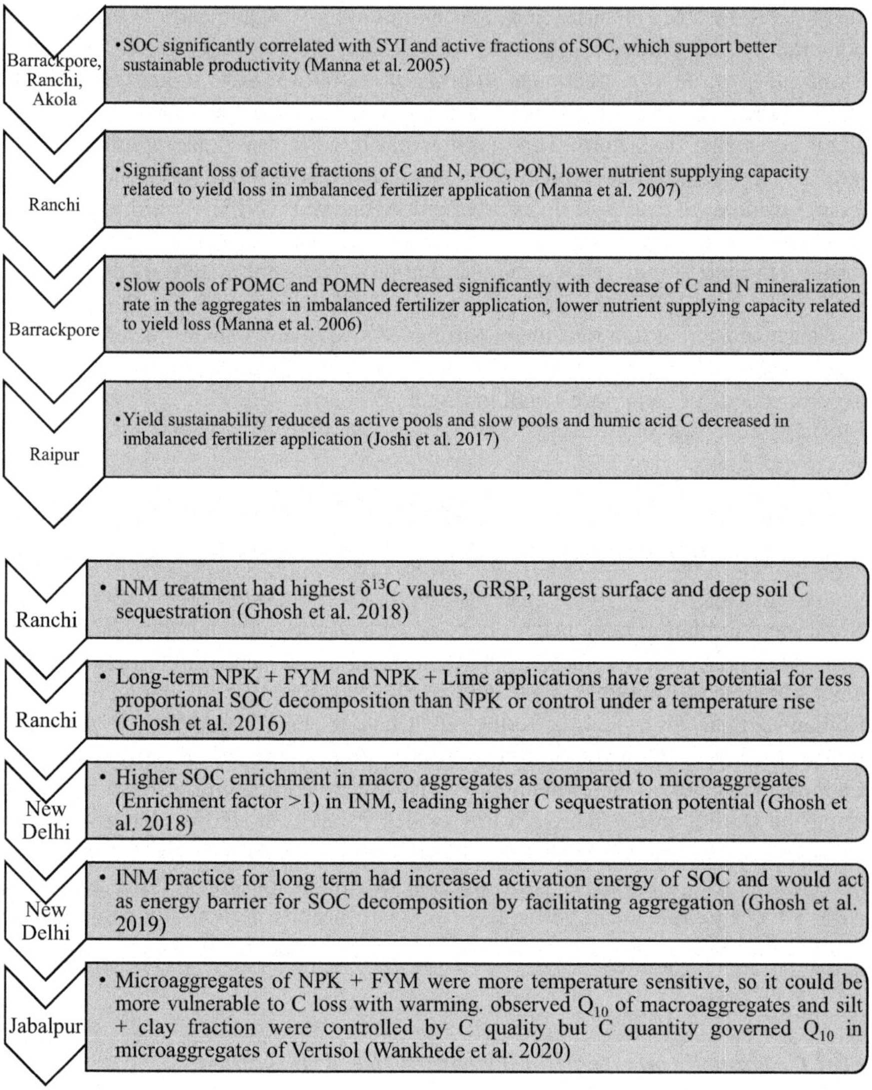


**Fig. 2** Experience from different long-term fertilizer experiments of India

sticky and heat-stable. Their long-term sequestration of carbon and aggregate stability are made possible by their recalcitrance, resistance to microbial diseases, and continuous retention in the soil. GRSPs make two types of contributions to the soil: direct and indirect.

According to Treseder and Holden (2013), the direct mechanism of GRSP and AMF in soil C sequestration is as follows: creation and placement of mycorrhizal debris and the mycorrhiza's release of oxidative and hydrolytic enzymes, which facilitates the degradation of SOM and leftover plant and microbial material. AMF boosting biomass output both above- and belowground, plant growth, and soil carbon input. The buildup of C is caused by the resistant chitinous amino sugars of the hyphal cell walls; nevertheless, the mean residual time of the C stored by the soil–GRSP complex is longer than the mean residual time of the C stored in the aggregates (Chen et al. 2012).

Though, GRSP plays a direct function in C accumulation, it also indirectly contributes to SOC sequestration through the following means:

- Fostering soil stability and aggregation: GRSP functions as a glue, holding soil particles cohesively to create aggregates. The SOC retained within these aggregates is better protected because of their increased resistance to erosion and breakdown.
- Improving nutrient cycling: GRSP allows helpful bacteria to thrive and encourages their activity. These microorganisms are crucial to the decomposition of organic materials and the release of nutrients into the soil. Plants develop more effectively when there are more nutrients available, which increases the amount of biomass that plants add to the soil. The soil microorganisms eventually transform this biomass into SOC.
- AMF spores resulted in improved plant growth and deeper root growth.
- Higher microbial activity by enhancing active C pools.

Improving nutrient cycling: GRSP allows helpful bacteria to thrive and encourages their activity. These microorganisms are crucial to the decomposition of organic materials and the release of nutrients into the soil. Plants develop more effectively when there are more nutrients available, which increases the amount of biomass that plants add to the soil. The soil microorganisms eventually transform this biomass into SOC. AMF spores resulted in improved plant growth and deeper root growth. Higher microbial activity by enhancing active C pools.

When compared to conventional farming systems, organic farming practices have demonstrated to increase the diversity, abundance, and species richness of Am fungi and Bradford reactive and immunoreactive soil proteins. As a result, in tandem with an increase in microbial biomass, organic carbon, and the GRSP components, organic inputs raise the percentage of macroaggregates and the mean weight diameter of aggregates (Agnihotri et al. 2022). In a comparable manner, organic orchards as opposed to conventional orchards had increased total glomalin and AMF species richness. The application of compost or farmyard manure, as well as the continuous application of manure from chickens, pigs, or cattle for long time, raises the GRSP content (Agnihotri et al. 2022).

## *Soil Carbon Sequestration Through the Microbial Necromass*

One of the primary constituents of soil organic matter (SOM) is microbial necromass. It builds up in soil as extracellular substances secreted by microorganisms or as the remnants and fragments of deceased microorganisms (Fig. 3). Because it is

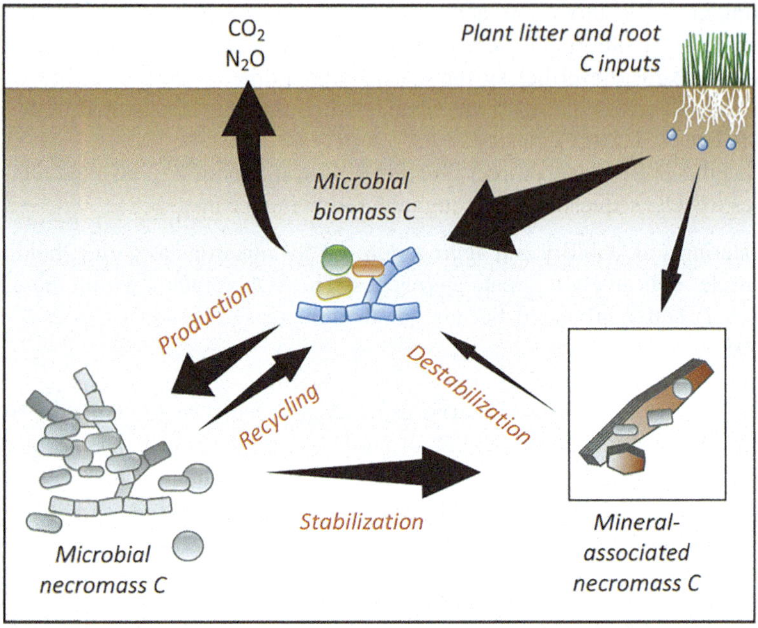


**Fig. 3** Microbial necromass persistence in soil (Source: Buckeridge et al. 2022)

shielded from microbial degradation due to its recalcitrant nature, it remains in soil for longer duration, and when recycled effectively, it contributes to the formation of microbial biomass (Buckeridge et al. 2022).

## *Microbial Necromass in Different Ecosystems*

According to Huang et al. (2019), microbial necromass is a significant part of SOC, making up over 50% of SOC sequestration in soil. Microbial necromass carbon was shown to be responsible for 55.6%, 61.8%, and 32.6% of sSOC in temperate farmland, grassland, and forestry soils, respectively (Liang et al. 2019). Necromass enters the nonliving SOC pool upon getting stabilized in microaggregates and linked to clay minerals. The quantity of bacterial, fungal, and total microbial necromass carbon increases with the quantity of labile organic carbon, nitrogen, and labile mineral phosphorus. This shows that greater levels of labile carbon, nitrogen, and phosphorus may increase microbial biomass, in turn increasing the accumulation of microbial necromass. According to Chen et al. (2020), there is a considerable positive correlation between microbial growth, biomass, and necromass and dissolved organic carbon (DOC) in soils. Microbial necromass development was facilitated by increased microbial biomass, quicker microbial growth and turnover, and greater root carbon intake and DOC (Prommer et al. 2019).

The amount of nitrogen in soil is significantly impacted by microbial necromass. This is due to the fact that proteins, nucleic acids, fungal chitin, glycoproteins, and bacterial peptidoglycans make up microbial necromass nitrogen, which accounts for over 60% of soil nitrogen (Warren 2021). For example, extracellular enzymes can break down microbial necromass and provide an alternate supply of nitrogen when easily available nitrogen is low or labile carbon is excessively prevalent. This is due to the fact that nitrogen obtained from bacteria repurposing necromass is a more efficient method than nitrogen obtained from resistant SOC fraction. Bacterial necromass influences microbial necromass in agriculture soil more significantly than forest soil. Numerous processes, notably N fertilization, contribute to this. In croplands, N fertilizers are frequently applied, and bacteria acquire more from it than fungi do. This is because bacteria, as opposed to fungi, tend to utilize greater N-rich resources and have smaller C:N ratios. However, fungi are the major player in regulating the decomposition of polymers such as lignin, hemicellulose, and cellulose. These polymers are less common in farmland soils than in forest soils. Furthermore, the availability of complex substrates, like plant roots or litter, encourages extensive bacterial necromass consumption toward N recycling in forests with a high C/N ratio.

Besides, tillage and other perturbations that significantly damage fungal hyphae in croplands could potentially be a factor behind the dominance of bacteria. When it comes to arable land, no-till fields have lower pH levels and higher soil microbial biomass than cultivated areas. In forests and grasslands, low soil pH promotes proliferation of fungi and, consequently, the accumulation of fungal necromass.

Fifty-one percent of the total SOC found in agricultural soil is composed of microbial necromass, compared to 47% in grassland soil. These percentages are higher than the proportion (35%) that is found in forest soil. This is caused by multiple variables, such as the following.

Root biomass: Compared to forests, grasslands have a higher amount of belowground plant residues. Greater amounts of belowground plant biomass promote microbial growth, proliferation, and necromass buildup, which raise the proportion of microbial necromass in the SOC in grasslands.

Decomposition of litter: Arable soil and grassland soils transform their litter through microbial action faster than forest soils, although the latter also have a greater quantity of partially degraded plant debris.

Lignin and cellulose: Microbial necromass in farmland and grasslands might be stored more efficiently; however, this is not true for forest soil due to the slow decomposition of lignin and cellulose found in forest land.

Organo-mineral interactions: The microbial necromass in croplands and grasslands forms enduring organo-mineral associations by being intimately connected to clay minerals. Recent direct results have shown that there is a substantial correlation between organic C linked to minerals and amino sugars (Samson et al. 2020). In agricultural land and grasslands, more carbon is stored in mineral-associated organic matter, which make up about 50% of the soil organic carbon content in most soils. Furthermore, this C storage has a minimum residence time of decades to centuries (Villarino et al. 2021). Wang et al. (2021) investigated the quantities of glucosamine and muramic acid as well as the composition of microbial necromass and its role in SOC buildup by conducting a meta-analysis of 1716 data points from cropland, grassland, and forest soil profiles worldwide. They found that the main source of SOC on farmland and grasslands is rhizodeposits that produce mineral-associated organic C; in forests, on the other hand, SOC is primarily created by the physical movement of persistent plant debris, which form particulate organic C. The contribution of microbial necromass to SOC varies depending on the habitat because of factors like physicochemical protection, environmental variables, and substrate availability (microbial in vivo conversion dominates in agricultural land and grasslands, whereas in forests, ex vivo modifications prevails). Furthermore, compared to forest areas, it has been proposed that microbial necromass contributes more to the buildup of SOC in croplands and grasslands.

Fungi are more adapted to break down plant biomass, whereas bacteria are better suited to break down dead microbial biomass. Consequently, it has been postulated that as compared to agricultural land and grasslands, forests had higher fungal:bacterial necromass ratios and fungal inputs to microbial necromass development. The quantity and type of vegetation are affected by temperature and average annual precipitation. This can then have an impact on the properties of the soil (pH, microbial biomass, etc.) and eventually necromass accumulation.

## Feasibility of Achieving the 4 per 1000 Target to Abate Climate Change

### *Lesson from Century Simulations of Long-Term Arable Experiments in European Conditions*

The goal of the 4 per 1000 program is to preserve and replenish soil organic carbon (SOC) stores in order to achieve food security, cope with climate change, and sustain soil fertility. To estimate the quantity of organic carbon inputs required to meet the aim of increasing SOC by 4 per 1000 annually for 30 years, Bruni et al. (2021) conducted a comparative analysis of 14 long-term agricultural locations in Europe. The aim was determined to be attainable using the Century model to simulate SOC stocks; however, experimental data indicated that treatments with greater carbon inputs were only raising SOC by 0.25% on average. This implies that the Century model might be overestimating the amount of carbon inputs required to meet the 4 per 1000 target and that the real quantity may be significantly greater. Furthermore, the study calculated that under 1 °C and 5 °C warming scenarios, respectively, annual carbon inputs would need to rise by 54% and 120% more to meet the target. The modelled carbon inputs necessary for achieving the aim depended linearly on the starting SOC stocks, which further raised doubts about the viability of the 4 per 1000 plan. This implies that large-scale SOC stock increases in a future with a warmer environment will be difficult.

### *Major Limitations to Achieving the 4 per 1000 Initiative for Increasing SOC Stock in Temperate Regions: Experience from Rothamsted Research, United Kingdom*

Through an analysis of the rates of SOC gain in treatments over 16 long-term studies in the southeast of the United Kingdom, Poulton et al. (2018) assessed the viability and potential benefits of the "4 per 1000" effort for raising SOC stocks. Over the course of 20 years, the effort aims to raise the SOC reserve by 4 per 1000 each year up to 40 cm soil depth. Three different soil types and comparative evaluation of 114 treatments over a period of 7–157 years comprised the experiments. The treatments included converting arable land into woods, applying N fertilizer, introducing pasture leys into uninterrupted arable systems, and incorporating organic matter using inversion plowing. SOC rise has been estimated to be >7 per 1000 annually up to 23 cm depth in about 65% of scenarios that is equivalent to 4 per 1000 annually up to 40 cm depth. The two longest fertilizer trials, which lasted for over 150 years, observed SOC increases of 18 per 1000 annually and 43 per 1000 annually in upper 23 cm depth over the first 20 years of yearly FYM addition at 35 t fresh material $ha^{-1}$ (around 3.2 t organic C $ha^{-1}$ $year^{-1}$). Increases beyond 7 per 1000 annually persisted for 40–60 years. SOC increase of only 3–8 per 1000 annually was observed across a number of decades in other tests when FYM was administered at lower rates and not every year. Over different time periods, SOC gains for different treatments ranged from 0 to 19 per 1000 per year. However, the current situation is still not adequate to reach this ambitious target. Factors that may limit the widespread adoption of practices to increase soil organic carbon (SOC) include the following.

Limited resources: Farmers may not have enough manure or crop residues to apply to their fields at the rates needed to increase SOC.

Existing adoption: Some practices, such as cover cropping, are already widely used, so there is limited scope for further adoption.

Negative impacts on food security: Converting agricultural land to forest or grassland could reduce global food production.

Economic costs: Implementing new management practices may be uneconomical for farmers, especially if they require changes to government policies or subsidies.

## Conclusion

Integrated nutrient management (INM) has been shown to be an effective way to sequester carbon in Indian agricultural soil over the long term. It can also help to mitigate greenhouse gas emissions and increase productivity and net returns without any additional costs in most cases. SOC has also been found to be a major driver of yield sustainability. Thus, compost/organic manure application has emerged as a "win–win" technology that can increase farm income and sequester carbon. Carbon sequestration under farmland with balanced fertilization and INM is now well recognized at the global level. In order to meet sustainable development goals and slow down climate change, policymakers are now pushing for greater financial assistance in the form of incentives or transferable carbon credits under the Clean Development Mechanism (CDM) to make the goals of C sequestration in soil more achievable.

## References

Agnihotri R, Sharma MP, Prakash A, Ramesh A, Bhattacharjya S, Patra AK, Manna MC, Kurganova I, Kuzyakov Y (2022) Glycoproteins of arbuscular mycorrhiza for soil carbon sequestration: Review of mechanisms and controls. Sci Total Environ 806(Pt-2):150571. https://doi.org/10.1016/j.scitotenv.2021.150571

Azeez G (2009) Soil carbon and organic farming: a review of the evidence of agriculture's potential to combat climate change summary of findings. http://www.nourishscotland.org/wp-content/uploads/2012/09/sa.pd

Bruni E, Guenet B, Huang Y, Clivot H, Virto I, Farina R, Kätterer T, Ciais P, Martin M, Chenu C (2021) Additional carbon inputs to reach a 4 per 1000 objective in Europe: feasibility and projected impacts of climate change based on Century simulations of long-term arable experiments. Biogeosciences 18:3981–4004. https://doi.org/10.5194/bg-18-3981-2021

Buckeridge KM, Creamer C, Whitaker J (2022) Deconstructing the microbial necromass continuum to inform soil carbon sequestration. Funct Ecol 36:1396–1410

Chen PH, Liu HL, Chen YJ, Cheng YH, Lin WL, Yeh CH, Chang CH (2012) Enhancing CO 2 bio-mitigation by genetic engineering of cyanobacteria. Energy Environ Sci 5(8):8318–8327

Chen X, Xia Y, Rui Y, Ning Z, Hu Y, Tang H, He H, Li H, Kuzyakov Y, Ge T, Wu J, Su Y (2020) Microbial carbon use efficiency, biomass turnover, and necromass accumulation in paddy soil depending on fertilization. Agric Ecosyst Environ 292:106816J

Davies J (2017) The business case for soil. Nature 543:309–311

Dutta D, Singh DK, Subash N, Ravisankar N, Kumar V, Meena AL, Panwar AS (2018) Effect of long-term use of organic, inorganic and integrated management practices on carbon sequestration and soil carbon pools in different cropping systems in Tarai region of Kumayun hills. Indian J Agric Sci 88(4):523–529

FAO and ITPS (2015) Status of the world's soil resources (SWSR) – Main report. Food and Agriculture Organization of the United Nations and Intergovernmental Technical Panel on Soils, Rome, Italy

FAO (2017) FAO and the SDGs Indicators: Measuring up to the 2030 Agenda for Sustainable Development. pp 1–40. content (fao.org)

Ghosh D (2017) Nutrient Dynamics Under Long Term Integrated Nutrient Management Practices Followed In Upland Inceptisols (Doctoral dissertation, Orissa University Of Agriculture And Technology)

Ghosh A, Bhattacharyya R, Dwivedi BS, Meena MC, Agarwal BK, Mahapatra P, Shahi DK, Salwani R, Agnihorti R (2016) Temperature sensitivity of soil organic carbon decomposition as affected by long-term fertilization under a soybean based cropping system in a sub-tropical Alfisol. Agric Ecosyst Environ 233:202–213

Ghosh A, Bhattacharyya R, Agarwal BK, Mahapatra P, Shahi DK, Singh G, Agnihotri R, Sawlani R, Sharma C (2018a) Long-term fertilization effects on 13C natural abundance, soil aggregation, and deep soil organic carbon sequestration in an Alfisol. Land Degrad Dev. https://doi.org/10.1002/ldr.3229

Ghosh A, Bhattacharyya R, Meena MC, Dwivedi BS, Singh G, Agnihotri R, Sharma C (2018b) Long-term fertilization effects on soil organic carbon sequestration in an Inceptisol. Soil Tillage Res 177:134–144. https://doi.org/10.1016/j.still.2017.12.006

Ghosh A, Bhattacharyya R, AbirDey BS, Dwivedi MC, Meena MC, Manna RA (2019) Long-term fertilisation impact on temperature sensitivity of aggregate associated soil organic carbon in a sub-tropical inceptisol. Soil Till Res 195:104369. ISSN 0167-1987. https://doi.org/10.1016/j.still.2019.104369

Huang Y, Liang C, Duan X, Chen H, Li D (2019) Variation of microbial residue contribution to soil organic carbon sequestration following land use change in a subtropical karst region. Geoderma 353:340–346. https://doi.org/10.1016/j.geoderma.2019.07.028

Jha P, Lakaria BL, Vishwakarma AK, Wanjari RH, Mohanty M, Sinha NK, Somasundaram J, Dheri GS, Dwivedi AK, Sharma RP, Muneshwar Singh RC, Dalal AKB, Patra AK, Chaudhari SK (2021) Modeling the organic carbon dynamics in long-term fertilizer experiments of India using the Rothamsted carbon model. Ecol Model 450:109562

Joshi SK, Bajpai RK, Kumar P, Tiwari A, Bachkaiya V, Manna MC, Asha S, Sudeshna B, Rahman MM, Wanjari RH, Singh M, Coumar V, Patra AK, Chaudhari SK (2017) Soil organic carbon dynamics in a Chhattisgarh vertisol after use of a rice–wheat system for 16 years. Agron J 109(6):1–14

Keesstra SD, Bouma J, Wallinga J, Tittonell P, Smith P, Cerdà A, Montanarella L, Quinton JN, Pachepsky Y, van der Putten WH, Bardgett RD, Moolenaar S, Mol G, Jansen B, Fresco LO (2016) The significance of soils and soil science towards realization of the United Nations Sustainable Development Goals. SOIL 2:111–128. https://doi.org/10.5194/soil-2-111-2016

Ladha JK, Dawe D, Pathak H, Padre AT, Yadav RL, Singh B, Singh Y, Singh Y, Singh P, Kundu AL, Sakal R, Ram N, Regmi AP, Gami SK, Bhandari AL, Amin R, Yadav CR, Bhattarai EM, Das S, Aggarwal HP, Gupta RK, Hobbs PR (2013) How extensive are yield declines in long-term rice–wheat experiments in Asia?. Field Crops Res 81(2–3):159–180. https://doi.org/10.1016/S0378-4290(02)00219-8

Lal R (2005) Carbon sequestration and climate change with specific reference to India. Proceedings of International Conference on Soil, Water and Environmental Quality–Issues and Strategies. IARI, New Delhi, pp 295–302

Lal R (2018) Digging deeper: a holistic perspective of factors affecting soil organic carbon sequestration in agroecosystems. Glob Chang Biol 24:1–17

Lal R, Smith P, Jungkunst HF, Mitsch WJ, Lehmann J, Ramachandran-Nair PK, McBratney AB, de Moraes-Sá JC, Schneider J, Zinn YL, Skorupa A, Zhang HL, Minasny B, Srinivasrao C, Ravindranat NH (2018) The carbon sequestration potential of terrestrial ecosystems. J Soil Water Conserv 73:145–152

Liang C, Amelung W, Lehmann J, Kästner M (2019). Quantitative assessment of microbial necromass contribution to soil organic matter. Glob Chang Biol 25:3578–3590. https://doi.org/10.1111/gcb.14781.Y

Lorenz K, Lal R (eds) (2018) Carbon sequestration in agricultural ecosystems. Springer International Publishing

Majumder B, Mandal B, Bandyopadhyay PK, Gangopadhyay A, Mani PK, Kundu AL, Mazumdar D (2008) Organic amendments influence soil organic carbon pools and rice–wheat productivity. Soil Sci Soc Am J 72:775–785

Manna MC, Swarup A, Wanjari RH, Ravankar HN, Mishra B, Saha MN, Singh YV, Sahi DK, Sarap PA (2005) Long-term effect of fertilizer and manure application on soil organic carbon storage, soil quality and yield sustainability under sub-humid and semi-arid tropical India. Field Crop Res 93:264–280

Manna MC, Swarup A, Wanjari RH, Mishra B, Shahi DK (2007) Long-term fertilization, manure and liming effects on soil organic matter and crop yields. Soil Tillage Res 94:397–409

Muntean M, Guizzardi D, Schaaf E, Crippa M, Solazzo E, Olivier JGJ, Vignati E (2018) Fossil $CO_2$ emissions of all world countries. Publications Office of the European Union, Luxembourg

Pathak H, Byjesh K, Chakrabarti B, Aggarwal PK (2011) Potential and cost of carbon sequestration in Indian agriculture: Estimates from long-term field experiments. Field Crops Res 120(1):102–111. https://doi.org/10.1016/j.fcr.2010.09.006

Paustian K, Parton WJ, Persson J (1992) Modeling soil organic matter in organic amended and nitrogen fertilizer long-term plots. Soil Sci Soc Am J 56:476–488

Poulton PR, Johnston AE, Macdonald AJ, White RP, Powlson DS (2018) Major limitations to achieving 4 per 1000 increases in soil organic carbon stock in temperate regions: evidence from long-term experiments at Rothamsted Research, UK. Glob Chang Biol 24(6):2563–2584

Prommer J, Walker TWN, Wanek W, Braun J, Zezula D, Hu Y, Hofhansl F, Richter A (2019) Increased microbial growth, biomass, and turnover drive soil organic carbon accumulation at higher plant diversity. Glob Chang Biol 26:669–681

Richards M, Bruun TB, Campbell BM, Le G, Huyer S, Kuntze V, Stn M, Oldvig M, Vasileiou I (2016) How countries plan to address agricultural adaptation and mitigation: an analysis of intended nationally determined contributions. CCAFS dataset

Samson M, Chantigny MH, Vanasse A, Menasseri-aubry S, Royer I, Angers DA (2020) Management practices differently affect particulate and mineral-associated organic matter and their precursors in arable Soils. Soil Biol Biochem 148:107867

Treseder KK, Holden SR (2013) Fungal carbon sequestration. Science 340:1528–1529. https://doi.org/10.1126/science.1236338

Trivedi A, Bhattacharyya R, Biswas DR, Das S, Das TK, Mahapatra P, Sharma C (2020) Long-term impacts of integrated nutrient management with equivalent nutrient doses to mineral fertilization on soil organic carbon sequestration in a sub-tropical Alfisol of India. Carbon Manag 11(5):483–497

UNEP (2015). The emissions gap report 2015. United Nations Environment Programme (UNEP), Nairobi

UNEP (2018) The emissions gap report 2018 [online]. [Cited 15 November 2019]. https://www.unenvironment.org/resources/bridginggapenhancing-mitigation-ambition-and-action-g20-leveland-globally

Villarino SH, Pinto P, Jackson RB, Piñeiro G (2021) Plant rhizodeposition: a key factor for soil organic matter formation in stable fractions. Sci Adv 7:3176

Wang C, Qu L, Yang L, Liu D, Morrissey E, Miao R, Liu Z, Wang Q, Fang Y, Bai E (2021) Large-scale importance of microbial carbon use efficiency and necromass to soil organic carbon. Glob Chang Biol 27:2039–2048

Wankhede M, Avijit G, Manna MC, Sukanya M, Sirothia P et al (2020) Does soil organic carbon quality or quantity govern relative temperature sensitivity in soil aggregates? Biogeochemistry 148:191–206. https://doi.org/10.1007/s10533-020-00653-y

Warren C (2021) What are the products of enzymatic cleavage of organic N? Soil Biol Biochem 154:108152

# Prediction of Crop Response to Atmospheric Greenhouse Gas Concentration and Climate Parameters

**Bharati Kollah, Rakesh Parmar, M. Homeshwari Devi, and Santosh R. Mohanty**

## Introduction

Human activities such as fossil fuel combustion, deforestation, and intensive agriculture have led to substantial increases in emissions of major greenhouse gases including carbon dioxide, methane, and nitrous oxides over the past century. According to the latest Intergovernmental Panel on Climate Change (IPCC) (2021) report, atmospheric concentrations of carbon dioxide have risen from pre-industrial levels of 280 ppm to over 410 ppm currently, the highest in at least two million years. This accumulation of greenhouse gases is causing significant changes in long-term climate patterns, including higher average temperatures, shifts in precipitation patterns, rising sea levels, ocean acidification, and an increased frequency of extreme weather events as highlighted by Wheeler and Von Braun (2013). These climatic changes are expected to negatively impact agricultural productivity and disrupt regional and global food systems that are critical for ensuring food and nutrition security worldwide (Challinor et al. 2018). Crop yields are affected in complex ways by various interacting climatic factors such as rising temperatures, changing rainfall profiles, elevated atmospheric carbon dioxide concentrations, increased climate variability and extreme events, as well as the incidence of pests, weeds, and diseases (Lobell and Field 2007). Therefore, developing robust projections of crop yield responses to changing environmental conditions under future climate change scenarios is imperative. Such projections can inform adaptation strategies, agricultural policies, technology investments, and management interventions aimed at increasing resilience. However, generating accurate predictions is challenging given the complex spatiotemporal dynamics involved. This chapter reviews different modeling techniques and integrated assessment approaches that are being used for developing projections of crop productivity under a range of

B. Kollah (✉) · R. Parmar · M. H. Devi · S. R. Mohanty
Indian Institute of Soil Science, Bhopal, MP, India

 
S. R. Mohanty, B. Kollah (eds.), *Greenhouse Gas Regulating Microorganisms in Soil Ecosystems*, https://doi.org/10.1007/978-3-031-70569-4_7

climate change scenarios. It analyzes their capabilities and limitations with a view to highlight areas where additional research efforts are needed to improve robustness, accuracy, and utility for supporting agricultural planning and policy worldwide.

## Predicting Crop Response to Elevated $CO_2$

Rising atmospheric carbon dioxide ($CO_2$) levels can have direct physiological effects on key crop processes like photosynthesis and transpiration. As demonstrated by Leakey et al. (2009), elevated $CO_2$ causes partial closure of leaf stomata which reduces crop water loss and improves water use efficiency. Enhanced $CO_2$ also stimulates the carboxylation reactions in photosynthesis, thereby increasing potential carbon fixation and carbohydrate production. However, the magnitude of these photosynthetic and transpiration responses depends on other factors like soil moisture availability, extreme temperature events, and nutrient limitations that interact with $CO_2$ effects under field conditions (Ainsworth and Rogers 2007). Well-watered and fertilized crops may show larger $CO_2$ fertilization effects compared to water or nutrient-stressed plants. Controlled environment studies under regulated $CO_2$ concentrations provide detailed insights into crop-specific physiological responses. However open-top field chambers and free air $CO_2$ enrichment (FACE) experiments offer more realistic assessments under actual production environments with interacting climatic variables (Ruiz-Vera et al. 2013). Statistical crop models rely on observed empirical relationships between historical yield data and $CO_2$ levels to develop correlative projections. Conversely, process-based models simulate mechanistic photosynthetic and metabolic responses to $CO_2$ increases based on scientific principles. But these models need robust validation with experimental data across genotypes, management regimes, and environments (Rötter et al. 2018a, b). Overall, each approach has intrinsic limitations in accurately predicting $CO_2$ fertilization effects across diverse cropping systems. Interactions with moisture, temperature, nutrients, and genetic factors make projections highly uncertain. More open-field experiments combined with enhanced crop modeling are needed to improve $CO_2$ response predictions and assess potential gains versus nullifying effects from other climate change factors. This can help adapt crop choice and management for optimizing future food production under higher $CO_2$ levels.

## Predicting Crop Response to Temperature

Increasing temperatures can alter crop developmental patterns and shift growing season lengths (Hatfield et al. 2011). Moderate warming may benefit some cereals like wheat, rice, and maize up to optimal temperature thresholds. However, further warming beyond ideal levels causes declines in yields and quality due to heat stress during critical growth stages as shown by Teixeira et al. (2013). High temperatures also decrease pollen viability, fertilization, grain filling, and seed quality for many crops. The impacts vary for annual versus perennial plants, with the latter facing greater risks from false springs and frost exposure (Bita and Gerats 2013). Controlled environment studies under set temperature ranges provide detailed crop-specific developmental and growth response curves. However, simulated field-level warming using infrared heaters, passive nighttime warming, and climate gradient studies offer better realism under actual production environments (Yin et al. 1995). Process-based models can simulate crop development processes and growth duration shifts affected by temperature changes. However, extreme temperature event impacts and responses require more robust integration with stochastic weather generators combined with refined model parameterization for key thresholds (Rötter et al. 2018a, b). There are also challenges in modeling potentially offsetting temperature effects on crop vegetative growth versus reproductive stages. Combined approaches are needed to improve crop-specific model calibration and yield response predictions under varying temperatures. This requires datasets from controlled warming experiments, climate analogues across spatial gradients, and long-term records to separate temperature-driven shifts from adaptive management. Understanding temperature change impacts on crops will be critical for adaptation through cultivar breeding, shifting planting dates, supplemental irrigation, and other measures. Studies should both assess transient warming effects on existing systems and consider long-term evolutionary adaptations of new crop genotypes and production systems.

## Predicting Crop Response to Changing Rainfall Patterns

Alterations in rainfall distribution patterns can directly affect soil moisture availability, which is critical for rain-fed crop productivity (Daryanto et al. 2016). Declining precipitation amounts and increased rainfall variability can limit yields, as water stress during sensitive crop growth stages impacts both productivity and quality. In contrast, excessive rainfall can damage crops through waterlogging, increase risks of fungal diseases, and disrupt field operations including planting, fertilizer application, and harvesting. Predicting crop responses requires differentiating between changes in total annual precipitation, within-season rainfall distribution, and daily extremes (Muller 2013). Local and regional hydrological models provide field-scale soil water balance estimates to complement broader climatic forecast models. However, there are challenges in integrating these approaches to properly capture both localized water dynamics and larger atmospheric drivers (Rivington and Koo 2010). Experimental rainfall manipulation studies using shelters, rainout shelters, and irrigation modulation offer useful validation data on soil and crop responses. But these have limited spatial coverage and may oversimplify extreme event impacts (Pryor et al. 2014). Detailed soil hydrological characterization considering infiltration, drainage, storage, and evapotranspiration dynamics is needed along with improved climatic inputs. Opportunities exist to couple hydrologic models with emerging remote sensing capabilities to better monitor soil moisture in near real-time over larger areas. Long-term studies comparing organic, conservation, and conventional tillage systems could elucidate adaptive mechanisms. Overall, predicting crop responses to altered rainfall remains a key challenge constrained by interacting soil, crop, and climatic uncertainties. Progress requires more integrative approaches, better hydrological understanding, increased spatial coverage, and longer-term studies. This can support adaptation strategies from supplemental irrigation to water harvesting, conservation tillage, residue management, and drought-tolerant cultivars.

## Integrated Assessment of Combined Climatic Effects

Individual climatic factors such as temperature, rainfall, and $CO_2$ interact in complex nonlinear ways to determine crop productivity (Lobell and Asseng 2017). For instance, elevated $CO_2$ can stimulate photosynthesis and water use efficiency, but higher evaporative demand under rising temperatures may limit benefits (Ruiz-Vera et al. 2013). Therefore, integrated assessments that combine $CO_2$, temperature, water availability, as well as other climatic variables are essential for realistic yield projections. Whole system approaches that also incorporate biotic interactions with pests, diseases, and weeds are relevant, as climate changes can alter pressure from these sources on crops (Deutsch et al. 2018). Experimental platforms like free air $CO_2$ enrichment (FACE) provide useful multifactor crop response data for model validation. Platforms that manipulate $CO_2$, temperature, water, and ozone factors simultaneously can offer additional insights into response surface contours. Crop models like APSIM, DSSAT, and DAYCENT have demonstrated some capability for integrated assessment of combined climatic impact pathways across diverse crops, soils, and environments (Ewert et al. 2015). However, further improvement is needed to better simulate extreme climate events and their nonlinear impacts on crop yields. There are also opportunities to incorporate crop genetics for predicting adaptation potential, as well as linking productivity projections with assessments of resource use sustainability. Additional capabilities for spatially explicit modeling are needed to capture microclimate effects and heterogeneities across agricultural landscapes. Furthermore, translating crop-level effects into wider food systems and economic outcomes remains a key challenge. Enhanced computing capabilities and stronger interdisciplinary linkages can facilitate more comprehensive climate impact assessments. Overall, integrated approaches that consider multiple interacting biophysical and socio-economic factors will be essential for developing robust projections to guide climate-smart agriculture policies and investments.

## Importance of Translating Crop Climate Impact Projections into Actionable Adaptations

While significant advances have been made in predicting crop productivity responses to climate change, translating these projections into practical on-farm adaptations remains a key challenge (Challinor et al. 2018). Without robust decision support systems and agricultural extension efforts, impact forecasts may not lead to actual adaptation outcomes and climate-resilient food systems (Vermeulen et al. 2013). Firstly, bridging the scale gap between broad projections and farm-level decisions is essential (Aggarwal et al. 2006). This requires developing user-friendly interfaces, mobile-based services, and digital tools to deliver customized climate information to farmers (Hansen et al. 2011). Secondly, communicating complex climate risks and uncertainties in an actionable way is critical for farmer decision-making (Lemos et al. 2012). Risk characterization methods that focus on overcoming cognitive and normative barriers can enhance trust and adoption (Steynor et al. 2019). Thirdly, economic assessments of adaptation options and enabling policies are needed alongside biophysical modeling to prioritize investments (Antle 2015). For instance, subsidy schemes and insurance can incentivize uptake of stress-tolerant cultivars, water-saving irrigation, and climate-smart practices. Fourthly, strengthening agricultural extension through participatory approaches and peer learning can sustain climate-resilient agriculture (Khatri-Chhetri et al. 2017). Farm advisors play a key role in supporting implementation, modifying interventions, and addressing site-specific barriers. Finally, vulnerability and equity assessments are imperative to target adaptation policies for resource-poor farmers most at risk from climate impacts (Vermeulen et al. 2013). In summary, a multidimensional approach is required spanning science, policy, institutions, and finance to enable farmers to adapt agricultural systems and livelihoods to progressive climate change (Jat et al. 2016).

## Key Challenges and Future Directions

Several key challenges remain in predicting climate change impacts on crops and translating projections to support on-farm decision-making. Many crop models operate at coarse spatial resolutions and temporal scales that miss critical farm-level nuances and extreme event impacts (Aggarwal et al. 2006). Cascading uncertainties across model ensembles and scenarios leads to yield projection ranges too wide for tactical adaptation planning (Asseng et al. 2014). The lack of quality field data for many regions also constrains model initialization, calibration, and validation (Ramirez-Villegas and Challinor 2016). For instance, sparse weather station networks provide inadequate inputs to capture microclimate variability across agricultural landscapes. Detailed field phenology observations are also needed to calibrate crop development simulations for specific cultivars. There are further challenges in representing the diversity of real-world farm management practices and responses.

Enhanced collaboration between model developers, agronomists, and farmers could help improve assumptions and adoption by end-users (Hutchings et al. 2012). Two-way communication is needed between the modeling community and practitioners to better represent on-farm decision dynamics and integrate projected climate risks into farming choices. Emerging technologies like remote sensing, genetics, and precision agriculture may provide valuable new datasets to support climate modeling and projections (White 2012). For example, remote sensing could help monitor crop stresses and soil moisture at finer spatial scales, while gene-based modeling may facilitate trait-driven impact assessments. However, significant data sharing, interoperability, and analytical challenges need addressing. Widespread investments are still required to upgrade modeling frameworks, expand climate observation networks, digitize historical records, and strengthen human capacity for research (Challinor et al. 2018). Improving access to climate projection services through digital platforms and agricultural extension can further support farmer adoption and use. Overall, an integrated agenda is essential across disciplines and stakeholder groups to enhance prediction capabilities, facilitate uptake, and enable locally relevant climate-smart solutions.

## Conclusion

This chapter has reviewed the range of techniques and approaches being used to predict crop yield responses under changing climatic conditions driven by rising atmospheric greenhouse gas concentrations. Crop models, controlled environment experiments, and climate gradient analogue analyses can all provide useful but partial insights into crop yield responses to isolated climate factors like $CO_2$, temperature, and rainfall. However, as discussed, integrated assessment frameworks that combine multiple interacting climatic and biological factors are essential for developing realistic projections of future crop productivity. Whole system approaches that also consider pests, diseases, and nutrient dynamics are thereby relevant. Several key challenges around spatial and temporal scaling, model uncertainties, lack of quality input data, and inadequate representation of farm management diversity need to be addressed to enhance real-world utility and adoption of crop climate impact projections. This requires stronger interdisciplinary collaboration between crop modelers, agronomists, climatologists, and social scientists to develop actionable climate-smart solutions for farmers. Continued research advances are also critically needed to improve model integration, refine simulations of extreme events, incorporate crop genetics and adaptation dynamics, and strengthen empirical validation. Investments in expanded observation networks, data digitization, computing capabilities, and human capital development will complement these research efforts. Overall, robust yet relevant crop productivity projections under climate change will be vital for identifying effective adaptation strategies, targeting technology investments, informing agricultural policies, and catalyzing on-farm responses. This can enable resilient food production systems despite rising climatic risks and uncertainties. However, realizing these practical benefits will require comprehensive initiatives that span across disciplines and better link science with on-ground decision-making. The coming decades will test our capacity to predict and adapt food systems to unprecedented environmental change through collective ingenuity, collaboration, and sustained commitments.

## References

Aggarwal PK, Banerjee B, Daryaei MG, Bhatia A, Bala A, Rani S et al (2006) InfoCrop: a dynamic simulation model for the assessment of crop yields, losses due to pests, and environmental impact of agroecosystems in tropical environments. I Model description. Agric Syst 89(1):1–25

Ainsworth EA, Rogers A (2007) The response of photosynthesis and stomatal conductance to rising [$CO_2$]: mechanisms and environmental interactions. Plant Cell Environ 30(3):258–270

Antle JM (2015) Climate change and agriculture: economic impacts. Choices 30(1):1–5

Asseng S, Ewert F, Martre P, Rötter RP, Lobell DB, Cammarano D et al (2014) Rising temperatures reduce global wheat production. Nat Clim Chang 5(2):143–147

Bita CE, Gerats T (2013) Plant tolerance to high temperature in a changing environment: scientific fundamentals and production of heat stress-tolerant crops. Front Plant Sci 4:273

Challinor AJ, Muller C, Asseng S, Deva C, Nicklin KJ, Wallach D et al (2018) Improving the use of crop models for risk assessment and climate change adaptation. Agric Syst 159:296–306

Daryanto S, Wang L, Jacinthe PA (2016) Global synthesis of drought effects on maize and wheat production. PLoS One 11(5):e0156362

Deutsch CA, Tewksbury JJ, Tigchelaar M, Battisti DS, Merrill SC, Huey RB, Naylor RL (2018) Increase in crop losses to insect pests in a warming climate. Science 361(6405):916–919

Ewert F, Rötter RP, Bindi M, Webber H, Trnka M, Kersebaum KC et al (2015) Crop modelling for integrated assessment of risk to food production from climate change. Environ Model Softw 72:287–303

Hansen JW, Mason SJ, Sun L, Tall A (2011) Review of seasonal climate forecasting for agriculture in sub-Saharan Africa. Exp Agric 47(2):205–240

Hatfield JL, Boote KJ, Kimball BA, Ziska LH, Izaurralde RC, Ort D et al (2011) Climate impacts on agriculture: implications for crop production. Agron J 103(2):351–370

Hutchings NJ, Reinds GJ, Leip A, Wattenbach M, Bienkowski JF, Dalgaard T et al (2012) A model for simulating the timelines of field operations at a European scale for use in complex dynamic models. Biogeosciences 9(11):4487–4496

Intergovernmental Panel on Climate Change (IPCC) (2021) Climate change 2021: the physical science basis. In: Contribution of working group I to the sixth assessment report of the Intergovernmental Panel on Climate Change. Cambridge University Press

Jat ML, Dagar JC, Sapkota TB, Singh Y, Govaerts B et al (2016) Climate change and agriculture: adaptation strategies and mitigation opportunities for food security in South Asia and Latin America. Adv Agron 137:127–235

Khatri-Chhetri A, Poudel B, Shirsath PB, Chaudhary P (2017) Assessment of climate-smart agriculture (CSA) options in Nepal. CGIAR Research Program on Climate Change, Agriculture and Food Security (CCAFS), New Delhi, India

Leakey AD, Ainsworth EA, Bernacchi CJ, Rogers A, Long SP, Ort DR (2009) Elevated $CO_2$ effects on plant carbon, nitrogen, and water relations: six important lessons from FACE. J Exp Bot 60(10):2859–2876

Lemos MC, Kirchhoff CJ, Ramprasad V (2012) Narrowing the climate information usability gap. Nat Clim Chang 2(11):789–794

Lobell DB, Asseng S (2017) Comparing estimates of climate change impacts from process-based and statistical crop models. Environ Res Lett 12(1):015001

Lobell DB, Field CB (2007) Global scale climate–crop yield relationships and the impacts of recent warming. Environ Res Lett 2(1):014002

Muller C (2013) African lessons on climate change risks for agriculture. Annu Rev Nutr 33:395–411

Pryor SC, Scavia D, Downer C, Gaden M, Iverson L, Nordstrom R et al (2014) Ch. 18: Midwest. In: Climate change impacts in the United States: the third national climate assessment, pp 418–440

Ramirez-Villegas J, Challinor A (2016) Towards a genotypic adaptation strategy for Indian groundnut cultivation using an ensemble of crop simulations. Clim Chang 138(1):223–238

Rivington M, Koo J (2010) Report on the meta-analysis of crop modelling for climate change and food security survey. CGIAR Research Program on Climate Change, Agriculture and Food Security (CCAFS), Copenhagen

Rötter RP, Palosuo T, Pirttioja NK, Dubrovsky M, Salo T, Fronzek S et al (2018a) What would happen to barley production in Finland if global warming exceeded 4 C? A model-based assessment. Eur J Agron 110:79–94

Rötter RP, Höhn J, Fronzek S (2018b) Projections of climate change impacts on crop production: a global and a Nordic perspective. Acta Agric Scand Sect B—Soil & Plant Sci 68(2):166–180

Ruiz-Vera UM, Siebers M, Gray SB, Drag DW, Rosenthal DM, Kimball BA et al (2013) Global warming can negate the expected $CO_2$ stimulation in photosynthesis and productivity for soybean grown in the Midwestern United States. Plant Physiol 162(1):410–423

Steynor A, Pasquini L (2019) Informing climate services in Africa through climate change risk perceptions. Clim Serv 15:100112

Teixeira EI, Fischer G, Van Velthuizen H, Walter C, Ewert F (2013) Global hot-spots of heat stress on agricultural crops due to climate change. Agric For Meteorol 170:206–215

Vermeulen SJ, Challinor AJ, Thornton PK, Campbell BM, Eriyagama N, Vervoort JM et al (2013) Addressing uncertainty in adaptation planning for agriculture. Proc Natl Acad Sci 110(21):8357–8362

Wheeler T, Von Braun J (2013) Climate change impacts on global food security. Science 341(6145):508–513

White S (2012) Climate change in global environmental history. A companion to global environmental history, pp 394–410

Yin X, Kropff MJ, McLaren G, Visperas RM (1995) A nonlinear model for crop development as a function of temperature. Agric For Meteorol 77(1–2):1–16

# Impact of Elevated $CO_2$ and Temperature on Greenhouse Gas Emission and Decomposition

**Upendra Kumar, Sudipta Dey, Megha Kaviraj, Snehasini Rout, and A. K. Nayak**

## Introduction

Climate change is currently one of the world's most critical concerns, and this major environmental issue is mostly linked to a rise in greenhouse gas (GHG) emissions (Ritchie and Roser 2020). China is currently among the top five GHGs generating countries on the list, emitting over twice as much as the second-highest emitter, the USA followed by India, Indonesia, and Russia. As per the Fifth Intergovernmental Panel on Climate Change (IPCC), global average land temperatures have increased roughly twice as much as ocean temperatures, and agricultural GHG emissions, which include crop and livestock production, forestry, and associated land use, account for up to 30% of anthropogenic emissions among all emitters (Tubiello et al. 2013). According to the latest data from the Consultative Group on International Agricultural Research (CGIAR), fertilizer production and applications in crops, storage, and packing of harvested produce account for up to one-third of all human GHG emissions (Gilbert 2012). Thus, GHG emissions in the atmosphere are rising, causing increasing global warming potential day by day.

Among the three GHGs, $CO_2$ is the most responsible for global warming. The cumulative warming effect of GHGs produced in the Earth's atmosphere from 1990 to 2019 increased by 45%, whereas $CO_2$ alone increased by 36% (NOAA 2020). $CO_2$ levels in the atmosphere have steadily risen from 280 ppm in pre-industrial times to 407.4 ppm in 2019 (Lindsey 2019), with 550–700 ppm expected by 2050 (Howard-Grenville et al. 2014). Agriculture, forest management, and other land

U. Kumar (✉) · S. Dey · S. Rout · A. K. Nayak
ICAR- National Rice Research Institute, Cuttack, Odisha, India

M. Kaviraj
ICAR- National Rice Research Institute, Cuttack, Odisha, India

The University of Birmingham, Edgbaston, Birmingham, UK

 
S. R. Mohanty, B. Kollah (eds.), *Greenhouse Gas Regulating Microorganisms in Soil Ecosystems*, https://doi.org/10.1007/978-3-031-70569-4_8

uses accounted for around 24% of direct $CO_2$ emissions, while industrial emissions accounted for 21%. $CO_2$ is also responsible for 20% of thermal adsorption, and its natural sources are organic degradation and respiration (Cassia et al. 2018). As a result, $CO_2$ is the most significant greenhouse gas emitted mostly by anthropogenic activity, and it also raises global temperatures and alters precipitation patterns (Bhattacharya et al. 2013).

The topmost meter of the global soil layer stores around 1500 Pg of total carbon and around 136 (92–140) Pg of total nitrogen, constituting the highest terrestrial carbon and nitrogen stocks. Important climate-relevant trace gases such as $CH_4$, $N_2O$, and $CO_2$ are very unequally distributed (by three to four orders of magnitude for carbon) (Nieder and Benbi 2008; Kutsch et al. 2009; Schaufler et al. 2010). Previous case studies have shown that the lowest (night) temperature has been expanding quicker than the outside temperature during the last 100 years. As a result, the concentrations of GHGs in the atmosphere have increased (Easterling et al. 1997). Agriculture is expected to account for 10–12% of total global anthropogenic GHG emissions, with 60% of global $N_2O$ and 50% of global $CH_4$ emissions, respectively. In comparison to $CO_2$ and $CH_4$, $N_2O$ is a more potent greenhouse gas, with a radiative forcing potential of roughly 12 times that of $CH_4$ and 298 times more potent concerning global warming than $CO_2$. Based on agricultural management, animal feeds and rice production were identified as the key sources of GHG emissions, while other cereals, fruits, and vegetables emit significantly less (Vetter et al. 2017).

Climate change as a global issue characterized by rising temperatures is having an impact on agricultural productivity and quality (DaMatta et al. 2010). Climate change could have major positive or negative effects on paddy cultivation and straw decomposition depending on the geographical location and the specific agricultural practices followed. However, there is a need for in-depth research to evaluate and improve our understanding of the effects of elevated $CO_2$ ($eCO_2$) and temperature on GHG emissions and decomposition. Hence, this chapter aims to summarize how $eCO_2$ and temperature affect GHG emissions, as well as to address decomposition mediated by microbes in the soil nutrient cycle.

## Biogeochemical Cycling of $CO_2$

Compared to $CH_4$ and $N_2O$, $CO_2$ is the dominant GHG emitted from agricultural cropping systems. Agriculture's net $CO_2$ release is minor in comparison to the total cycle, and it is mostly used for energy, on-farm energy, and the manufacture and transportation of agricultural products (Snyder et al. 2009). Rice is a major staple food that responds to $eCO_2$ levels by increasing its carbon absorption rates; consequently, the higher the $CO_2$ levels in the atmosphere, the greater the grain production (Sasaki et al. 2007). The $CO_2$ fertilization effect is frequently diminished as a result of reduced photosynthetic capability under $eCO_2$ (Reich et al. 2006). Rising temperatures are projected to increase plant respiration and soil N mineralization, which can release more $CO_2$ into the atmosphere.

## Biogeochemical Cycling of $CH_4$

The concentration of methane in the atmosphere is 1.77 μmol $mol^{-1}$ at the moment, and it accounts for 16% of global warming (Serrano-Silva et al. 2014). The concern needs to be addressed over the atmospheric $CH_4$ concentrations, which are increasing at a rate of 0.003 μmol $CH_4$ $mol^{-1}$air $year^{-1}$ (Khalil et al. 2007; Kvenvolden 2002). The terrestrial ecosystem serves as a source for increased rate of methane emissions which is primarily caused by methanogenesis and anaerobic decomposition. The main causes of methane emissions include an array of conditions, including aerobic and anaerobic environments, pH, salinity, redox potential, soil organic matter (SOM), and $O_2$ concentration (Dalal et al. 2008). The two primary groups of microorganisms in the process of methanogenesis and methane oxidation, respectively, are methanogens and methanotrophs (Fig. 1). According to Thauer et al. (2008), methanogens can be found in peat bogs, ruminants, freshwater habitats, hydrothermal vents, petroleum reservoirs, anaerobic sewage digesters, rice paddies, and sediments. Methanogens, however, are strict anaerobes that can only develop and operate in environments with low redox potential, such as moist soils or anoxic sediments (Smith et al. 2003). Conversely, methanotrophs can thrive in oxic environments and can be readily activated in the presence of moist anoxic conditions (Angel et al. 2012).

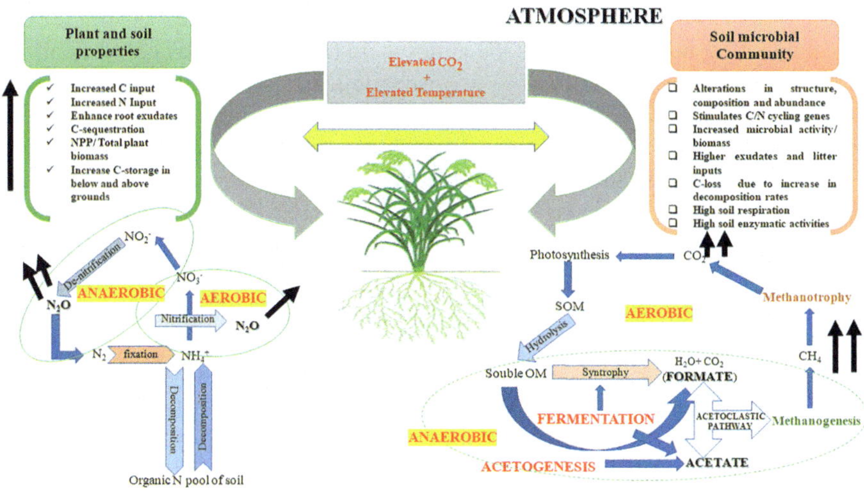


**Fig. 1** A conceptual framework of coupling soil microbial community and ecosystem functioning in response to elevated $CO_2$ and temperature.

## Microbial Biochemistry of Methane Oxidation

Biological methane oxidation, a metabolic process for energy generation, may be carried out aerobically by methanotrophs or anaerobically by anaerobic archaea (Fig. 1) (Ettwig et al. 2012). Methanotrophs are capable of preventing two-thirds of the $CH_4$ produced annually and hence are considered as only biological sink for methane. Hydroxyl radicals (OH) oxidize 90% of atmospheric methane, with methanotrophs re-mineralizing the remaining 5%. Methanotrophs can be found in a wide range of environments where oxic and anoxic conditions co-exist. Optimal conditions for methanotroph activity are neutral pH, mesophilic temperatures, and low salinities. Aerobic methanotrophs account for the bulk of biological methane oxidation in oxic conditions, with 600 Tg $CH_4$ oxidized each year on land compared to 300 Tg per year $CH_4$ in anaerobic habitats. Anaerobic $CH_4$ oxidation can also occur in marine environments, with anaerobic bacteria accounting for about 75% of marine methane oxidation through reverse methanogenesis, acetogenesis, and methylogenesis (Strous and Jetten 2004; Colwell et al. 2008). Comprehensive research on upland soil found that methanotrophs like *Methylocaldum*, *Methylosinus*, and *Methylocystis* account up to 70% microbial community (Knief et al. 2003). Methanotrophs are found at a wide range of temperatures, ranging from 0 °C to 72 °C. The methanotrophic bacterial clade *Methylacidiphilales* includes thermophilic and acidophilic bacteria, all of which have been isolated from geothermally active acidic environments (Dunfield et al. 2007).

## Effect of Elevated Carbon Dioxide ($eCO_2$) and Temperature on $CH_4$ Emission

The amount of methane ($CH_4$) in the atmosphere has been increasing by about 1% per year (Cheng et al. 2000). The atmospheric $CH_4$ are microbially driven processes occurring in anaerobic habitats, such as natural and farmed wetlands, sediments, sewage, and landfills. Studies have shown that elevated $CO_2$ increases $CH_4$ emissions from paddy soils by an average of 43%, while elevated soil temperatures (2 °C) increase $CH_4$ emissions by 42% (Liu et al. 2016). Under flooded conditions, a significant increase in $CH_4$ production has been seen in rice paddies. Methanogenesis in tropical rice soils was enhanced in response to $eCO_2$ concentrations and high temperatures with minimal $CH_4$ oxidation (Das and Adhya 2012). Agricultural soils other than rice are a sink for atmospheric $CH_4$, but only to a lesser extent than native soils. Rice is responsible for 11% of anthropogenic $CH_4$ emissions. The study reported interactive effects of $eCO_2$ and temperature could increase methane emissions from tropical paddy fields in an open-top chambers experiment (OTCs) (Ziska et al. 1997). Rice cultivation has a global warming potential more than two times than that of wheat or maize, owing mostly to high $CH_4$ emissions (Carlson et al. 2017). The net $CH_4$ emitted from rice fields has been estimated by setting equilibrium between production and consumption (Conrad 2007). Under anaerobic circumstances, methanogenic archaea produce methane, which is primarily determined by the availability of carbon (Tokida et al. 2011). While anaerobic $CH_4$ oxidation has been reported, aerobic methanotrophic bacteria are responsible for the majority of $CH_4$ oxidation in rice fields (Conrad 2007). The availability of $CH_4$ and $O_2$ is a major determinant of this process (Cai et al. 2016). In some tropical soils, rising temperatures not only boost methanogenesis but also suppress the $CH_4$ oxidation pathway, due to a reduction of soil redox potential and an increase in substrates such as acetate, which favor the growth of methanogens (Das and Adhya 2012). The type of methanogenesis that occurs in soil is affected by temperature. Low temperatures favor acetoclastic methanogenic activity, while higher temperatures favor hydrogenotrophy.

## Nitrogen Cycling and $N_2O$ Emission

The nitrogen (N) cycle is a set of essential biogeochemical pathways such as N mineralization, immobilization, and a variety of redox reactions that result in N transformation that is carried out by several microbial communities (Ishii et al. 2009; Gruber and Galloway 2008). The N cycle is a complex biogeochemical cycle with multiple transformative phases that can be observed and studied through different laboratory and field based approaches. In addition $NO_3^-NH_4^+$ from the atmosphere, the plant root also absorbs amino acids as N sources (Nelson et al. 2016). The International Fertilizer Association (IFA) and the Food and Agriculture Organization of the United Nations (FAO) (2001) have listed climate change, soil organic carbon (SOC) content, soil texture, drainage, availability of $NO_3^-$-N, soil pH, and N application rate as some of the major key factors which may affecting $N_2O$ emissions from fertilized field crops. Soil microorganisms also play a crucial role in regulating $N_2O$ emissions. Nitrification and denitrification are two microbial soil processes that result in $N_2O$ emissions (Snyder et al. 2009). Although there are many factors that combine to regulate nitrification and denitrification, the range of oxygen ($O_2$) concentrations in the soil ultimately determines how much $N_2O$ is produced. The rate of nitrification or denitrification by soil microorganisms is influenced by temperature. Rising temperature directly correlates with the rate of bacterial nitrification and denitrification of soil.

## Nitrifying Microbes

Nitrogen (N) fertilization reduces the relative abundance of rhizospheric bacteria, particularly type II methanotrophs and *Bradyrhizobium*, which in turn results in a reduction of biological $N_2$ fixation, which directly affects the soil $N_2$-fixing bacterial community (Kumar et al. 2018). $N_2$-fixing bacteria convert atmospheric $N_2$ gas to ammonium, which is available for plant use (Yoneyama et al. 2019). Elevated $CO_2$, which favors $N_2$ fixation, can increase plant N requirements (Kumar et al. 2020, 2024; Luo et al. 2004). Besides, $eCO_2$ may induce exudation of organic acids from roots that act as carbon sources for diazotrophs (Bhattacharyya et al. 2013) which may also result in higher abundance of $N_2$-fixing bacteria. A previous study found that soil diazotrophic abundance increased significantly in two different rice cultivars under $eCO_2$ to meet the increased N demand in paddy fields (Yu et al. 2018). Plant roots interact with a variety of soil-derived microorganisms, including $N_2$-fixing bacteria, which influences nutrient uptake (Berendsen et al. 2012; Shinoda et al. 2019). Root-associated bacteria in rice paddy fields also vary significantly between growth stages (Chaparro et al. 2014; Okubo et al. 2014). Higher plant growth can indirectly influence soil microorganisms by depleting the available N pool for microbial absorption over time, potentially making microorganisms N limited (Séneca et al. 2020). Long-term exposure to $eCO_2$ levels reduces the amount of $NH_4^+$ and $NO_3^-$ in the soil. $CO_2$ enrichment benefits plant growth and development, nitrogen uptake, soil microbial population and activity, and nitrogen mineralization rate. These factors influence the availability of $NH_4^+$ and $NO_3^-$, which accelerates N cycling (Waqas et al. 2021). Ammonia-oxidizing bacteria (AOB) and archaea (AOA) populations control soil nitrification, which helps plant to obtain enough N. AOA belong to the *Thaumarchaeota* phylum, while AOB belong to the *Betaproteobacteria* and *Gammaproteobacteria* phyla (Nguyen et al. 2019). Interestingly, temperature and $CO_2$ levels have little or no effect on the composition and activity of nitrifier communities in soil (Liu et al. 2021).

## Denitrifying Microbes

Denitrification is an anaerobic process in which denitrifiers reduce nitrate ($NO_3^-$) to nitrite ($NO_2^-$) and then subsequently to different nitrogen gases (NO, $N_2O$, and $N_2$), which can lead to nitrogen loss (Philippot et al. 2007). Cleveland (2007) found that microorganisms contain more nitrogen than soil, suggesting that rising $CO_2$ levels in the atmosphere could increase nitrogen availability or alter the process of nitrogen turnover in response to changing soil substrates. Root exudation can decrease nitrate-N levels, which can affect the types of denitrification end products produced (Diao et al. 2020). At higher temperatures (52 °C or above), total denitrification by the microbial community was observed. Metagenomic sequencing has shown that temperature differences can shift the structure and richness of the microbial community, revealing 84 bacterial genera and 7 phyla. Temperature stress inhibited *Firmicutes* and *Bacteroidetes* but favored *Proteobacteria.* At lower temperatures, the predominance of *Halomonas* and a significant decrease in *Azoarcus* suggested that these two genera may be more involved in denitrification (Liao et al. 2018).

## Effect of $eCO_2$ and Temperature on $N_2O$ Emission

$N_2O$ emissions from N-fertilized soils increased significantly under $eCO_2$ treatments compared to non-fertilized soils. This increase was due to the effects of $eCO_2$ on soil moisture, labile carbon, or both (Kammann et al. 2008; Niboyet et al. 2011). Denitrification and $N_2O$ emissions are favored by anaerobic conditions, and labile carbon uptake is a primary source of energy for denitrifiers (Firestone 1982). Plants' need for N increased when they were not fertilized. Higher C availability and increased anaerobic situation due to an increase in soil moisture favored denitrification. Photosynthetic C fixation by plants exposed to $eCO_2$ increased the carbon availability. However, increased N absorption by both plants and microorganisms reduces the amount of $NO_3^-$ availability to denitrifiers (Luo et al. 2004).

Under $eCO_2$ and high temperatures, $N_2O$ emissions from paddy fields increased by 24%, resulting in a 25% increase in global warming potential (GWP) (Bhattacharya et al. 2016). Reports indicated that continuous maize cultivation has a lower GWP than maize–soybean systems due to fertilizer-induced $N_2O$ emissions ranging from 1.9% to 3.5% in 2003, 0.8% to 1.5% in 2004, and 0.4% to 0.5% in 2005, with no consistent differences between these systems (Adviento-Borbe et al. 2007) (Table 1). Under $eCO_2$, N availability can be increased by increasing N uptake and efficiently using N or strictly regulating the internal soil N cycle. Increased gross N immobilization or decreased gross N mineralization can both lead to a decrease in net N mineralization under $eCO_2$.

## Soil Organic Matter and Its Decomposition

Carbon content in soil is nearly three times higher than that in the atmosphere (Houghton 2007), and interestingly soil C cycle is much slower than that of $CO_2$ in the atmosphere (Houghton 2007). The carbon balance between intake via photosynthesis and losses via respiration is indicative of whether a soil is a source or sink of C, which is likely a key aspect of soil functioning (Freeman et al. 2004). Factors like moisture, temperature, plant residue composition, and the ability of soil to stabilize soil organic matter (SOM) can all influence the decomposition rate of the above- and below-ground vegetative matter and litter into SOM, as well as their subsequent decomposition/mineralization to dissolved organic carbon (DOC) and $CO_2$, respectively (Lal 2004). Plants include similar C compounds such as starches, celluloses, lipids, proteins, and phenolic compounds, but each of their proportions varies depending on the species and age of the plant. The predominant components of SOM are humic compounds, which are defined as "complex mixtures of biologically altered organic waste". They are insoluble but contain non-humic components like polysaccharides, hydrocarbons, esters, and acids as well as soil minerals and abrasive plant remains. Environmental parameters like low temperatures, anaerobic conditions, poor quality of substrates, low pH, and lower nutrient availability limit

**Table 1** Effect of $eCO_2$ and temperature on GHG emissions from different crops and their significant outcome

| Crop | Location | GHG | Rate of emission | | Significance | References |
|---|---|---|---|---|---|---|
| Rice | NRRI, Cuttack, India | $CH_4$, $N_2O$ | 26%, 24% | | $eCO_2$ + -elevated temperature increased the GWP up to 25% | Bhattacharya et al. (2013) |
| Rice | California, USA | $CH_4$ | 0.3–2.2 mg $m^{-2}$ $h^{-1}$ | | Increased $CH_4$ emission both in vegetated plots with burnt straw and bare flooded plots | Redeker et al. (2000) |
| Maize | University of Nebraska, Lincoln, USA | $CO_2$, $N_2O$, $CH_4$ | 5612 kg $CO_2$-C $ha^{-1}$ $yr^{-1}$, 1.87 kg $N_2O$-N $ha^{-1}$ $yr^{-1}$, −3.36 kg $CH_4$-C $ha^{-1}$ $yr^{-1}$ | | Increased soil carbon sequestration in continuous maize system had lower GWP than maize–soybean system | Adviento-Borbe et al. (2007) |
| *Miscanthus*<br>Willow<br>ACA (wheat–wheat–oilseed system)<br>ACB (oilseed–wheat–oilseed system) | England, UK | $N_2O$, $CH_4$ | µg $N_2O$-N $m^{-2}$ $h^{-1}$<br>3.7<br>0.2<br>8.3<br>32.3 | µg $CH_4$-C $m^{-2}$ $h^{-1}$<br>−2.5<br>−6.5<br>−5<br>−4 | Perennial bioenergy crops emit less GHGs than annual crops with no or less N fertilizer | Drewer et al. (2012) |
| Soybean<br>Maize<br>Soybean–maize intercropping | INTA Agricultural Experimental Station Balcarce, Argentina | $N_2O$<br>$CO_2$ | µg $N_2O$-N $m^{-2}$ $h^{-1}$<br>33.2<br>33.1<br>30.7 | mg $CO_2$-C $m^{-2}$ $h^{-1}$<br>36.6<br>9.6<br>10.1 | Greenhouse gas fluxes were equivalent between ambient and elevated $CO_2$+ Temperature | Oelbermann et al. (2021) |
| Natural bog<br>Permanent grassland<br>Oat–potato<br>Oat–spring barley<br>Potato–spring barley | Denmark | $N_2O$<br>$CH_4$ | kg $N_2O$ $ha^{-1}$<br>0.3<br>3.0<br>3.7<br>1.5<br>37.7 | Kg $CH_4$ $ha^{-1}$<br>172<br>−1.5<br>1.5<br>−1.4<br>−0.6 | The pattern of $N_2O$ emissions suggested an increased risk of pollution from potato cultivation before and during the period of potato growth | |
| Soybean | Argentina | $N_2O$, $CO_2$ | 373.5 kg C $ha^{-1}$<br>−1127 kg C $ha^{-1}$ | | The global balance resulted in a carbon loss of 1328.5 kg C eq $ha^{-1}$ per year | Lewczuk et al. (2017) |
| – | Germany | $N_2O$, $CH_4$ | Kg $N_2O$-N $ha^{-1}$ $yr^{-1}$<br>≤0.5–1.5 | Kg $CH_4$-C $ha^{-1}$ $yr^{-1}$<br>0.4–1.2 | $N_2O$ emission was the lowest for SRC in both regions, Gierstadt (low $N_2O$ emissions) and Reiffenhausen (medium | Walter et al. (2015) |

microbial-mediated decomposition in peatland and wetland ecosystems, which tends to maintain SOM levels relatively constant (Hobbie et al. 2001). DOC is the primary end product of anoxic habitats, although its quantity fluctuates as conditions become more aerobic (Freeman et al. 2004).

## Microbial Decomposition of SOM

Biological and non-biological processes, as well as natural and human activities, all contribute to greenhouse gas (GHG) emissions. Soils can contribute a significant amount of GHG emissions, depending on the circumstances. Native soils are predominantly methane and nitrous oxide sinks, and they can store as much carbon as they release. Agriculture emitted 5100–6100 Tg $CO_2$ equivalents per year, accounting for 10–12% of global $CO_2$ equivalent emissions (Smith 2012). The soil microbial population is regulated by a variety of indirect plant-mediated processes, such as rhizodeposition, root exudates, and variations in their growth as a result of increasing $CO_2$ concentrations (Carney et al. 2007; Kandeler et al. 2008). Higher C:N ratios can affect microbial breakdown, suggesting that increased plant biomass could exacerbate nitrogen limitation. Progressive nitrogen shortage could impede plant responses to elevated $CO_2$ ($eCO_2$). It could also affect the structural abundance and functional capabilities of microbial communities (Reich et al. 2006; Tu et al. 2016).

## Effect of $eCO_2$ and Temperature on Soil Microbial Communities

Soil microbiota play a dynamic role in nutrient cycling in terrestrial ecosystems, but the impact of environmental factors can shift the structural and functional characteristics of soil microbial communities (Lee and Kang 2016). Microbial nutrient cycling also influences the biological activities, soil C sequestration, and GHG emissions in agricultural soil. Elevated $CO_2$ increases the plant biomass and root exudates, resulting in higher C and N additions to the soil (Phillips et al. 2011: Morgan et al. 2011). A rise in substrate availability and microenvironmental changes under $eCO_2$ have revealed highly varied patterns across different terrestrial ecosystems (He et al. 2015). However, no consistent patterns have emerged (Kelley et al. 2011; Smith and Myers 2018), making ecosystem functioning in the presence of $eCO_2$ challenging to predict (Drigo et al. 2013). In terms of nitrogen metabolism, the rice rhizosphere had greater structural diversity and functional activity, such as nitrogen fixation, denitrification, and assimilatory and dissimilatory nitrate reduction (Table 2). *Methanosphaera*, *Methanothermococcus*, *Methanobacterium*, and *Methanothermus* were not found under ambient $CO_2$, but showed their

**Table 2** Effect of $eCO_2$ and temperature on microbial communities responsible for decomposition

| Study area | C-N pools | Targeted pathway | Microbes involved | Enzymatics | Reference |
|---|---|---|---|---|---|
| Paddy fields, NRRI, Cuttack, India | Soil microbial biomass (MBN: 20%) | Acetoclastic pathway | *Methanobacterium, Methanosphaera, Methanothermus, Methanothermococcus* | DHA: 35% | Bhattacharya et al. (2016) |
| Temperate grasslands, Tasmania | Increased labile C pools | Photosynthetic pathway (C mineralization) | Heterotrophic bacteria | – | Osanai et al. (2015) |
| Wetland type fens, Finland | – | Cellobiose degradation | *Firmicutes, Proteobacteria, Acidobacteria* | Increased enzymatic activities | Juottonen et al. (2017) |
| Coastal estuary (restored peat), California, USA | Increased C inputs C assimilation | Methanogenesis (acetoclastic pathway) | *Proteobacteria, Chloroflexi, Bacteroidetes, Euryarchaeota, Methanomicrobia* | Increased phenol oxidase activity | He et al. (2015) |
| Mires, Germany | NPOC (79 mg $L^{-1}$) | Methanogenesis (acetoclastic pathway) Acetogenesis | *Methanoregulaceae, Methanosarcinaceae, Methanosaetaceae* | – | Hunger et al. (2015) |
| Freshwater marsh, China | – | – | *Methylobacter, Methylocystis* | – | Yun et al. (2015) |

prominent abundance under $eCO_2$ (Bhattacharyya et al. 2016). Soil respiration, mediated by heterotrophic bacteria, was a major determinant of decomposition speed. Elevated temperatures, in combination with $eCO_2$, alter the labile carbon pool in the soil, resulting in increased soil C outflow. Carbon generated from cellobiose breakdown is primarily taken by Firmicutes and Proteobacteria. When two fens (mesotrophic and oligotrophic) with different nutritional status and $CO_2$ emission were evaluated, the results showed that mesotrophic fen released three times higher $CH_4$ compared to $CO_2$ emitted by oligotrophic peat. However, the data showed that anaerobic decomposition can use different sets of bacterial degraders depending on the amount of $CH_4$ generated. Members of the Phyla *Proteobacteria* and *Chloroflexi* are primarily responsible for the decomposition of soil C inputs in coastal estuaries (Acetoclastic pathway), followed by *Bacteroidetes*, *Euryarchaeota*, and *Methanomicrobia* (He et al. 2015).

## Conclusion

The majority of greenhouse gases (GHGs) are emitted or fluxed from soil biogeochemical carbon and nitrogen cycles. Soil microbes play a vital role in the mineralization, decomposition, and assimilation of GHGs. Agriculture alone contributes about 10–12% of the global anthropogenic GHG emissions, accounting for around 50% of global methane ($CH_4$) and 60% of global nitrous oxide ($N_2O$) emissions. The combined effects of $eCO_2$ and temperature increased $CH_4$ emissions, which are largely driven by the availability of carbon. This was accompanied by an increase in methanogenic bacteria and archaea. Similarly, $eCO_2$ treatments in N-fertilized fields lead to a significant rise in $N_2O$ emissions. The $N_2O$ emission rate is largely determined by the range of oxygen concentrations in soil; however, the interplay of regulatory factors complicates the nitrification and denitrification process. Soil microbes are actively involved in biogeochemical cycling, and their interactions modify the abundance, composition, and functional characteristics of the microbial community. Microbially production of GHG is influenced by the interplay between soil C and soil organic matter storage capacity. Sudden changes in flooding frequency and $eCO_2$ can lead to the loss of C supplies and could alter the microbial community, which need further investigation. Furthermore, $eCO_2$-induced fluctuations in soil microenvironments may induce prolonged consequences on terrestrial biogeochemical cycles and climate change. Further research is required in this context to assess the implications of various mitigation measures such as adjusting sowing and harvesting dates, pulse cropping, intercropping, and fertilizer management related to the agricultural C and N cycle.

## References


Adviento-Borbe MAA, Haddix ML, Binder DL, Walters DT, Dobermann A (2007) Soil greenhouse gas fluxes and global warming potential in four high-yielding maize systems. Glob Chang Biol 13(9):1972–1988

Angel R, Claus P, Conrad R (2012) Methanogenic archaea are globally ubiquitous in aerated soils and become active under wet anoxic conditions. ISME J 6(4):847–862

Berendsen RL, Pieterse CM, Bakker PA (2012) The rhizosphere microbiome and plant health. Trends Plant Sci 17(8):478–486

Bhattacharyya P, Roy KS, Neogi S, Manna MC, Adhya TK, Rao KS, Nayak AK (2013) Influence of elevated carbon dioxide and temperature on belowground carbon allocation and enzyme activities in tropical flooded soil planted with rice. Environ Monit Assess 185(10):8659–8671

Bhattacharyya P, Roy KS, Das M, Ray S, Balachandar D, Karthikeyan S, Nayak AK, Mohapatra T (2016) Elucidation of rice rhizosphere metagenome in relation to methane and nitrogen metabolism under elevated carbon dioxide and temperature using whole genome metagenomic approach. Sci Total Environ 542:886–898

Cai Y, Zheng Y, Bodelier PL, Conrad R, Jia Z (2016) Conventional methanotrophs are responsible for atmospheric methane oxidation in paddy soils. Nat Commun 7(1):1–10

Carlson KM, Gerber JS, Mueller ND, Herrero M, MacDonald GK, Brauman KA, Havlik P, O'Connell CS, Johnson JA, Saatchi S, West PC (2017) Greenhouse gas emissions intensity of global croplands. Nat Clim Chang 7(1):63–68

Carney KM, Hungate BA, Drake BG, Megonigal JP (2007) Altered soil microbial community at elevated $CO_2$ leads to loss of soil carbon. Proc Natl Acad Sci 104(12):4990–4995

Cassia R, Nocioni M, Correa-Aragunde N, Lamattina L (2018) Climate change and the impact of greenhouse gasses: $CO_2$ and NO, friends and foes of plant oxidative stress. Front Plant Sci 9:273

Chaparro JM, Badri DV, Vivanco JM (2014) Rhizosphere microbiome assemblage is affected by plant development. ISME J 8(4):790–803

Cheng W, Chander K, Inubushi K (2000) Effects of elevated $CO_2$ and temperature on methane production and emission from submerged soil microcosm. Nutr Cycl Agroecosyst 58(1):339–347

Cleveland CC, Nemergut DR, Schmidt SK, Townsend AR (2007) Increases in soil respiration following labile carbon additions linked to rapid shifts in soil microbial community composition. Biogeochemistry 82(3):229–240

Colwell FS, Boyd S, Delwiche ME, Reed DW, Phelps TJ, Newby DT (2008) Estimates of biogenic methane production rates in deep marine sediments at Hydrate Ridge, Cascadia Margin. Appl Environ Microbiol 74(11):3444–3452

Conrad R (2007) Microbial ecology of methanogens and methanotrophs. Adv Agron 96:1–63

Dalal RC, Allen DE, Livesley SJ, Richards G (2008) Magnitude and biophysical regulators of methane emission and consumption in the Australian agricultural, forest, and submerged landscapes: a review. Plant Soil 309(1):43–76

DaMatta FM, Grandis A, Arenque BC, Buckeridge MS (2010) Impacts of climate changes on crop physiology and food quality. Food Res Int 43(7):1814–1823

Das S, Adhya TK (2012) Dynamics of methanogenesis and methanotrophy in tropical paddy soils as influenced by elevated $CO_2$ and temperature interaction. Soil Biol Biochem 47:36–45

Diao T, Peng Z, Niu X, Yang R, Ma F, Guo L (2020) Changes of soil microbes related with carbon and nitrogen cycling after long-term $CO_2$enrichment in a typical Chinese maize field. Sustain For 12(3):1250

Drewer J, Finch JW, Lloyd CR, Baggs EM, Skiba U (2012) How do soil emissions of $N_2O$, $CH_4$ and $CO_2$ from perennial bioenergy crops differ from arable annual crops? GCB Bioenergy 4(4):408–419

Drigo B, Kowalchuk GA, Knapp BA, Pijl AS, Boschker HT, Van Veen JA (2013) Impacts of 3 years of elevated atmospheric $CO_2$ on rhizosphere carbon flow and microbial community dynamics. Glob Chang Biol 19(2):621–636

Dunfield PF, Yuryev A, Senin P, Smirnova AV, Stott MB, Hou S, Ly B, Saw JH, Zhou Z, Ren Y, Wang J (2007) Methane oxidation by an extremely acidophilic bacterium of the phylum Verrucomicrobia. Nature 450(7171):879–882

Easterling DR, Horton B, Jones PD, Peterson TC, Karl TR, Parker DE, Salinger MJ, Razuvayev V, Plummer N, Jamason P, Folland CK (1997) Maximum and minimum temperature trends for the globe. Science 277(5324):364–367

Ettwig KF, Speth DR, Reimann J, Wu ML, Jetten MS, Keltjens JT (2012) Bacterial oxygen production in the dark. Front Microbiol 3:273

Firestone MK (1982) Biological denitrification. In: Nitrogen in agricultural soils, vol 22, pp 289–326

Freeman C, Fenner N, Ostle NJ, Kang H, Dowrick DJ, Reynolds B, Lock MA, Sleep D, Hughes S, Hudson J (2004) Export of dissolved organic carbon from peatlands under elevated carbon dioxide levels. Nature 430(6996):195–198

Gilbert N (2012) One-third of our greenhouse gas emissions come from agriculture. Nature News 31:10–12

Gruber N, Galloway JN (2008) An Earth-system perspective of the global nitrogen cycle. Nature 451(7176):293–296

He S, Malfatti SA, McFarland JW, Anderson FE, Pati A, Huntemann M, Tremblay J, Glavina del Rio T, Waldrop MP, Windham-Myers L, Tringe SG (2015) Patterns in wetland microbial community composition and functional gene repertoire associated with methane emissions. MBio 6(3):00066–00015

Hobbie EA, Weber NS, Trappe JM (2001) Mycorrhizal vs saprotrophic status of fungi: the isotopic evidence. New Phytol:601–610

Houghton RA (2007) Balancing the global carbon budget. Annu Rev Earth Planet Sci 35:313–347

Howard-Grenville J, Buckle SJ, Hoskins BJ, George G (2014) Climate change and management. Acad Manag J 57(3):615–623

Hunger S, Gößner AS, Drake HL (2015) Anaerobic trophic interactions of contrasting methane-emitting mire soils: processes versus taxa. FEMS Microbiol Ecol 91(5)

Ishii S, Ikeda S, Minamisawa K, Senoo K (2009) Nitrogen cycling in rice paddy environments: past achievements and future challenges. Microbes Environ:1110170328

Juottonen H, Eiler A, Biasi C, Tuittila ES, Yrjälä K, Fritze H (2017) Distinct anaerobic bacterial consumers of cellobiose-derived carbon in boreal fens with different $CO_2/CH_4$ production ratios. Appl Environ Microbiol 83(4):e02533–e02516

Kammann C, Müller C, Grünhage L, Jäger HJ (2008) Elevated $CO_2$ stimulates $N_2O$ emissions in permanent grassland. Soil Biol Biochem 40(9):2194–2205

Kandeler E, Mosier AR, Morgan JA, Milchunas DG, King JY, Rudolph S, Tscherko D (2008) Transient elevation of carbon dioxide modifies the microbial community composition in a semi-arid grassland. Soil Biol Biochem 40(1):162–171

Kelley AM, Fay PA, Polley HW, Gill RA, Jackson RB (2011) Atmospheric $CO_2$ and soil extracellular enzyme activity: a meta-analysis and $CO_2$ gradient experiment. Ecosphere 2(8):1–20

Khalil MAK, Butenhoff CL, Rasmussen RA (2007) Atmospheric methane: trends and cycles of sources and sinks. Environ Sci Technol 41(7):2131–2137

Knief C, Lipski A, Dunfield PF (2003) Diversity and activity of methanotrophic bacteria in different upland soils. Appl Environ Microbiol 69(11):6703–6714

Kumar U, Kaviraj M, Rout S, Chakraborty K,Swain P, Nayak PK, Nayak AK (2024) Combined effects of elevated CO2 and nitrogen fertilizers on plant physiological variables in two contrasting Azolla spp. Sci Horticu 331:113152

Kumar U, Nayak AK, Shahid M, Gupta VV, Panneerselvam P, Mohanty S, Kaviraj M, Kumar A, Chatterjee D, Lal B, Gautam P (2018) Continuous application of inorganic and organic fertilizers over 47 years in paddy soil alters the bacterial community structure and its influence on rice production. Agric Ecosyst Environ 262:65–75

Kumar U, Nayak AK, Sahoo S, Kumar A, Kaviraj M, Shahid M (2020) Combined effects of elevated CO2, N fertilizer and water deficit stress on diazotrophic community in sub-humid tropical paddy soil. Appl Soil Ecol155:1–12

Kutsch WL, Bahn M, Heinemeyer A (eds) (2009) Soil carbon dynamics: an integrated methodology. Cambridge University Press

Kvenvolden KA (2002) Methane hydrate in the global organic carbon cycle. Terra Nova 14(5):302–306

Lal R (2004) Soil carbon sequestration impacts on global climate change and food security. Science 304(5677):1623–1627

Lee SH, Kang H (2016) Elevated $CO_2$ causes a change in microbial communities of rhizosphere and bulk soil of salt marsh system. Appl Soil Ecol 108:307–314

Lewczuk NA, Posse G, Richter K, Achkar A (2017) $CO_2$ and $N_2O$ flux balance on soybean fields during growth and fallow periods in the Argentine Pampas—a study case. Soil Tillage Res 169:65–70

Liao R, Miao Y, Li J, Li Y, Wang Z, Du J, Li Y, Li A, Shen H (2018) Temperature dependence of denitrification microbial communities and functional genes in an expanded granular sludge bed reactor treating nitrate-rich wastewater. RSC Adv 8(73):42087–42094

Lindsey R (2019) Climate change: atmospheric carbon dioxide. https://www.climate.gov/news-features/understanding-climate/climate-change-atmospheric-carbon-dioxide

Liu Y, Liu X, Cheng K, Li L, Zhang X, Zheng J, Zheng J, Pan G (2016) Responses of methanogenic and methanotrophic communities to elevated atmospheric $CO_2$ and temperature in a paddy field. Front Microbiol 7:1895

Liu J, Han J, Zhu C, Cao W, Luo Y, Zhang M, Zhang S, Jia Z, Yu R, Zhao J, Bao Z (2021) Elevated atmospheric $CO_2$ and nitrogen fertilization affect the abundance and community structure of rice root-associated nitrogen-fixing bacteria. Front Microbiol 12:628108

Luo Y, Su BO, Currie WS, Dukes JS, Finzi A, Hartwig U, Hungate B, McMurtrie RE, Oren RAM, Parton WJ, Pataki DE (2004) Progressive nitrogen limitation of ecosystem responses to rising atmospheric carbon dioxide. Bioscience 54(8):731–739

Morgan JA, LeCain DR, Pendall E, Blumenthal DM, Kimball BA, Carrillo Y, Williams DG, Heisler-White J, Dijkstra FA, West M (2011) $C_4$ grasses prosper as carbon dioxide eliminates desiccation in warmed semi-arid grassland. Nature 476(7359):202–205

Nelson MB, Martiny AC, Martiny JB (2016) Global biogeography of microbial nitrogen-cycling traits in soil. Proc Natl Acad Sci 113(29):8033–8040

Nguyen LT, Broughton K, Osanai Y, Anderson IC, Bange MP, Tissue DT, Singh BK (2019) Effects of elevated temperature and elevated $CO_2$ on soil nitrification and ammonia-oxidizing microbial communities in field-grown crop. Sci Total Environ 675:81–89

Niboyet A, Brown JR, Dijkstra P, Blankinship JC, Leadley PW, Le Roux X, Barthes L, Barnard RL, Field CB, Hungate BA (2011) Global change could amplify fire effects on soil greenhouse gas emissions. PLoS One 6(6):e20105

Nieder R, Benbi DK (2008) Carbon and nitrogen in the terrestrial environment. Springer Science & Business Media

NOAA (2020) Climate change indicators: climate forcing. https://www.epa.gov/climate-indicators/climate-change-indicators-climate-forcing

Oelbermann M, Morgan S, Echarte L (2021) Elevated carbon dioxide and temperature effects on soil properties from sole crops and intercrops. Soil Use Manag 38:435–447

Okubo T, Tokida T, Ikeda S, Bao Z, Tago K, Hayatsu M, Nakamura H, Sakai H, Usui Y, Hayashi K, Hasegawa T (2014) Effects of elevated carbon dioxide, elevated temperature, and rice growth stage on the community structure of rice root—associated bacteria. Microbes Environ 29:184–190

Osanai Y, Janes JK, Newton PC, Hovenden MJ (2015) Warming and elevated $CO_2$ combine to increase microbial mineralisation of soil organic matter. Soil Biol Biochem 85:110–118

Philippot L, Hallin S, Schloter M (2007) Ecology of denitrifying prokaryotes in agricultural soil. Adv Agron 96:249–305

Phillips RP, Finzi AC, Bernhardt ES (2011) Enhanced root exudation induces microbial feedbacks to N cycling in a pine forest under long-term $CO_2$ fumigation. Ecol Lett 14(2):187–194

Redeker KR, Wang NY, Low JC, McMillan A, Tyler SC, Cicerone RJ (2000) Emissions of methyl halides and methane from rice paddies. Science 290(5493):966–969

Reich PB, Hungate BA, Luo Y (2006) Carbon-nitrogen interactions in terrestrial ecosystems in response to rising atmospheric carbon dioxide. Annu Rev Ecol Evol Syst 37:611–636

Ritchie H, Roser M (2020) $CO_2$ and greenhouse gas emissions. Our world in data

Sasaki H, Hara T, Ito S, Uehara N, Kim HY, Lieffering M, Okada M, Kobayashi K (2007) Effect of free-air $CO_2$ enrichment on the storage of carbohydrate fixed at different stages in rice (*Oryza sativa* L.). Field Crop Res 100(1):24–31

Schaufler G, Kitzler B, Schindlbacher A, Skiba U, Sutton MA, Zechmeister-Boltenstern S (2010) Greenhouse gas emissions from European soils under different land use: effects of soil moisture and temperature. Eur J Soil Sci 61:683–696

Séneca J, Pjevac P, Canarini A, Herbold CW, Zioutis C, Dietrich M, Simon E, Prommer J, Bahn M, Pötsch EM, Wagner M (2020) Composition and activity of nitrifier communities in soil are unresponsive to elevated temperature and $CO_2$, but strongly affected by drought. ISME J 14(12):3038–3053

Serrano-Silva N, Valenzuela-Encinas C, Marsch R, Dendooven L, Alcántara-Hernández RJ (2014) Changes in methane oxidation activity and methanotrophic community composition in saline alkaline soils. Extremophiles 18(3):561–571

Shinoda R, Bao Z, Minamisawa K (2019) $CH_4$ oxidation-dependent $^{15}N_2$ fixation in rice roots in a low-nitrogen paddy field and in *Methylosinus* sp. strain 3S-1 isolated from the roots. Soil Biol Biochem 132:40–46

Smith P (2012) Agricultural greenhouse gas mitigation potential globally, in Europe and in the UK: what have we learnt in the last 20 years? Glob Chang Biol 18(1):35–43

Smith MR, Myers SS (2018) Impact of anthropogenic $CO_2$ emissions on global human nutrition. Nat Clim Chang 8(9):834–839

Smith KA, Ball T, Conen F, Dobbie KE, Massheder J, Rey A (2003) Exchange of greenhouse gases between soil and atmosphere: interactions of soil physical factors and biological processes. Eur J Soil Sci 54(4):779–791

Snyder CS, Bruulsema TW, Jensen TL, Fixen PE (2009) Review of greenhouse gas emissions from crop production systems and fertilizer management effects. Agric Ecosyst Environ 133(3–4):247–266

Strous M, Jetten MS (2004) Anaerobic oxidation of methane and ammonium. Ann Rev Microbiol 58:99–117

Thauer RK, Kaster AK, Seedorf H, Buckel W, Hedderich R (2008) Methanogenic archaea: ecologically relevant differences in energy conservation. Nat Rev Microbiol 6(8):579–591

Tokida T, Adachi M, Cheng W, Nakajima Y, Fumoto T, Matsushima M, Nakamura H, Okada M, Sameshima R, Hasegawa T (2011) Methane and soil $CO_2$ production from current-season photosynthates in a rice paddy exposed to elevated $CO_2$ concentration and soil temperature. Glob Chang Biol 17(11):3327–3337

Tu Q, Zhou X, He Z, Xue K, Wu L, Reich P, Hobbie S, Zhou J (2016) The diversity and co-occurrence patterns of $N_2$-fixing communities in a $CO_2$-enriched grassland ecosystem. Microb Ecol 71(3):604–615

Tubiello FN, Salvatore M, Rossi S, Ferrara A, Fitton N, Smith P (2013) The FAOSTAT database of greenhouse gas emissions from agriculture. Environ Res Lett 8(1):015009

Vetter SH, Sapkota TB, Hillier J, Stirling CM, Macdiarmid JI, Aleksandrowicz L, Green R, Joy EJ, Dangour AD, Smith P (2017) Greenhouse gas emissions from agricultural food production to supply Indian diets: implications for climate change mitigation. Agric Ecosyst Environ 237:234–241

Walter K, Don A, Flessa H (2015) Net $N_2O$ and $CH_4$ soil fluxes of annual and perennial bioenergy crops in two central German regions. Biomass Bioenergy 81:556–567

Waqas MA, Ashraf MN, Ahmed W, Wang B, Sardar MF, Ma P, Li R, Wan Y, Kuzyakov Y (2021) Long-term warming and elevated $CO_2$ increase ammonia-oxidizing microbial communities and accelerate nitrification in paddy soil. Appl Soil Ecol 166:104063

Yoneyama T, Terakado-Tonooka J, Bao Z, Minamisawa K (2019) Molecular analyses of the distribution and function of diazotrophic rhizobia and methanotrophs in the tissues and rhizosphere of non-leguminous plants. Plan Theory 8(10):408

Yu Y, Zhang J, Petropoulos E, Baluja MQ, Zhu C, Zhu J, Lin X, Feng Y (2018) Divergent responses of the diazotrophic microbiome to elevated $CO_2$ in two rice cultivars. Front Microbiol 9:1139

Yun J, Zhang H, Deng Y, Wang Y (2015) Aerobic methanotroph diversity in Sanjiang wetland, Northeast China. Microb Ecol 69(3):567–576

Ziska LH, Namuco O, Moya T, Quilang J (1997) Growth and yield response of field-grown tropical rice to increasing carbon dioxide and air temperature. Agron J 89(1):45–53

# Carbon Speciation by Microbes in Soil: Indicators of Carbon Sequestration and Soil Function

**Yumnam Bijilaxmi Devi, Thounaojam Thomas Meetei, and Thounaojam Thorny Chanu**

## Introduction

Effective management of soil carbon storage is imperative as it constitutes a crucial ecosystem service which is pivotal for efficiently sequestering carbon (Ontl and Schulte 2012). The amount of carbon stored in soil exceeds the total amount of carbon present in both the atmosphere and the biotic pool (Stockmann et al. 2013). Therefore, soil is a significant reservoir and source of carbon, making it important to understand carbon speciation in soil. Soil plays an important role in the global carbon cycle (Adamu et al. 2021). Anthropogenic activities have caused heightened disturbance of originally stable carbon forms in soil. For the record, human activities have altered the global carbon cycle, resulting in a 31% increase in atmospheric $CO_2$ levels from 280 ppmv in 1800 to 368 ppmv in 2000 (Houghton et al. 2001). The use of mineral fertilizers, manure, and other amendments in agricultural systems contributes to global anthropogenic greenhouse gas emissions (GHGs) (Smith et al. 2007; Snyder et al. 2009). Increasing nitrogen (N) fertilizer use has led to high rates of $N_2O$ emissions. Employing a combination of organic and inorganic fertilization can aid in mitigating these emissions and protect the environment (Efthimiadou et al. 2010). However, incorporating organic materials can also increase $CO_2$ emissions (Huang et al. 2004; Johnson et al. 2007). Lowering the level of $CO_2$ in atmosphere is a major challenge in mitigating climate change. The ability

Y. B. Devi (✉)
College of Horticulture and Forestry, Rani Lakshmi Bai Central Agricultural University, Jhansi, UP, India

T. T. Meetei
College of Agriculture, Rani Lakshmi Bai Central Agricultural University, Jhansi, UP, India

T. T. Chanu
Department of Botany, School of Life Science, Assam Don Bosco University, Sonapur, Assam, India

 
S. R. Mohanty, B. Kollah (eds.), *Greenhouse Gas Regulating Microorganisms in Soil Ecosystems*, https://doi.org/10.1007/978-3-031-70569-4_9

of carbon to change its form increases the potential for carbon sequestration and climate change mitigation (Jackson et al. 2017; Stockmann et al. 2013). There is limited information on how N fertilization affects carbon speciation and transformation. Therefore, understanding how soil responds to the addition of organic and inorganic materials is important for assessing soil sustainability and productivity (Olaleye et al. 2020). Carbon speciation in soil, microbes, plants, and water plays an important role in carbon sequestration. Human activities are impeding this process and sequestration. This chapter focuses on carbon speciation in soil by soil microbes in particular. Soil is an important reservoir of carbon in the form of soil organic carbon (SOC), which has the potential to store more carbon than the atmosphere. Carbon is stored in soil through the primary producers of an ecosystem, where carbon fixed during plant photosynthesis is transferred to the soil as soil organic matter (SOM) (Horwath 2007). SOM consists of diverse components, which may differ depending on the plant source. These components typically encompass cellulose, hemicellulose, lignin, proteins, and simple sugars generally in decreasing order of concentration. The fractions of SOM that can be easily decomposed by soil organisms are considered the non-humic fraction of SOM (Lehmann and Kleber 2015). The humic fractions of SOM are those fractions that are very stable in soil and cannot be further decomposed. In other words, the humic fractions of SOM are generated from the non-humic fractions after repeated decomposition by different groups of organisms in the soil. During the plant life cycle, at least all 17 essential nutrients required for proper growth and metabolism are assimilated from the soil. These essential nutrients are then returned to the soil through nutrient transformation process (nutrient cycle) after the crop life cycle ends. The non-humic fractions of SOM are crucial for any nutrient transformation process (Tate 1984). This is because these fractions can be decomposed and transformed to inorganic forms by soil microorganisms within a short period of time. Therefore, the non-humic fractions of SOM are associated with soil quality. However, the humic fractions of SOM are considered the reserved fractions of SOM due to their high polymerization and stability.

## Carbon Speciation

Organic matter serves as the primary source of carbon in soil and constitutes a substantial portion of terrestrial carbon storage. Carbon can exist in different species, including inorganic, organic, and elemental forms, each with its own chemical properties and environmental effects (Kumpiene et al. 2011). The decomposition of organic matter in soil produces different species of carbon. Humic and non-humic substances are formed during this process (Kaushal et al. 2021). Humic substances are constituents of soil organic matter characterized by their low molecular weight (Fig. 1). They are associated with various compounds of biological origin and participate in reactions that occur in the soil (Gautam et al. 2021). Humic substances are composed of three components: humic acid, fulvic acid, and humin. Sugars,

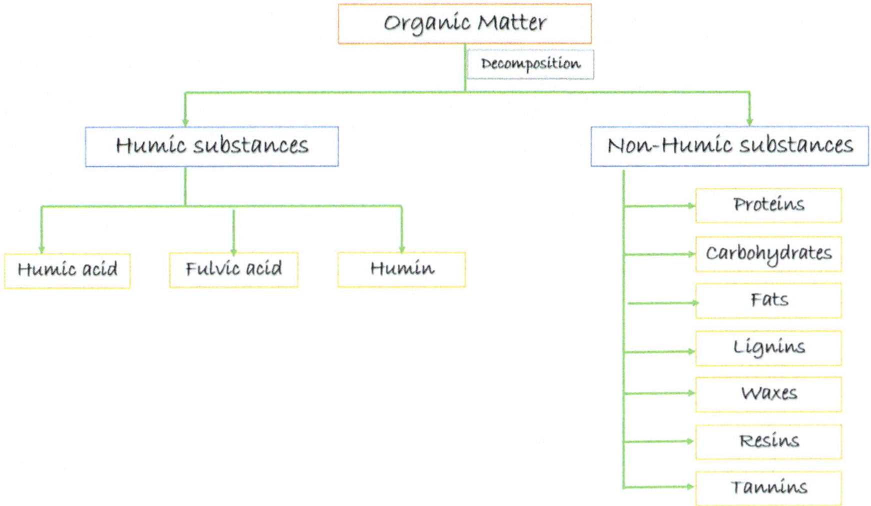


**Fig. 1** Forms of organic matter after decomposition

fats, and carbohydrates make up the non-humic portion of soil organic matter, contributing about 5–25%. Over time, carbohydrates are converted to sugars, hemicellulose, and cellulose. These compounds play a crucial role as a binding agent in maintaining aggregate stability.

Soil organic carbon fractions can be broadly divided into two pools: active/labile and passive/non-labile. These pools are highly influenced by land use management practices.

The labile form of carbon includes soil microbial biomass carbon (SMBC) and other fractions such as total organic carbon (TOC), dissolved organic matter (DOC), particulate organic matter (POM), and water-soluble carbon (WSC). Labile pools undergo quick decomposition and cycling of carbon (Ghosh et al. 2012). The passive carbon pool (POC) consists of more complex carbonaceous materials that are resistant to decomposition. POC can be further divided into less labile and non-labile pools (Fig. 2). POC is a recalcitrant fraction that contributes around 60% of organic carbon and is not easily influenced by land use management practices (Sainepo et al. 2018). These portions are protected from microbial activity and are therefore least decomposed (Dwivedi et al. 2019). In the case of bamboo systems, POC may be comprised of phytolith-occluded carbon (PhytOC) (Huang et al. 2014). Carbon may also be present in functional groups such as carboxylic and phenolic groups.

Different species of soil organic carbon (SOC) are produced from the decomposition of organic compounds in the soil. The primary and secondary sources of soil organic compounds are plants and animals, respectively. Soil organic compounds consist of living (edaphon) and nonliving (soil organic matter) parts. The nonliving part, soil organic matter (SOM), is composed of different components (cellulose, hemicellulose, lignin, proteins, and simple sugars) that vary from one organic

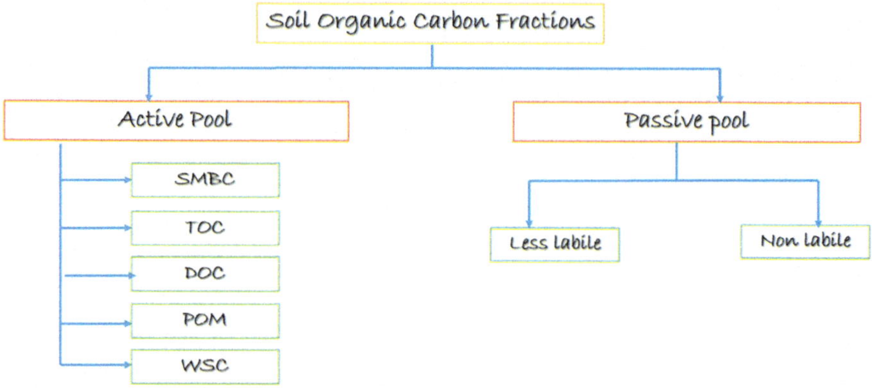


**Fig. 2** Fractions of soil organic carbon making up species of carbon. SMBC soil microbial biomass carbon, TOC total organic carbon, DOC dissolved organic matter, WSC water-soluble carbon

matter type to another. SOM consists of unaltered and altered materials. The altered materials are the decaying and decomposed products of organic compounds. Therefore, the altered material of SOM can be basically classified into humic and non-humic substances. Non-humic substances are those fractions of SOM that are freshly released from crop residues. They are considered as active fractions that can be easily decomposed by soil microorganisms.

Humic substances are those fractions of SOM that are formed from non-humic substances after repeated decomposition and polymerization by soil microorganisms. During the decomposition of non-humic substances, different nutrients are released into the soil. However, once humus is formed, further decomposition is restricted.

As a result, non-humic substances play an important role in soil quality, while humic substances serve as a C reservoir and promote soil aggregate stability. During the decomposition of non-humic fractions of SOM to the final stage of decomposition (the formation of humic substances), the material passes through different intermediate stages. Therefore, different species of SOC are produced during the conversion of organic compounds to humic substances, which is mediated by different groups of soil microorganisms. Soil is an important reservoir of carbon (C), containing about 1500 Pg C, which is more than the total contributions of both the biotic pool (560 PgC) and the atmospheric pool (760 PgC). Therefore, the role of soil in reducing the impact of climate change is well-established. However, our understanding of the specific functions of different soil organic carbon (SOC) species in climate and soil health sustainability is lacking. This is because SOC composition varies widely across terrestrial ecosystems, depending on factors such as climate, biosphere, topography, parent material, and time. Anthropogenic activities, such as land use change and land cover change to increase agricultural production, are disrupting the global carbon cycle.

Since 1850, these activities have led to a 450% increase in global cultivated area. Significant changes have been observed in tropical and subtropical regions since

1950. These anthropogenic activities have resulted in less biomass being returned to the soil and more $CO_2$ being released into the atmosphere, exacerbating climate change. However, not all SOC species are responsible for $CO_2$ release. Some fractions of SOC are stable and remain in the soil, supporting aggregate stability. All SOC species have different roles in the soil. The concentration of each species depends on the stage of microbial decomposition of organic matter. The functioning capacity of the soil also varies with different SOC species. Therefore, it is important to identify and understand the specific roles of different SOC species. However, some SOC species cannot be easily determined using traditional organic carbon determination methods. Sophisticated instruments and principles are required to characterize and classify these species. Many researchers have conducted research on SOC speciation and have suggested various management practices to improve soil health and reduce $CO_2$ emissions. Some of the examples are cited from which, one can have an idea of carbom speciation and its management process (Table 1).

## Microbial Communities and C Speciation

Carbon is transformed into various species after it is fixed in the soil by autotrophic organisms, especially photoautotrophic and chemoautotrophic microbes that synthesize atmospheric carbon into organic material (Gougoulias et al. 2014). Root-driven carbon storage is mainly attributed to the exchange and transport of nutrients between plant roots and microbes in the rhizosphere. These microbes may be bacteria (such as Rhizobiales) or fungi (such as Russulales). The bacterial order Rhizobiales plays an important role in nutrient exchange, and the fungal order Russulales acts as a stability promoter that can retain and stabilize microbial-derived organic matter. It also acts as a nutrient transporter by transporting nutrients between roots and soil (Song et al. 2020).

Processes such as decomposition, methanogenesis, methane oxidation, bacterial carbon dioxide fixation, and excretion of elements from dead and decaying organic matter change the forms of carbon in soil and play a significant role in carbon speciation (Fig. 3). The terrestrial carbon cycle demonstrates how microbial communities actively differentiate various carbon species over time. Carbon cycling is an important process that keeps carbon in check in the environment. Without proper carbon cycling, carbon speciation will be incomplete, and without microbial communities, the potential for storing carbon in soil will decline. The makeup of soil microbial communities governs a multitude of soil ecosystem services, impacting soil structure, decomposition rates, nutrient transformations, and other essential processes. (Bhattacharyya et al. 2022). A thorough comprehension and effective management of microbial communities will enhance the carbon cycling process, thereby ensuring the stability of carbon speciation. Many studies have been conducted on the potential of microbes in carbon sequestration from which one can understand the importance of microbes in C speciation (Table 2).

**Table 1** Review of carbon speciation studies and their suggestions

| Place studied | Problems addressed | Management practices suggested | References |
|---|---|---|---|
| South Africa | The persistent effects of human-induced disturbances on the dynamics and differentiation of organic carbon in tropical forest and subtropical grassland ecosystems | Alterations in the inherent molecular structures of the SOM | Solomon et al. (2007) |
| Forest soils of the USA | Distribution of organic carbon, its various forms, and relationship with elemental content within soil microaggregates | The effectiveness of a sample preparation method increases by involving the selection of particles with optimal thicknesses, rather than creating thin sections from soil microaggregates | Wan et al. (2007) |
| Sweden | The categorization of carbon types present in ash, residual waste, and contaminated soil | Speciation of total organic carbon determined by chemical degradation is highest in ash compared to residual waste and contaminated soil. Contaminated soil followed by residual waste is more stable in terms of carbon | Kumpiene et al. (2011) |
| China | Dynamics and speciation of organic carbon during decomposition of leaf litter and fine roots in four subtropical plantations | Site microclimate and litter chemistry played more critical roles in regulating fine root mass loss for different C fractions than leaf litter during decomposition. Thus, the main factors controlling decomposability of belowground litter C fractions were environmental and initial litter quality, whereas physical properties and site conditions may impact the decomposability of aboveground litter C fractions | Wang et al. (2013) |
| Columbus, USA | Effects of the different types of organic carbon on the biogeochemistry and microbial ecology of hyporheic corridors | Mixing of groundwater and river water in a hyporheic corridor can stimulate biogeochemical activity | Stegen et al. (2018) |
| Nigeria | Effect of manure and urea fertilization on yield, carbon speciation, and greenhouse gas emissions | The low organic matter sandy soils of SSA can be significant sources of $CO_2$ and $N_2O$, and fertility managements to optimize yield, build/maintain soil productivity, and mitigate greenhouse gas emission will differ by ecoregion, soil type, and maybe even crop | Olaleye et al. (2020) |
| Goa, India | Soil and water conservation measures and its impact on soil carbon sequestration and soil quality | The long-term implementation of soil and water conservation measures on sloping terrain serves as a promising carbon sink and enhances soil quality over time | Mahajan et al. (2021) |

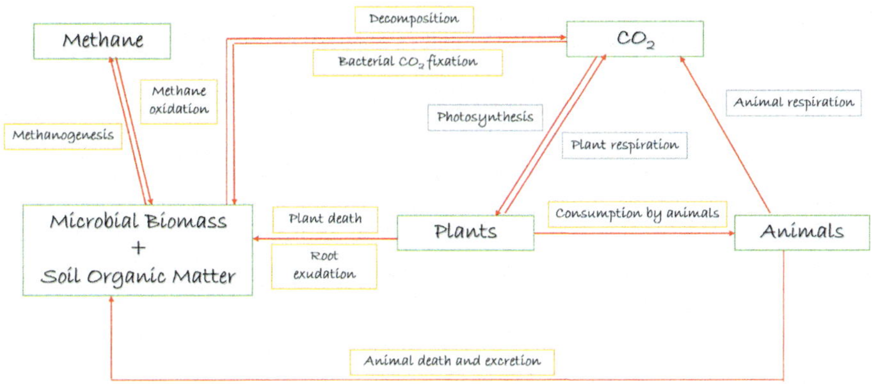


**Fig. 3** Terrestrial carbon cycle where soil microorganisms mediate the processes (Gougoulias et al. 2014)

**Table 2** Benefits provided by microbes in C speciation

| Microbe associated | Benefits/functions provided | References |
|---|---|---|
| Bacteria, actinomycetes, Fungi | Microbial mineralization of heavy metals as microbes integrate the effect of soil properties | Dumestre et al. (1999) |
| Bacteria and fungi | Bacterial and fungal contributions to carbon sequestration in agroecosystems | Six et al. (2006) |
| Mycorrhizal fungi | Microbial contributions to carbon storage | King (2011) |
| Microbes | Microbial carbon pump and its contribution in carbon sequestration in the ocean | Jiao and Azam (2011) |
| Microbes | The stabilization of plant carbon into soil carbon | Schimel (2013) |
| Bacteria and fungi | Developed a mechanistic model to simulate and investigate the microbial control over SOC cycling | Fan and Liang (2015) |
| Bacteria and cyanobacteria | To determine the microbial composition and abundance and measurement of organic pools | Kheirfam (2020) |

## Mechanism Followed by Microbes in C Speciation

**Decomposition of organic matter** Organic matter decomposition releases various compounds, including carbon dioxide ($CO_2$), energy, plant nutrients, water, and resynthesized organic carbon compounds (Khatoon et al. 2017). Microbes are responsible for decomposing these resynthesized carbon compounds, leading to the formation of various carbon species.

**Methanogenesis** Methanogenesis is an anaerobic process carried out by archaea, a group of microorganisms known as methanogens. They degrade organic carbon by fermenting acetate to $CO_2$ and $CH_4$ or by oxidizing $H_2$ to $H_2O$ (Bryant 1979). These processes also affect the various species of carbon in the soil.

**Methane oxidation** Methane ($CH_4$) is oxidized to carbon dioxide ($CO_2$) with molecular oxygen ($O_2$) by organisms called methanotrophs. This process assimilates carbon, and methanotrophs use methane as their sole source of carbon and energy, helping to sequester carbon (Sahoo et al. 2021). Bacterial carbon dioxide fixation is done by a group of bacteria, including cyanobacteria and other photosynthetic organisms which uses carbon dioxide to grow. During this process, $CO_2$ is converted into organic building blocks (Kumar et al. 2018). Carbon fixation pathways occur in the chloroplasts of plants and in the cells of cyanobacteria. Through this process, inorganic carbon is converted to organic carbon aiding in the sequestration of carbon.

Excretion of exudates: Root exudates are a source of soluble organic carbon in soil, known as rhizodeposition. Plants also contribute to carbon speciation by altering the form of carbon in soil. These exudates largely influence the microbes in the rhizosphere, which in turn influence the forms of carbon in soil (Wen et al. 2022). For example, root exudate chemistry affects soil carbon mobilization. Carboxylic acids led to an increase in actinobacteria populations and promoted carbon mobilization in the soil, whereas exudate amino acids resulted in elevated proteobacteria populations, thereby hindering carbon release in the soil. These examples show the importance of root exudates in mobilizing soil carbon.

$CO_2$ released by microbial respiration: Microbial respiration releases carbon dioxide gas into the soil, which can hamper carbon speciation (Nanda et al. 2016). The released carbon dioxide is vulnerable to various microbial processes, such as carbon fixation and methanogenesis.

## Soil Functions

Over the past two decades, increasing soil carbon (C) storage through C sequestration has been considered a measure to mitigate global climate change. However, $CO_2$ is not the only greenhouse gas (GHG) emitted from soil that contributes to the greenhouse effect. Methane ($CH_4$) and nitrous oxide ($N_2O$) are other GHGs that are emitted from soil and have a greater potential to increase the greenhouse effect. $CH_4$ is 30 times more potent than $CO_2$ at increasing atmospheric temperature, and $N_2O$ is 300 times more potent (Al-Ghussain 2019). Therefore, the full extent of C sequestration's ability to mitigate climate change is still unknown.

In addition, considering the increasing human population, industrialization, and urbanization globally, the $CO_2$ emissions that are currently being released cannot be controlled solely through C sequestration. This is one reason why policymakers are hesitant to invest in soil research for environmental sustainability. Therefore, GHG reduction from the atmosphere to soil may not be a compelling factor to attract funding agencies for soil organic matter (SOM) initiatives. However, emphasizing the importance of SOM for soil function and soil ecosystem services could be a more effective approach.

Without a doubt, SOM improves soil functioning capacity, which is essential for ensuring food production to meet the needs of the growing population (Roy et al. 2006). However, mismanagement of SOM can lead to increased C flux from soil to atmosphere, which reduces soil functional capacity. Soil offers humans a range of ecosystem services, encompassing provisioning, regulating, and supporting functions. Of these services, provisioning, regulating, and supporting services all depend on the availability of SOM. Therefore, proper land use management practices are essential to increase SOM content and improve soil function.

Soil organic carbon (SOC) is a vital component that regulates the proper functioning of soil (Ontl and Schulte 2012). Its functions can be broadly discussed in terms of soil health, which includes the physical, chemical, and biological properties of soil.

Since the start of the green revolution, the focus has been on crop production through the application of inorganic nutrients, with soil fertility being the primary priority. However, soil physical properties, which are essential for proper soil function, have been neglected.

SOC plays a vital role in soil physical stability by cementing primary soil particles (sand, silt, and clay) together. This results in a proper ratio of solid, liquid, and gas (50:25:25), which supports life on Earth (Peth et al. 2008). The solid phase provides support for organisms, the liquid phase facilitates nutrient movement and uptake, and the gaseous phase enhances certain chemical reactions.

The chemical properties of soil depend on the colloidal properties of inorganic (clay) and organic (humus) colloids. Soil colloids have different colloidal properties, which make them chemically active. One of the most important properties of colloids is their surface charge (positive or negative). This surface charge is responsible for important chemical properties such as cation exchange capacity (CEC), anion exchange capacity (AEC), buffering capacity, soil acidity, and basicity. Therefore, the presence of humus (a fraction of SOM) is essential for stimulating all of these chemical properties.

The biological properties of soil are entirely dependent on the availability of SOC. Both the substrate and energy for organism growth are derived from the decomposition of SOC (Kibblewhite et al. 2008). SOC improves both the physical and chemical properties of soil, providing support and the necessary chemical reactions for soil organisms. Carbon (C) in soil is an important food source for organisms to build their biomass. As the availability of SOC in soil increases, the microbial population increases, leading to a wider diversity of soil microbes.

Plants and animals constitute the main contributors of soil organic matter (SOM) or soil organic carbon (SOC), while decaying organic matter represents a secondary source. SOM contains a variety of substances (cellulose, hemicellulose, lignin, proteins, and simple sugars) that soil organisms rely on for their livelihood. Depending on the extent of SOM decomposition, it can be classified into humic and non-humic fractions. Non-humic fractions have a very short mean residence time (MRT) in soil, while humic fractions have a long MRT and are not easily decomposed (Wattel-Koekkoek et al. 2003). The non-humic fraction with its short MRT can be easily decomposed by soil organisms, which improves soil quality. On the other hand, the stable humic fraction with its colloidal properties is important for soil aggregate stability and C storage in soil.

Soil organic matter (SOM) is a heterogeneous mixture of organic substances originating from both living plant and microbial cells (simple organic compounds) and more complex products of plant, animal, and microorganism residues with different sizes, complexities, and stabilities (Weng et al. 2022). Due to this heterogeneity, SOM constituents decompose at different rates, resulting in different species of carbon in the soil with different functions. This variation in SOM decomposition, due to the differential size, complexity, and stability of its constituents, results in different organic carbon species, and this needs to be analyzed separately in order to study the geochemical cycling of carbon.

## Factors Affecting C Speciation

Combustion of fossil fuel: Fossil fuels are largely made up of carbon and hydrogen. When they are burned, carbon is released into the atmosphere, disrupting atmospheric stability. The consequences of burning fossil fuels are still uncertain, but it is clear that it influences soil respiration and carbon speciation. Microbial contributions to climate change through the carbon cycle also strongly affect carbon speciation (Crutzen and Andreae 1990).

Conversion of natural ecosystems to agricultural land: Natural ecosystems help to maintain a balance in nature by storing all forms of carbon in optimal amounts. Disturbing these ecosystems for artificial land use can lead to consequences such as over-decomposition of organic matter and removal of the primary source of production, i.e., plant biomass during harvesting (Gougoulias et al. 2014). It is estimated that 42–78 Gt of carbon has been lost due to intensive agriculture, and regaining lost soil through cultivation may be one potential solution to mitigate climate change (Lal 2004).

Deforestation: Forests and trees store carbon in their biomass and in the soil. When trees are cut down, the sequestered carbon in the soil and vegetation is released into the atmosphere as carbon dioxide. On average, global loss of tropical forests contributes 4.8 billion tons of carbon dioxide per year, which is 8–10% of annual human emissions of carbon dioxide (Singh 2007).

## Conclusion

This chapter emphasizes the role of microbes in carbon sequestration through the process of carbon speciation. Carbon speciation, the process of creating various forms of carbon, is crucial in soil dynamics. Microbial activity plays a significant role in shaping these carbon species which serves as an indicator of a soil's carbon sequestration potential. By comprehending how microbes influence carbon speciation and actively managing these mechanisms, we can enhance carbon sequestration in soil. This enhancement will not only improve soil functions but also contribute to providing valuable ecosystem services to humans.

## References


Adamu H, Umar YA, Akanang H, Sabo A (2021) Evaluation of carbon sequestration potential of soils—what is missing? J Geosci Environ Protect 9(8):39–47

Al-Ghussain L (2019) Global warming: review on driving forces and mitigation. Environ Prog Sustain Energy 38(1):13–21

Bhattacharyya SS, Ros GH, Furtak K, Iqbal HM, Parra-Saldívar R (2022) Soil carbon sequestration–an interplay between soil microbial community and soil organic matter dynamics. Sci Total Environ 152928

Bryant MP (1979) Microbial methane production—theoretical aspects. J Anim Sci 48(1):193–201

Crutzen PJ, Andreae MO (1990) Biomass burning in the tropics: impact on atmospheric chemistry and biogeochemical cycles. Science 250(4988):1669–1678

Dumestre A, Sauve S, McBride M, Baveye P, Berthelin J (1999) Copper speciation and microbial activity in long-term contaminated soils. Arch Environ Contam Toxicol 36(2):124–131

Dwivedi D, Tang J, Bouskill N, Georgiou K, Chacon SS, Riley WJ (2019) Abiotic and biotic controls on soil organo–mineral interactions: developing model structures to analyze why soil organic matter persists. Rev Mineral Geochem 85(1):329–348

Efthimiadou A, Bilalis D, Karkanis A, Froud-Williams B (2010) Combined organic/inorganic fertilization enhance soil quality and increased yield, photosynthesis and sustainability of sweet maize crop. Aust J Crop Sci 4:722

Fan Z, Liang C (2015) Significance of microbial asynchronous anabolism to soil carbon dynamics driven by litter inputs. Sci Rep 5(1):1–7

Gautam RK, Navaratna D, Muthukumaran S, Singh A, More N (2021) Humic substances: its toxicology, chemistry and biology associated with soil, plants and environment. In: Humic substances. IntechOpen, London, p 97

Ghosh PK, Venkatesh MS, Hazra KK, Kumar N (2012) Long-term effect of pulses and nutrient management on soil organic carbon dynamics and sustainability on an inceptisol of indo-gangetic plains of India. Exp Agric 48(4):473–487

Gougoulias C, Clark JM, Shaw LJ (2014) The role of soil microbes in the global carbon cycle: tracking the below-ground microbial processing of plant-derived carbon for manipulating carbon dynamics in agricultural systems. J Sci Food Agric 94(12):2362–2371

Horwath W (2007) Carbon cycling and formation of soil organic matter. In: Soil microbiology, ecology and biochemistry. Academic Press, pp 303–339

Houghton JT, Ding YDJG, Griggs DJ, Noguer M, van der Linden PJ, Dai X et al (2001) Climate change 2001: the scientific basis: contribution of working group I to the third assessment report of the Intergovernmental Panel on Climate Change. Cambridge University Press

Huang Y, Zou J, Zheng X, Wang Y, Xu X (2004) Nitrous oxide emissions as influenced by amendment of plant residues with different C: N ratios. Soil Biol Biochem 36(6):973–981

Huang ZT, Li YF, Jiang PK, Chang SX, Song ZL, Liu J, Zhou GM (2014) Long-term intensive management increased carbon occluded in phytolith (PhytOC) in bamboo forest soils. Sci Rep 4(1):1–5

Jackson RB, Le Quéré C, Andrew RM, Canadell JG, Peters GP, Roy J, Wu L (2017) Warning signs for stabilizing global $CO_2$ emissions. Environ Res Lett 12(11):110202

Jiao N, Azam F (2011) Microbial carbon pump and its significance for carbon sequestration in the ocean. In: Microbial carbon pump in the ocean, vol 10, pp 43–45

Johnson JMF, Franzluebbers AJ, Weyers SL, Reicosky DC (2007) Agricultural opportunities to mitigate greenhouse gas emissions. Environ Pollut 150(1):107–124

Kaushal R, Tewari S, Thapliyal SD, Kumar A, Roy T, Islam S et al (2021) Build-up of labile, non-labile carbon fractions under fourteen-year-old bamboo plantations in the Himalayan foothills. Heliyon 7(8):e07850

Khatoon H, Solanki P, Narayan M, Tewari L, Rai JPN, HinaKhatoon C (2017) Role of microbes in organic carbon decomposition and maintenance of soil ecosystem. Int J Chem Stud 5(6):1648–1656

Kheirfam H (2020) Increasing soil potential for carbon sequestration using microbes from biological soil crusts. J Arid Environ 172:104022

Kibblewhite MG, Ritz K, Swift MJ (2008) Soil health in agricultural systems. Philos Trans R Soc B: Biol Sci 363(1492):685–701

King GM (2011) Enhancing soil carbon storage for carbon remediation: potential contributions and constraints by microbes. Trends Microbiol 19(2):75–84

Kumar M, Sundaram S, Gnansounou E, Larroche C, Thakur IS (2018) Carbon dioxide capture, storage and production of biofuel and biomaterials by bacteria: a review. Bioresour Technol 247:1059–1068

Kumpiene J, Robinson R, Brännvall E, Nordmark D, Bjurström H, Andreas L et al (2011) Carbon speciation in ash, residual waste and contaminated soil by thermal and chemical analyses. Waste Manag 31(1):18–25

Lal R (2004) Soil carbon sequestration impacts on global climate change and food security. Science 304(5677):1623–1627

Lehmann J, Kleber M (2015) The contentious nature of soil organic matter. Nature 528(7580):60–68

Mahajan GR, Das B, Manivannan S, Manjunath BL, Verma RR, Desai S et al (2021) Soil and water conservation measures improve soil carbon sequestration and soil quality under cashews. Int J Sediment Res 36(2):190–206

Nanda S, Reddy SN, Mitra SK, Kozinski JA (2016) The progressive routes for carbon capture and sequestration. Energy Sci Eng 4(2):99–122

Olaleye A, Peak D, Shorunke A, Dhillon G, Oyedele D, Adebooye O, Akponikpe PI (2020) Effect of manure and urea fertilization on yield, carbon speciation and greenhouse gas emissions from vegetable production systems of Nigeria and Republic of Benin: a phytotron study. Agronomy 10(3):400

Ontl TA, Schulte LA (2012) Soil carbon storage. Nat Educ Knowledge 3(10)

Peth S, Horn R, Beckmann F, Donath T, Fischer J, Smucker AJM (2008) Three-dimensional quantification of intra-aggregate pore-space features using synchrotron-radiation-based microtomography. Soil Sci Soc Am J 72(4):897–907

Roy RN, Finck A, Blair GJ, Tandon HLS (2006) Plant nutrition for food security. A guide for integrated nutrient management. FAO Fertilizer Plant Nutr Bull 16:368

Sahoo KK, Goswami G, Das D (2021) Biotransformation of methane and carbon dioxide into high-value products by methanotrophs: current state of art and future prospects. Front Microbiol 12:636486

Sainepo BM, Gachene CK, Karuma A (2018) Assessment of soil organic carbon fractions and carbon management index under different land use types in Olesharo Catchment, Narok County, Kenya. Carbon Bal Manage 13(1):1–9

Schimel J (2013) Microbes and global carbon. Nat Clim Chang 3(10):867–868

Singh SP (2007) Himalayan forest ecosystem services. Central Himalayan Environment Association, Nainital, p 53

Six J, Frey SD, Thiet RK, Batten KM (2006) Bacterial and fungal contributions to carbon sequestration in agroecosystems. Soil Sci Soc Am J 70(2):555–569

Smith P, Martino Z, Cai D (2007) 'Agriculture', in climate change 2007: mitigation. Cambridge University Press

Snyder CS, Bruulsema TW, Jensen TL, Fixen PE (2009) Review of greenhouse gas emissions from crop production systems and fertilizer management effects. Agric Ecosyst Environ 133(3–4):247–266

Solomon D, Lehmann J, Kinyangi J, Amelung W, Lobe I, Pell A et al (2007) Long-term impacts of anthropogenic perturbations on dynamics and speciation of organic carbon in tropical forest and subtropical grassland ecosystems. Glob Chang Biol 13(2):511–530

Song W, Tong X, Liu Y, Li W (2020) Microbial community, newly sequestered soil organic carbon, and δ15N variations driven by tree roots. Front Microbiol 11:314

Stegen JC, Johnson T, Fredrickson JK, Wilkins MJ, Konopka AE, Nelson WC et al (2018) Influences of organic carbon speciation on hyporheic corridor biogeochemistry and microbial ecology. Nat Commun 9(1):1–11

Stockmann U, Adams MA, Crawford JW, Field DJ, Henakaarchchi N, Jenkins M et al (2013) The knowns, known unknowns and unknowns of sequestration of soil organic carbon. Agric Ecosyst Environ 164:80–99

Tate KR (1984) The biological transformation of P in soil. In: Biological processes and soil fertility. Springer, Dordrecht, pp 245–256

Wan J, Tyliszczak T, Tokunaga TK (2007) Organic carbon distribution, speciation, and elemental correlations within soil microaggregates: applications of STXM and NEXAFS spectroscopy. Geochimica et Cosmochimica Acta 71(22):5439–5449

Wang H, Liu S, Wang J, Shi Z, Lu L, Guo W et al (2013) Dynamics and speciation of organic carbon during decomposition of leaf litter and fine roots in four subtropical plantations of China. For Ecol Manag 300:43–52

Wattel-Koekkoek EJW, Buurman P, Van Der Plicht J, Wattel E, Van Breemen N (2003) Mean residence time of soil organic matter associated with kaolinite and smectite. Eur J Soil Sci 54(2):269–278

Wen T, Yu GH, Hong WD, Yuan J, Niu GQ, Xie PH et al (2022) Root exudate chemistry affects soil carbon mobilization via microbial community reassembly. Fundamental Res 2(5):697–707

Weng Z, Lehmann J, Van Zwieten L, Joseph S, Archanjo BS, Cowie B et al (2022) Probing the nature of soil organic matter. Crit Rev Environ Sci Technol 52(22):4072–4093

# Nitrogen Cycle, $N_2O$ Emission, and Its Decomposition Processes in Soil

**Mark Coyne, Yawen Huang, and Wei Ren**

## Introduction

Nitrogen (N, molecular weight 14) is a critical biotic element. After carbon, hydrogen, and oxygen, it dominates biomass, being a critical building block of proteins, genetic information (RNA and DNA), cell wall and cell membrane components, and numerous secondary metabolites. In most cases, it is the most critical nutrient required for agricultural productivity (Smil 2001). Nitrogen fixation by microbiota is the single most important means of improving soil fertility. Nitrogen absence through immobilization, volatilization, and leaching is an essential contributor to low fertility. Its presence in surface and groundwater leads to eutrophication, impaired water quality, and significant human health concerns (Tomich et al. 2016). Nitrogen is also one of the most chemically and biologically dynamic plant nutrients. Depending on the environmental aeration status and whether nitrogen is present as a free nutrient or assimilated into organic compounds, nitrogen undergoes oxidation and reduction reactions, transforming it from reduced forms (oxidation

M. Coyne (✉)
Department of Plant and Soil Sciences, University of Kentucky, Lexington, KY, USA
e-mail: mark.coyne@uky.edu

Y. Huang
Department of Plant and Soil Sciences, University of Kentucky, Lexington, KY, USA

Key Laboratory of Low-Carbon Agriculture in Southeastern China, Nanjing Agricultural University, Nanjing, China

W. Ren
Department of Plant and Soil Sciences, University of Kentucky, Lexington, KY, USA

Department of Natural Resources and the Environment, University of Connecticut, Storrs, CT, USA



S. R. Mohanty, B. Kollah (eds.), *Greenhouse Gas Regulating Microorganisms in Soil Ecosystems*, https://doi.org/10.1007/978-3-031-70569-4_10

**Table 1** Global reservoirs of atmospheric and terrestrial N

| Source | N content (Tg) |
|---|---|
| Lithosphere | $164 \times 10^9$ |
| Atmosphere | $3.7 \times 10^9$ |
| Litter and soil N | $1.0 \times 10^5$ |
| Vegetation | $3.5 \times 10^3$ |
| Inorganic soil N | $9.1 \times 10^1$ |

Sources: Soderlund and Svensson (1976), Ward (2012), Zaehle (2013)
Tg = Teragram = $10^9$ kg

state −3) to oxidized forms (oxidation state +5), which vary in their ability to return to atmospheric, terrestrial, or aquatic pools.

The major terrestrial nitrogen pool consists of inorganic nitrogen within the matrix of parent material and minerals (Table 1). This is fundamentally unavailable to plant growth. The dominant available pools are the atmosphere (78% by volume $N_2$) and the terrestrial soil pool, which consists of soil organic matter (including biomass) and inorganic nitrogen (dominated by $NH_4$+ and $NO_3$–) adsorbed to soil minerals or in soil solution. Nitrogen cycling between these pools is critical to life processes and environmental quality.

## Nitrogen Cycle

The nitrogen cycle can be best understood as a series of transformations that convert nitrogen between organic and inorganic forms, and between oxidized and reduced states, depending on the environment. As a general rule, nitrogen occurs in its most reduced state (−3) as a component of organic molecules, its most oxidized state (+5) in well-aerated soil environments, and its most stable state (0) as atmospheric nitrogen gas ($N_2$). Eight key biochemical processes transform nitrogen between these various states: assimilation, mineralization, nitrification, anaerobic ammonia oxidation (ANAMMOX), nitrate respiration, denitrification, dissimilatory nitrate reduction to ammonium (DNRA), and nitrogen fixation (Fig. 1). Lazcano et al. (2021) have summarized the key genes and relevant enzymes involved in nitrification and denitrification.

Nitrogen cycling processes that transform nitrogen into gaseous products and intermediates at the expense of soil fertility replenish atmospheric nitrogen that was initially converted to organic nitrogen by biological nitrogen fixation (and to a limited extent into inorganic nitrogen by abiotic transformation, e.g., lightning). Intermediate products such as nitric oxide (NO) and nitrous oxide ($N_2O$) play a significant role in atmospheric chemistry. The extent to which human activity, such as increased legume planting, anthropogenic combustion, industrial nitrogen fixation, and fertilization, increases nitrogen cycling and the likelihood that these intermediates reach the atmosphere is a major environmental concern, particularly with respect to the key gaseous intermediate nitrous oxide ($N_2O$).

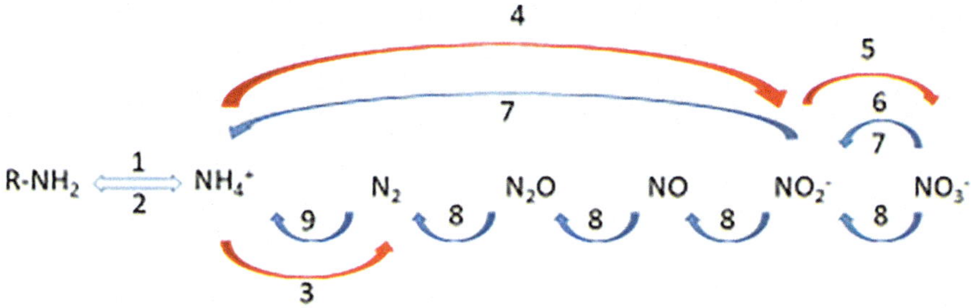


**Fig. 1** Critical biological transformations in the N cycle. Clear arrows do not change oxidation state; red arrows are oxidative; blue arrows are reductive. 1 Assimilation, 2 Mineralization, 3 ANAMMOX, 4 $NH_3$ oxidation (nitrification), 5 $NO_2^-$ oxidation (nitrification), 6 $NO_3^-$ respiration, 7 DNRA, 8 Denitrification, 9 Nitrogen fixation. ANAMMOX anaerobic ammonium oxidation, DNRA dissimilatory nitrate reduction to ammonium

## $N_2O$ as an Important Atmospheric Gas

Nitrous oxide ($N_2O$) is an important atmospheric gas. It has a mixing ratio of approximately 333 parts per billion by volume (ppbv) in 2020 and is increasing in the atmosphere at an annual rate of approximately 0.9 ppb per year (Dlugokencky 2021). Most $N_2O$ (70%) derives from soil, and fertilizers and manures are estimated to account for 23–31% of global anthropogenic $N_2O$ emissions from soil (Syakila and Kroeze 2011). $N_2O$ is significant for two critical reasons: it interferes with stratospheric chemistry by depleting ozone ($O_3$), and it promotes the greenhouse effect, contributing 298 times more than $CO_2$ on a per molecule basis (IPCC 2013). It is considered the largest stratospheric ozone-depleting substance in the twenty-first century (Ravishankara et al. 2009). Three biological processes dominate $N_2O$ production in soil: denitrification, nitrification, and to a lesser extent DNRA. In addition, several abiotic processes (collectively referred to as chemodenitrification) also produce trace quantities of $N_2O$. Denitrification and DNRA generally (but not exclusively) occur in wet or reducing environments. Nitrification generally (but not exclusively) occurs in well-aerated or oxidizing environments. This presents a challenge for controlling $N_2O$ emissions from terrestrial environments to reduce its environmental effects, particularly its contribution as a greenhouse gas. $N_2O$ is an inevitable byproduct of biological N cycling, and mitigating one formation pathway by changing the environment can accentuate $N_2O$ formation by other pathways.

## Detection and Sourcing of $N_2O$

Detecting nitrous oxide ($N_2O$) in the environment is challenging due to its low atmospheric concentrations and its sporadic fluxes with significant spatiotemporal variations due to the multiple processes that regulate $N_2O$ production and consumption (Rapson and Dacres 2014). Nevertheless, detecting $N_2O$ fluxes is critical to help understand the complex processes that govern $N_2O$ fluxes and enable accurate greenhouse gas accounting at multiple scales. The quality of experimental data largely influences the accuracy of models used for national and global greenhouse gas inventories.

## *Sampling Methods*

Chambers are the most common method for measuring trace gas fluxes from soil (Denmead 2008). Chamber size can range from less than 1 $m^2$ to greater than 100 $m^2$. The general principle is to determine the change in $N_2O$ concentration within a specific volume over a period of time. Chambers can be classified as flow-through or closed, depending on whether they are open to the air, and as manual or automatic, depending on their operating system. One of the strengths of chambers is that they are simple to operate and do not require fast or high-precision analytical techniques (Denmead 2008). However, chambers with smaller surface areas (< 1 $m^2$) may require more replicates to capture large spatial variability in soil $N_2O$ emissions. It is important to note that the sampling window and frequency can have a significant impact on data quality, especially for manual chambers, where readings are often taken at a single time of day.

Micrometeorological techniques, such as eddy covariance, eddy accumulation, flux-gradient methods, integrated horizontal flux, backward Lagrangian stochastic dispersion technique, and moving platforms, are often used to measure gas fluxes over large areas (e.g., landscape scales). These techniques require gas sensors (often fast-response sensors) to be installed to measure wind, temperature, and gas concentrations at one or more points above the soil or vegetation canopy (Rapson and Dacres 2014).

## *Analytical Techniques*

Current analytical techniques used to analyze $N_2O$ can be divided into three categories: chromatographic, optical, and amperometric (Rapson and Dacres 2014). Gas chromatography with an electron-capture detector (GC-ECD) is the most widely used technique due to its low cost and precision of 0.18–0.4 ppb. However, it requires frequent calibration and has a long sample run time (4–6 minutes per sample), which limits its use for continuous measurements.

Optical techniques, such as Fourier-transform infrared spectroscopy (FTIR), can provide continuous measurements with fewer calibration requirements. Laser absorption spectroscopy is another optical technique that enables rapid and highly sensitive measurements, but it is expensive and can only be used to measure one or two trace gases at a time.

Amperometric methods are sensitive, simple, and economical, but they can only be used to measure dissolved $N_2O$.

## Sourcing

Being able to distinguish the source (or sink) of $N_2O$ in the environment is useful in determining the mechanisms of its production (or consumption) and to develop strategies to mitigate its production and accentuate its uptake.

## Pathway-Specific Inhibition

General commercial nitrification inhibitors target $NH_3$ oxidation by $NH_3$ monooxygenase (AMO), which initiates the steps that form $N_2O$ later in the nitrification pathway. They include dicyandiamide (DCD), thiourea, 2-chloro-6-(trichloromethyl) pyridine (nitrapyrin), and 4-amino-1,2,4-triazole·HCl (ATC) (Table 2).

Methane ($CH_4$) is a competitive inhibitor of $NH_3$ monooxygenase (AMO) and is oxidized to methanol. Compounds that release acetylene ($C_2H_2$) are suicide inhibitors of AMO. Acetylene does not distinguish between bacterial and archaeal AMO, but octyne ($C_8H_{14}$) is a specific inhibitor of bacterial AMO, which enables laboratory studies to distinguish the prokaryotic source of nitrification (Taylor et al. 2013). Denitrification by either prokaryotes or fungi is not amenable to field-scale inhibition except by nonspecific bactericidal or fungicidal agents. Individual steps in the prokaryotic denitrification pathway have been inhibited at the laboratory scale. Tungsten (W) is an analog for molybdenum (Mo) in molybdoproteins, such as nitrate reductase, and prevents the initial step in nitrate reduction, which also applies to dissimilatory nitrate reduction to ammonium (DNRA). The chelating agent

**Table 2** Nitrification inhibitors

| Mode of action | Chemical | Formula |
|---|---|---|
| Cu-chelating agents | Hydrazinecarbothiamide | $NH_2CSNNH_2$ |
| | Diethyldithiocarbamate | $(C_2H_5)_2NS_2Na$ |
| | Thiourea | $NH_2CSNH_2$ |
| $e^-$ transport by cytochromes | Sodium azide | $NaN_3$ |
| | p-450 inhibitors | |
| Pyridine compounds (interact with active site of AMO) | N-Serve | 2-Chloro-6-(trichloromethyl) pyridine |
| | AM | 2-Amino-4-chloro-6-methylpyrimidine triazoles |
| | MT | 1-Mercapto-1,2,4-triazole |
| | ATC | 4-Amino-1,2,4-6-triazole-HCl |
| Miscellaneous | Hydrazine | $NH_2$-$NH_2$ |
| | Methane (competitive substrate) | $CH_4$ |
| | Dicyandiamide (DCD) (cyanoquinidine) | $C_2N_4H_4$ |

diethyl-dithiocarbamate (DDC) can be used to distinguish between denitrifiers that have a cytochrome or copper-based nitrite reductase (Coyne and Tiedje 1990a). The final reduction of nitrous oxide ($N_2O$) to nitrogen ($N_2$) is inhibited by the suicide inhibitor acetylene ($C_2H_2$). P450 inhibitors have been used to generically prevent fungal activity and their denitrification.

## *Isotopomers*

When using $^{15}N$ tracers to evaluate $N_2O$ emission, the site of the $^{15}N$ in $N_2O$ is revealing as to the pathway involved. The various $^{15}N$-labled $N_2O$ isotopes can be distinguished by mass spectrometry and are referred to as isotopomers (Stoll 2007), not merely isotopes differing from 14-$N_2O$ but differing in the location of the $^{15}N$. Nitrogen atoms in the linear $N_2O$ molecule are distinguished as α (central) or β (terminal). In labeling studies, the distribution of $^{15}N$ between α and β positions will differ depending on the pathway. It can be distinguished because $N_2O$ fragmentation in the ion source during mass spectrometry will show where the label is (Sutka et al. 2003). Site preference is the difference between $\delta^{15}N$ α and $\delta^{15}N$ β, where δ refers to delta value or atom %$^{15}N$ enrichment. Site preference is positive (+) when the α or central N atom is labeled and negative (−) when the β or terminal N atom is labeled. If there is a relatively uniform and statistically significant discrimination for one isotope over another in a given position, then you can infer the source. The site preference for denitrification and nitrifier denitrification is −10 to 0 $^0/_{00}$ and that of nitrification and fungal denitrification is 33–37 $^0/_{00}$ (Sutka et al. 2003). The limitation of this procedure obviously is the ability to clearly distinguish beyond gross comparisons of the major $N_2O$-producing pathways.

## Sites of $N_2O$ Emission in the Nitrogen Cycle

It is wrong to conceive of the individual components of the N cycle as being either spatially or temporally segregated or even wholly independent. Processes that produce and consume $N_2O$ can be closely integrated in the soil. In a pasture system, Robertson et al. (1988), for example, clearly show the dependence of denitrification rates, which both consume and produce $N_2O$, on the extent of nitrification, which produces $N_2O$ and $NO_3^-$. Lazcano et al. (2021) reviewed the consequences for $N_2O$ emission on the type of organic fertilizers added to soil. Fertilizer choice influenced the amount of $NH_4^+$ and $NO_3^-$ added, the amount of C substrate for heterotrophic denitrifiers, and even soil moisture content and redox potential. Philippot et al. (2011) report that the denitrifiers in rhizospheres are more likely than counterparts in the bulk soil to lack the *nosZ* gene for $N_2O$ reduction and therefore have a greater propensity to emit $N_2O$. So, in understanding $N_2O$ emission from soil as a consequence of the N cycle, it is important to realize that no process or group of processes, and consequently the net effect on $N_2O$ emission, is entirely isolated even though in this reading they are treated as though they were.

## Nitrification

Autotrophic nitrification is the sequential oxidation of $NH_3$ to $NO_2^-$ and $NO_3^-$ by distinct taxonomic groups of lithotrophic bacteria and archaea. Heterotrophic nitrification is mediated by heterotrophic bacteria and fungi with the resulting oxidation of both organic and inorganic N. The autotrophic prokaryotes are well known to produce NO (e.g., Remde and Conrad 1990) and $N_2O$ (e.g., Poth and Focht 1995) through various mechanisms, including enzymatic pathways similar to denitrifying prokaryotes. One distinction is that a greater proportion of NO is produced by these pathways than by classical denitrification.

### *Environmental Regulators*

The dominant regulators of nitrification in the soil environment are $O_2$, the availability of $NH_3$ and $NO_2^-$, and pH. Although much early literature attributed $N_2O$ production in otherwise well-aerated soils to anaerobic microsites (Sexstone et al. 1985), it is abundantly clear that autotrophic nitrification is also a unique source as has been established by isotope analysis (e.g., Park et al. 2011). It has also been observed that with increasing moisture content in aerated soil, the proportion of $N_2O$ derived from nitrification increases until about 60% water filled the pore space at which time $N_2O$ produced during denitrification becomes steadily more important (Davidson 1991). Nitrite, which in high enough concentration will inhibit autotrophic $NH_3$ oxidation, is thought to be removed from the cell environment by conversion to NO and $N_2O$ as a detoxification mechanism. Likewise, in microaerophilic environments, utilizing $NO_2^-$ for respiratory denitrification to produce $N_2O$ liberates molecular $O_2$ for use in $NH_3$ oxidation.

Ammonia-oxidizing archaea (AOA) generally outnumber ammonia-oxidizing bacteria (AOB) in most nonagricultural soils by ratios exceeding 10:1 (Leininger et al. 2006). This ratio decreases as fertility levels increase, particularly fertilization with N (Shen et al. 2008). Because AOB contribute more to potential nitrification than AOA as fertility increases, they are regarded as the functionally dominant group responsible for nitrification in agricultural soils (Jia and Conrad 2009). Liu et al. (2017) demonstrated that potential nitrification in soils fertilized with 336 kg N $ha^{-1}$ were far more responsive to increasing substrate $NH_4^+$ than unfertilized soils.

Autotrophic nitrification is pH sensitive because the initial deprotonation of $NH_4^+$ to produce $NH_3$ for oxidation is impeded by increasing soil acidity. Nitrification can occur in culture at lower pH than is typically observed in soil because the micro-environment surrounding soil particles is more acid than bulk pH measurements indicate. The difficulty in isolating autotrophic nitrifiers from acid forest soils was attributed to this property, and subsequent nitrification in such environments was attributed to heterotrophic nitrification instead. This is probably incorrect, because abundant populations of AOA exist in such environments, indicating that low pH is less an issue for this group. High pH is likewise an issue. The subsequent oxidation of $NO_2^-$ to $NO_3^-$ by $NO_2^-$ oxidizing bacteria (NOB) is inhibited by high pH because NOB are sensitive to $NH_3$ in solution (Pearce et al. 1998). Sufficiently high $NO_2^-$ concentrations will ultimately also inhibit AOA and AOB.

## *Mechanisms of Evolution*

Two distinct pathways are invoked for $N_2O$ production by nitrifiers. In one, the incomplete oxidation of $NH_2OH$ leaves reactive NO that is quickly transformed by NO reductase to $N_2O$. This is attributed to a metabolic imbalance between the rates of $NH_2OH$ formation and its subsequent oxidation (Wunderlein et al. 2012). In the other, the $NO_2^-$ formed during $NH_3$ oxidation is reduced to $N_2O$ by conventional $NO_2^-$ and NO reductase, albeit at considerably higher $O_2$ concentration than typical denitrifying bacteria (nitrifier denitrification). Recently, a direct pathway for conversion of $NH_2OH$ to $N_2O$ by cytochrome $P_{460}$ was discovered in *Nitrosomonas europaea* (Caranto et al. 2016). Nitrous oxide production is also attributed to simple chemical reactions such as spontaneous $NH_2OH$ decomposition or chemical oxidation by high $NO_2^-$ concentrations (chemodenitrification) (Wunderlein et al. 2012).

## *Organisms of Origin*

The key organisms responsible for $N_2O$ production during autotrophic nitrification are the $NH_3$-oxidizing prokaryotic bacteria best typified by *Nitrosomonas europaea*. Nitrifying archaea seem not to be involved in $N_2O$ production, possibly because the $K_m$ for $NH_3$ by this group is considerably lower than for $NH_3$-oxidizing bacteria. Nitrite-oxidizing bacteria are likewise not involved in $N_2O$ production. Comammox bacteria, typified by *Nitrosospira* spp., which are capable of the complete oxidation of $NH_3$ to $NO_3^-$ have not been sufficiently studied to determine their contributions to $N_2O$ production (Daims et al. 2015; Lu et al. 2020). However, Shaw et al. (2006) demonstrated that some *Nitrosospira* spp. do produce $N_2O$ via nitrifier denitrification (in which the reduction of $NO_2^-$ proceeds through NO and $N_2O$ to $N_2$) (Schmidt et al. 2004).

## Nitrate Respiration and DNRA

It has been recognized for a considerable period that $N_2O$ is produced by various bacteria and fungi that are not identified as either nitrifiers or respiratory denitrifiers (Yoshida and Alexander 1970). This includes microbes regarded as $NO_3^-$ respirers (converting $NO_3^-$ to $NO_2^-$ anaerobically, but no further) and dissimilatory $NO_3^-$ reducers producing $NH_4^+$ as the end product (DNRA). Some, but not all, of the $N_2O$ produced by fungi can be associated with fungal denitrification, which has correspondence with many aspects of respiratory denitrification. Among the two groups, it appears that $NO_3^-$ respirers have a much greater propensity to make $N_2O$ than will be produced during DNRA.

### *Environmental Regulators*

Nitrate, carbon, aeration, and growth stage will all influence $N_2O$ within these organisms. DNRA is a metabolically competitive $NO_3^-$ dissimilation process when $NO_3^-$ concentrations are low and C concentrations are high (Rehr and Klemme 1988). Kaspar (1982) suggested that $NO_2^-$ reduction to $N_2O$ was a detoxification mechanism in Propionibacteria. In C-rich and strictly anaerobic environments, such as the animal rumen, DNRA is the dominant mechanism of $NO_3^-$ reduction. Obligate anaerobes such as *Clostridium* that perform DNRA in the rumen do not produce $N_2O$ (Bleakley and Tiedje 1982). Among other dominant soil groups, facultatively anaerobic enterobacteria characterized by prokaryotes such as *Escherichia coli*, *Klebsiella*, and *Citrobacter* that are capable of respiring $NO_3^-$ can produce significant amounts of $N_2O$—up to 36% of the $NO_3^-$ reduced (Bleakley and Tiedje 1982). This production is most pronounced during stationary phase growth, and it does not necessarily have to occur in anaerobic environments, although the induction of dissimilatory $NO_3^-$ reductase is an anaerobic process.

### *Mechanisms of Emission*

The mechanism by which non-denitrifying prokaryotes produce $N_2O$ was investigated by Smith in a series of papers (Smith 1982, 1983; Smith and Zimmerman 1981). The process is enzymatic, and it does not involve a $NO_2^-$ reductase. Autoclaved cells do not produce $N_2O$ (Bleakley and Tiedje 1982). Nitrate dissimilators with inactivated $NO_2^-$ reductase produce $N_2O$ (Smith 1983). On the other hand, mutants in $NO_3^-$ reductase or cells grown in the presence of tungsten (W) to produce inactive $NO_3^-$ reductase lack $N_2O$ production (Smith 1983). The likely source is therefore $NO_2^-$ reduction by a functioning $NO_3^-$ reductase, particularly when cells are in stationary growth (Bleakley and Tiedje 1982), $NO_2^-$ concentrations are high (Smith observed a $K_m$ for $NO_2^-$ reduction of 0.9 mM) (Smith 1983), or because the further reduction of $NO_2^-$ to $NH_4^+$ is impaired by the lack of oxidizable C (Smith 1982).

## *Organisms of Origin and Points of Control*

Based on culturable approaches to enumeration (anaerobic growth on $NO_3^-$), Smith and Zimmerman (1981) observed that $NO_3^-$ respirers were the numerically dominant group. Of 214 isolates, 209 produced $N_2O$, and 155 accumulated $NO_2^-$ as the end product of growth. The most common genera were *Bacillus* and *Enterobacter*. Therein lies a further complexity in $N_2O$ production from soil. At redox potentials in the soil environment that permit $NO_3^-$ reduction but not further reduction of $NO_2^-$, it is possible for both nitrification and $NO_3^-$ respiration to contribute to $N_2O$ production. That $NO_2^-$ does not accumulate in such soils is undoubtedly due to the activity of $NO_2^-$ oxidizing bacteria and/or the further reduction of $NO_2^-$ via dissimilative processes to $NH_4^+$ when abundant C is present. It illustrates the difficulty of entirely controlling $N_2O$ evolution from soil environments. Nitrous oxide evolution would appear to be possible even in only modestly wet soils, unable to fully develop an active denitrifying population.

## Prokaryotic Denitrification

Classical denitrification is an anaerobic or facultatively anaerobic process in which nitrogen oxides are reduced sequentially during respiratory metabolism to produce gaseous intermediates and end products. The complete pathway consists of nitrate, nitrite, nitric oxide, and nitrous oxide reductases (Nar, Nir [heme or Cu], Nor, and Nos) (Fig. 2). The pathway can be complete or truncated. Zumft (1997) reviewed the cellular and molecular aspects of prokaryotic denitrification, while Firestone (1982) and Coyne (2008, 2018) reviewed denitrification in soil environments.

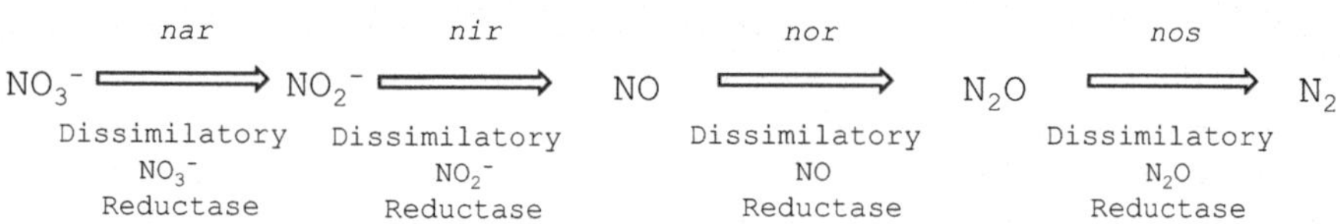


**Fig. 2** The sequential pathway of prokaryotic respiratory denitrification

## Denitrifying Organisms

Denitrification is polyphyletic among bacteria and archaea. Thus, while it is a good physiological descriptor, it is not very useful for developing phylogenetic relationships because the capacity is broadly distributed among numerous physiological groups such as aerobic heterotrophs (fermentative, $N_2$-fixing, and oligotrophic members), anaerobic heterotrophs (e.g., *Shewanella denitrificans*), lithotrophs (Fe-oxidizing, $H_2$-oxidizing, S-oxidizing, and nitrifying), and phototrophs (e.g., *Rhodopseudomonas*).

## Environmental Regulators

Oxygen, C, and $NO_3^-$ are the dominant regulators of denitrification in the soil environment. Tiedje (1988) indicates that when $O_2$ concentration in soil is less than 10 μMol $L^{-1}$, denitrification is largely derepressed. The sequential induction of enzyme synthesis has been demonstrated by Hochstein et al. (1984) and Coyne and Tiedje (1990b) with $N_2O$ reductase the last enzyme to be induced and the first enzyme to be repressed when $O_2$ returns to a system. As a consequence of sequential induction and activity, intermediate denitrification products are observed to occur in flooded soil with time (Cooper and Smith 1963). Denitrification is also a hysteretic process in soil (Groffman and Tiedje 1988).

## Hole in the Pipe

Nitrous oxide fluxes are considerably greater from well-aerated soils than wetland soils probably because a greater proportion of $N_2O$ produced in wetland soils is reduced completely to $N_2$ as it diffuses through the saturated soil (Conrad 1996). Completely flooded soils can act as a sink for $N_2O$, while upland soils will generally be a net source of $N_2O$ (Conrad 1996). Firestone and Davidson (1989) devised a conceptual model to explain why denitrification intermediates such as $N_2O$ appear in various environments. Called the "Hole-in-the-Pipe" model, it visualizes denitrification as a throughput process in which the efficient conversion of $NO_3^-$ to the final end product $N_2$ is determined by the size of holes in each step of the process through which intermediates may leak. Anything that interferes with the efficient transfer of substrate ($NO_3^-$), electron supply (C source), or electron delivery (enzyme synthesis or reduction) can slow product flow (Fig. 3).

For example, Betlach and Tiedje (1981) demonstrated with a kinetic model that if the rate of $NO_3^-$ reduction in denitrifiers exceeds the rate of $NO_2^-$ reduction (a product of excess $NO_3^-$ availability in soil), then a denitrifying culture can transiently accumulate $NO_2^-$ (similarly NO production rate relative to reduction rate

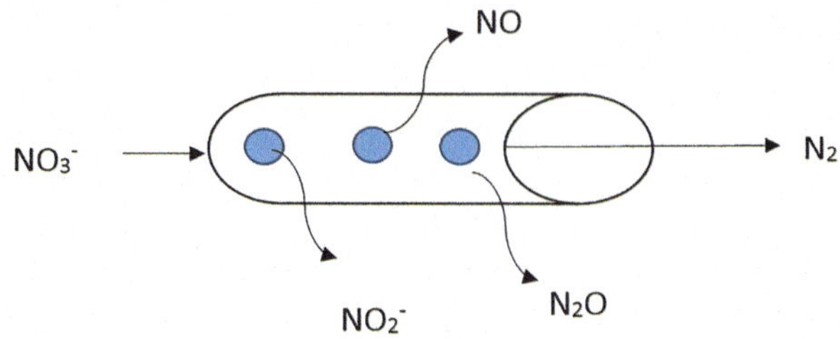


**Fig. 3** The "Hole-in-the-Pipe" model (Firestone and Davidson 1989). Anything that impedes the flow of intermediates through the denitrification "pipeline" leads to intermediate loss because the pipeline contains holes at each step in the sequential process

and $N_2O$ production rate relative to reduction). Likewise, Firestone and Tiedje (1979) demonstrated the transient accumulation of $N_2O$ in saturated soils during the onset of anaerobiosis because $N_2O$ is formed before $N_2O$ reductase is fully synthesized and active.

## Fungal Denitrification

In its broadest sense, denitrification is the transformation of inorganic ammonia and soluble N oxides into gaseous products. Some fungi, therefore, are denitrifiers because they produce trace quantities of NO and $N_2O$ (Shoun et al. 1992). Mechanistically, the process is different than that found in prokaryotes, as are some of the enzymes (Lazcano et al. 2021).

### *Mechanisms of Emission*

The metabolic pathway of fungal denitrification is similar to, but not entirely like, that in prokaryotic denitrifiers (Fig. 4). Some of the key differences are the roles of formate and NADH oxidation in various reductive steps, the association of $NO_2^-$ reductase with the mitochondria, and the localization of NO reductase in the cytoplasm. Although it appears that some fungi are able to reduce $N_2O$ to $N_2$, fungi dominantly appear to accumulate $N_2O$ as the terminal product of denitrification.

### *Organisms of Emission*

Increasing numbers of surveys have shown that fungal denitrification is a much more common and widespread phenomenon than previously imagined (Maeda et al. 2015). *Aspergillus*, *Fusarium* (particularly *Fusarium oxysporum*), and *Penicillium* are three genera that are frequently isolated in surveys of denitrifying fungi. It is a

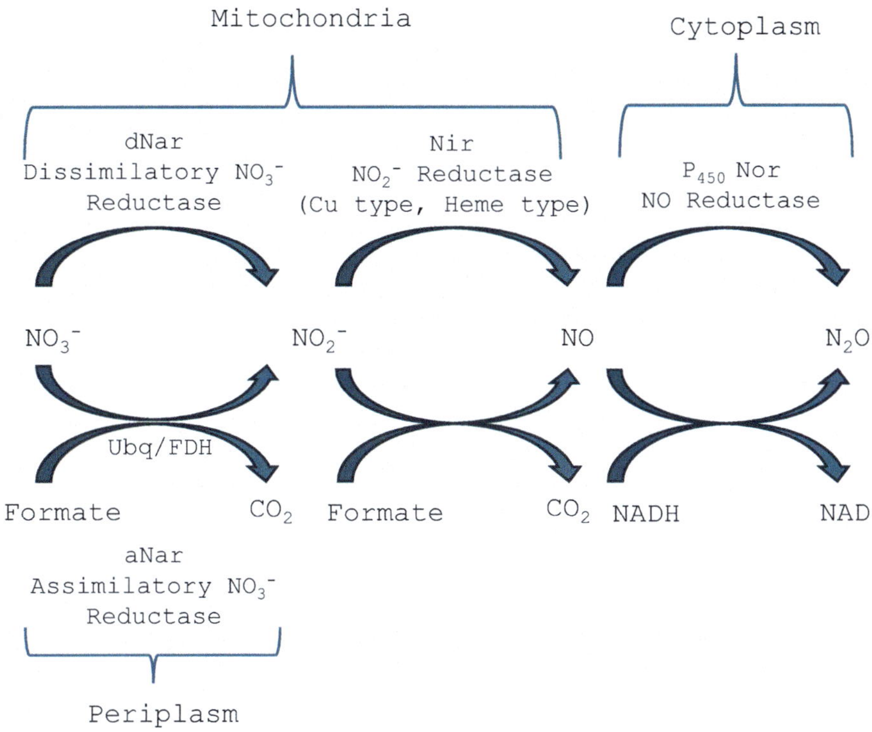


**Fig. 4** Model pathway of fungal denitrification (after Coyne 2018)

fairly common trait now that it is being actively investigated. For example, Maeda et al. (2015) found over 70 denitrifying strains in a survey of 207 isolates representing a diversity of fungal groups.

## Control Points

By means of differential inhibitors, it is estimated that fungi contribute up to 18% of the $N_2O$ flux from soil (Herold et al. 2012). One of the interesting characteristics of fungal denitrification is that $NO_2^-$ rather than $NO_3^-$ is the preferred substrate based on the amount of $N_2O$ formed (Maeda et al. 2015). Compared to prokaryotes, fungi also appear to be less sensitive to the inhibition of $O_2$. There is a significant effect of pH on the contribution of fungi to $N_2O$ production. Fungi are significant $N_2O$ producers in acid soils (Huang et al. 2017), and as soil pH increases, the relative contribution of fungi to $N_2O$ production decreases (Shutan et al. 2019). In acid soil environments, where active fungal biomass increases relative to prokaryotic biomass and the tendency for prokaryotic $N_2O$ reductase activity is to decline, the denitrifying fungi play an increasingly significant role in $N_2O$ evolution (Shutan et al. 2019).

## Abiotic Processes

### *Chemodenitrification*

There are various chemical mechanisms in soil as part of the abiotic N cycle that produce $N_2O$ in trace amounts. These are collectively called "chemodenitrification." The Van Slyke reaction will produce $N_2O$ when organic N is exposed to acid conditions. Hydroxylamine ($NH_2OH$) decomposition in soil will produce $N_2O$. Although hydroxylamine is known as a transient intermediate in the first step of nitrification ($NH_3$ oxidation to hydroxylamine by ammonia monooxygenase), the chemical decomposition of $NH_2OH$ in soil is considered relatively minor (Minami and Fukushi 1986). Nitrite can also be converted to $N_2O$ by metal-catalyzed chemical reduction in acidic conditions.

### *Fe Reduction and $N_2O$ Production*

A key point is that highly acidic environments conducive to chemodenitrification are not conducive to agriculture or the use of N fertilizers that could be chemically transformed. In circumneutral soils, however, there is growing evidence of significant abiological $N_2O$ production during Fe(III) reduction (Dhakal et al. 2021; Matocha and Coyne 2007; Rakshit et al. 2008). Siderite ($FeCO_{3(s)}$) in anaerobic conditions will abiologically reduce $NO_2^-$ to $N_2O$. During this process, siderite is oxidized to lepidocrocite ($\gamma$-$FeOOH_{(s)}$) (Rakshit et al. 2008). Similarly, goethite ($FeOOH_{(s)}$) will react with $NO_2^-$ in Fe-reducing environments to produce $N_2O$. The consequence of these observations is that iron-containing secondary materials in Fe-reducing soils with abundant $NO_3^-$ and where $NO_3^-$ and Fe-reducing zones overlap, or across aerobic–anaerobic interfaces, create the potential for $N_2O$ emission (Rakshit et al. 2008).

### *$N_2O$ Decomposition in Soil*

Nitrous oxide emission from soil reflects the net effect of production vs. consumption processes. Processes that form $N_2O$ faster than it can be reduced or strip $N_2O$ from the environment of the cell faster than it can diffuse back will cause $N_2O$ to be emitted from soil in measurable amounts (Chapuis-Lardy et al. 2007). Nitrous oxide consumption in soil is a simple reflection of its use, particularly in anaerobic environments, as a terminal electron acceptor in respiratory processes, which leads to the final production of $N_2$. Relative to $N_2O$ production, $N_2O$ consumption seems far less appreciated, and when it occurs, it has often been regarded as an artifact of research collection methods (Chapuis-Lardy et al. 2007). Like $N_2O$ production, consumption processes can be described by a combination of chemical, physical, enzymatic, and microbiological processes that produce either more (e.g., $N_2$) or less (e.g., NO, $NO_2^-$) reduced N compounds.

Huang et al. (2018) report an overall significant effect of tillage (no-tillage vs. conventional tillage) on $N_2O$ production (a relative effect ca. 10) in a meta-analysis of almost 300 field experiments. This response can be routinely attributed to (1) higher $H_2O$ content, (2) anaerobic microsites/denitrification hotspots, and (3) higher organic matter content. It is also attributable to greater diffusivity (greater pore diameter and conductivity) in undisturbed relative to disturbed environments (Kreba et al. 2017). So, for a gas like $N_2O$ with high solubility, its ability to escape the soil environment readily after production plays a critical role in its evaluation. That capacity exists more in aerobic nitrifying environments and diminishes in wetter denitrifying environments. On the other hand, the likelihood of $N_2O$ uptake would be greater in the latter as long as the $N_2O$ had the capacity to diffuse through overlying aerobic soil into the saturated environments beneath.

## *$N_2O$ Uptake by Soils*

Blackmer and Bremner (1976) noted the capacity for some soils to have greater $N_2O$ uptake than production. Vedenina and Zavarzin (1977) described the biological removal of $N_2O$ in oxidizing conditions that Vedenina et al. (1980) later ascribed to the co-metabolism of $N_2O$ by catalase during hydrogen peroxide transformation. The abiotic activity by either NO reductase or $N_2O$ reductase is probably minimal because both enzymes are membrane bound and require biologically active cell membranes for function in natural conditions (Coyne 2018).

Other biological consuming processes could occur, but they should be minor. Nitrous oxide is a substrate of the nitrogenase complex in $N_2$ fixation, being reduced to $N_2$ and $NH_3$. Denitrifying rhizobia, for example, will preferentially convert $N_2O$ to $N_2$ as measured by $^{15}N$ labeling rather than $^{15}NH_3$ due to the reducing environment within a legume nodule (Coyne and Focht 1987). Technically, any $N_2$-fixing prokaryote should therefore be able to consume $N_2O$, but the kinetics of this process are unknown, and it is unlikely to be significant given the low concentration of $N_2O$ in the soil environment relative to $N_2$, the principle substrate of nitrogenase.

As a reduced compound, $N_2O$ reflects a potential source of energy for microbial growth in aerobic environments and technically in anaerobic environments as well if a suitable electron acceptor were present. Anaerobic ammonia oxidation (ANAMMOX), for example, couples $NH_3$ oxidation and $NO_2^-$ reduction in anaerobic environments, which forms $N_2$. It does not appear to consume or produce $N_2O$, however, and its significance in terrestrial soils is unknown (although the microbes associated with ANAMMOX—*Planctomyces*—are present in soil).

## *Uptake Rates*

Schlesinger (2013) surveyed 100 studies reporting $N_2O$ uptake by soil. The median uptake was 4 μg $m^{-2}$ $h^{-1}$ ranging from 1 to 207 μg $m^{-2}$ $h^{-1}$ and was log-normally distributed with uptake ranging from 1 to 10 μg $m^{-2}$ $h^{-1}$. Peatlands and wetlands had the greatest uptake, which overall represented just 2% of $N_2O$ production rates. There was an inverse relationship between potential production and uptake: high $N_2O$ emission meant limited $N_2O$ uptake, and low emission meant reflected high uptake. A literature review by Chapuis-Lardy et al. (2007) across a wide range of agricultural and nonagricultural soils indicated uptake rates ranging from 0.0014 to 484 μg $N_2O$-N $m^{-2}$ $h^{-1}$.

As with many trace gases, the kinetics of $N_2O$ uptake by soil do not match the $K_m$ for uptake by cultures (the concentration being far lower in soil). The threshold for soil uptake, for example, is <0.8 nM (Blackmer and Bremner 1976), while the $K_m$ for microbial $N_2O$ uptake is between 0.5 and 100 μM (Conrad 1996) in culture and 700–1000 μM for soil incubations (Blackmer and Bremner 1976). The implication is that culturable prokaryotes (fungi do not appear to reduce $N_2O$) do not reflect the populations active in soil.

## *Microbial Basis for $N_2O$ Uptake*

Philippot et al. (2011) noted that up to one-third of denitrifiers have truncated pathways lacking the *nosZ* gene for $N_2O$ reduction. They attempted to show the consequences of this truncation by inoculating soil with a truncated *Agrobacterium tumefaciens* culture and observing $N_2O$ production. To their surprise, the increase in $N_2O$ emission was not as great in some soils as it ought, based on potential denitrification rates, which implied a microbial population actively consuming the $N_2O$ that was produced.

Bazylinski et al. (1986) had investigated a *Pseudomonas aeruginosa* isolate that was unusual in being able to grow solely on $N_2O$ as the terminal electron acceptor, and both *Wolinella* and *Campylobacter* spp. have the same phenotype (Payne et al. 1982). Environmental studies show microbes with genes producing $N_2O$ (*nirS*/*nirK*/*nor*) are 2-10 fold greater than those with genes reducing $N_2O$ (*nosZ*). So, why isn't more $N_2O$ evolved from soil irrespective of physicochemical barriers such as diffusion and solubility? Jones et al. (2013) subsequently performed a phylogenetic analysis for *nosZ* that revealed two distinct clades (I and II) that primarily differed in the signal proteins inserting $N_2O$ reductase into the membrane and periplasmic space. Clade I was associated with classical denitrifying species and Clade II with prokaryotes reducing $N_2O$ only. Clade II appeared to be as abundant or more abundant than Clade I in terrestrial soils. Domeignnoz-Horra et al. (2016) determined that the diversity of *nosZ* Clade II was positively related to the soil $N_2O$ sink capacity. *Dyadobacter fermentans* (which has no other denitrification genes but *nosZ*) reduced $N_2O$ emission from soil into which it was inoculated, indicating active consumption. Hallin et al. (2018) reviewed the molecular basis for enhanced $N_2O$ uptake by prokaryotes in soil. Clade I primarily consists of α, β, and ɣProteobacteria with 83% of the genomes having *nosZ* also possessing either *nirS* or *nirK*, the genes responsible for $NO_2^-$ reduction that irreversibly transform the soluble products of nitrification back to a gaseous state. Clade II consists of 51% non-denitrifiers (both bacteria and archaea) with the *nosZ* gene product more closely associated with the cytoplasmic membrane than the periplasmic space. Clade II predominates in Bacteroidetes (one of the most common phyla in soil), Gemmatimonadetes, and δProteobacteria. Induced expression of Clade II *nosZ* appears to occur at low $NO_3^-$ concentration (Hallin et al. 2018) which suggests an interesting parallel to the effects of $NH_4^+$ concentration on the relative activity of autotrophic nitrifiers—low $NH_4^+$ concentration favors AOA, which do not produce $N_2O$, while high $NH_4^+$ favors AOB that do. In the case of $N_2O$ uptake in soils, low $NO_3^-$ concentration favors the expression of non-denitrifying $N_2O$ consumers, while high $NO_3^-$ concentrations favors the activity of denitrifiers, which we realize "leak" significant amounts of $N_2O$ into the soil atmosphere. One consequence of these new observations is that the adverse consequences of temperature and pH for enhanced $N_2O$ evolution in soil may not so much reflect an influence on Clade I denitrifier *nosZ* as on those genera with Clade II *nosZ* that are inhibited in $N_2O$ uptake.

## Conclusion

Nitrous oxide evolution has been described as a "splendidly complicated interplay of different genera and species in changing environments." The extent of $N_2O$ production and consumption has genetic and environmental controls. As with all things microbiological in soils, the complexity and interaction of organisms and processes associated with $N_2O$ have only become more complicated with time as new techniques in analysis and discovery have been used. In isolation, mechanisms for either producing or consuming $N_2O$ seem manageable. Collectively, management reflecting strategies to mitigate evolution may also accentuate production. Cutting $N_2O$ production at its source (the initial oxidation of $NH_3$) would be nearly impossible in natural environments and impractical in agricultural environments and might have untoward consequences overall for environmental quality and plant growth. Several questions seem pertinent for future research: (1) What is an appropriate spatiotemporal scale over which to measure $N_2O$ flux to effectively and efficiently model its contribution to global climate change? (2) What is the differential control of fertility levels on $N_2O$ evolution by nitrifiers, and can it be managed by fertilizer choice? (3) How does aeration status in soil affect the partitioning of $N_2O$ production between prokaryotes and fungi, and how will that be influenced by changing rainfall patterns? (4) Does the propensity of denitrifiers lacking *nosZ* around rhizospheres influence the concentration of clade II $N_2O$ reducers in those locations and lead to a spatial distribution of phenotype that might be used? (5) What is the significance of precision soil fertility management, including liming, on minimizing $N_2O$ flux from soil? Developments to control of $N_2O$ production and consumption in soil would therefore seem to have a bright research future as there are clearly answers and strategies yet to be investigated.

**Acknowledgements** The preparation of this review was made possible by the following funding agencies:

USDA National Institute of Food and Agriculture, Hatch project 1024123; the Northeast Sustainable Agriculture Research and Education program under sub-award number 21-276-UK; USDA National Institute of Food and Agriculture, Hatch project 2352437000; and NSF project 2045235.

## References

Bazylinski DA, Soohoo CK, Hollocher TC (1986) Growth of *Pseudomonas aeruginosa* on nitrous oxide. Appl Environ Microbiol 51:1239–1246

Betlach MR, Tiedje JM (1981) Kinetic explanation for accumulation of nitrite, nitric oxide, and nitrous oxide during bacterial denitrification. Appl Environ Microbiol 42:1074–1084

Blackmer AM, Bremner JM (1976) Potential of soil as a sink for atmospheric nitrous oxide. Geophys Res Lett 3:739–742

Bleakley BH, Tiedje JM (1982) Nitrous oxide production by organisms other than nitrifiers or denitrifiers. Appl Environ Microbiol 44:1342–1348

Caranto JD, Vilbert AC, Lancaster KM (2016) *Nitrosomonas europaea* cytochrome $P_{460}$ is a direct link between nitrification and nitrous oxide emission. Proc Natl Acad Sci USA 113:14704–14709

Chapuis-Lardy L, Wrage N, Metay A, Chottes J-L, Bernoux M (2007) Soils, a sink for $N_2O$? a review. Glob Chang Biol 13:1–17. https://doi.org/10.1111/j.1365-2486.2006.01280x

Conrad R (1996) Soil microorganisms as controllers of atmospheric trace gases ($H_2$, CO, $CH_4$, OCS, $N_2O$, NO). Microbiol Rev 60:609–640

Cooper GS, Smith RL (1963) Sequence of products formed during denitrification in some diverse western soils. Soil Sci Soc Am Proc 27:659–662

Coyne MS (2008) Biological denitrification. In: Schepers JS, Raun W (eds) Nitrogen in agricultural systems. Agronomy Monograph 49. ASA-CSSA-SSSA, Madison, pp 197–249

Coyne MS (2018) Denitrification in soil. In: Lal R, Stewart BA (eds) Advances in soil science: soil nitrogen uses and environmental impacts. CRC Press, Boca Raton, pp 95–139

Coyne MS, Focht DD (1987) Nitrous oxide reduction in nodules: denitrification or nitrogen fixation? Appl Environ Microbiol 53:1168–1170

Coyne MS, Tiedje JM (1990a) Distribution and diversity of dissimilatory $NO_2^-$ reductases in denitrifying bacteria. In: Revsbech NP, Sørensen J (eds) Denitrification in soil and sediment: FEMS Symposium No. 56. Plenum Press, New York, pp 21–35

Coyne MS, Tiedje JM (1990b) Induction of denitrifying enzymes in oxygen-limited *Achromobacter cycloclastes* in continuous culture. FEMS Microbiol Ecol 73:263–270

Daims H, Lebedeva EH, Pjevac P, Han P, Herbold C, Albertson M, Jehmlich N, Palattinszky M, Vierheilig J, Bulaev A, Kirkegaard RH, von Bergen M, Rattei T, Bendinger B, Nielson PH, Wagner M (2015) Complete nitrification by *Nitrospira* bacteria. Nature 528:504–509

Davidson EA (1991) Fluxes of nitrous oxide and nitric oxide from terrestrial ecosystems. In: Rogers JE, Whitman WB (eds) Microbial production and consumption of greenhouse gases: methane, nitrogen oxides, and halomethanes. American Society for Microbiology, Washington DC, pp 219–235

Denmead OT (2008) Approaches to measuring fluxes of methane and nitrous oxide between landscapes and the atmosphere. Plant Soil 309:5–24

Dhakal P, Coyne MS, McNear DH, Wendroth O, Vandiviere MM, D'Angelo EM, Matocha CJ (2021) Reactions of nitrite with goethite and surface Fe(II)-goethite complexes. Sci Total Environ 782:146406. https://doi.org/10.1016/j.scitotenv.2021.146406

Dlugokencky E NOAA/GML (2021) Trends in atmospheric nitrous oxide. https://gml.noaa.gov/ccgg/trends_$N_2$o/. Accessed 28 October 2021

Domeignoz-Horta LA, Putz M, Spor A, Bru D, Breuil MC, Hallin S, Philppot L (2016) Non-denitrifying nitrous oxide-reducing bacteria—an effective $N_2O$ sink in soil. Soil Biol Biochem 103:376–379

Firestone MK (1982) Biological denitrification. In: Stevenson FJ (ed) Nitrogen in agricultural soils. American Society for Agronomy, Madison, pp 289–326

Firestone MK, Davidson EA (1989) Microbiological basis of NO and $N_2O$ production and consumption in soil. In: Andreae MO, Schimel DS (eds) Exchange of trace gases between terrestrial ecosystems and the atmosphere. Wiley, New York, pp 7–21

Firestone MK, Tiedje JM (1979) Temporal change in nitrous oxide and dinitrogen from denitrification following onset of anaerobiosis. Appl Environ Microbiol 38:673–679

Groffman PM, Tiedje JM (1988) Denitrification hysteresis during wetting and drying cycles in soil. Soil Sci Soc Am J 52:1626–1629

Hallin S, Philippot L, Löffler FE, Sanford RA, Jones CM (2018) Genomics and ecology of novel $N_2O$-reducing microorganisms. Trends Microbiol 26:43–55

Herold MB, Baggs EM, Daniell TJ (2012) Fungal and bacterial denitrification are differently affected by long term pH amendment and cultivation of arable soil. Soil Biol Biochem 54:25–35

Hochstein LI, Betlach M, Kritikos G (1984) The effect of oxygen on denitrification during steady-state growth of *Paracoccus halodenitrificans*. Arch Microbiol 137:74–78

Huang Y, Xiao X, Long XE (2017) Fungal denitrification contributes significantly to $N_2O$ production in a highly acid tea soil. J Soils Sediments 17:1599–1606

Huang Y, Ren W, Wang L, Hui D, Grove JH, Yang X, Tao B, Goff B (2018) Greenhouse gas emissions and crop yield in no-tillage systems: a meta-analysis. Agric Ecosyst Environ 268:144–153. https://doi.org/10.1016/j.agee.2018.09.002

IPCC (2013) Climate change 2013: the physical science basis. In: Stocker TL et al (eds) Contribution of working Group I to the Fifth Assessment Report if the Intergovernmental Panel on Climate Change, Cambridge/New York

Jia Z, Conrad R (2009) Bacteria rather than Archaea dominate microbial ammonia oxidation in an agricultural soil. Environ Microbiol 11:1658–1671

Jones CM, Graf DRH, Bru D, Philippot L, Halin S (2013) The unaccounted yet abundant nitrous oxide-reducing microbial community: a potential nitrous oxide sink. ISME J 7:417–426

Kaspar H (1982) Nitrite reduction to nitrous oxide by propionibacteria: detoxification mechanism. Arch Microbiol 133:126–130

Kreba S, Wendroth O, Coyne MS, Walton J (2017) Soil gas diffusivity, air-filled porosity, and pore continuity: land use and spatial patterns. Soil Sci Soc America 81:477–489. https://doi.org/10.2136/sssaj2016.10.0344

Lazcano C, Zhu-Barker X, Decock C (2021) Effects of organic fertilizers on the soil microorganisms responsible for $N_2O$ emissions: a review. Microorganisms 9:983. https://doi.org/10.3390/microorganisms9050983

Leininger S, Urich T, Schloter M, Schwark L, Qi J, Nicol GW, Prosser JL, Schuster SC, Schleper C (2006) Archaea predominate among ammonia-oxidizing prokaryotes in soil. Nature 442:806–809

Liu S, Coyne MS, Grove JH (2017) Long-term tillage and nitrogen fertilization: consequences for nitrifier density and activity. Appl Soil Ecol 120:121–127. https://doi.org/10.1016/j.apsoil.2017.07.034

Lu X, Taylor AE, Myrold DD, Neufeld JD (2020) Expanding perspectives of soil nitrification to include ammonia-oxidizing archaea and comammox bacteria. Soil Sci Soc Am J 84:287–302. https://doi.org/10.1002/saj2.20029

Maeda K, Spor A, Edel-Hermann V, Heraud C, Breuil M-C, Bizouard F, Toyoda S, Yoshida N, Steinberg C, Philippot L (2015) $N_2O$ production, a widespread trait in fungi. Sci Rep 5:9697. https://doi.org/10.1038/srep09697

Matocha C, Coyne MS (2007) Short-term response of soil iron to nitrate addition. Soil Sci Soc Am J 71:108–117. https://doi.org/10.2136/sssaj2005.0170

Minami K, Fukishu S (1986) Emissions of nitrous oxide from a well-aerated andosol treated with nitrite and hydroxylamine. Soil Sci Plant Nutr 32:233–237

Park S, Perez T, Boering KA, Trumbore SE, Gil J, Marquina S, Tyler SC (2011) Can $N_2O$ stable isotopes and isotopomers be useful tools to characterize sources and microbial pathways of $N_2O$ production and consumption in tropical soils? Glob Biogeochem Cycles 25:BB1001

Payne WJ et al (1982) Nitrogen-oxide reduction in *Wolinella succinogenes* and *Campylobacter* spp. J Bacteriol 152:915–918

Pearce RC, Zhang Y, Coyne MS (1998) Nitrogen transformations in the tobacco float system. Tob Sci 42:82–88

Philippot L, Andert J, Jones CM, Bru D, Hallin S (2011) Importance of denitrifiers lacking the genes encoding the nitrous oxide reductase for $N_2O$ emissions from soil. Glob Chang Biol 17:1497–1504. https://doi.org/10.1111/j.1365-2486.2010.02334.x

Poth M, Focht DD (1995) 15N kinetic analysis of $N_2O$ production by *Nitrosomonas europaea*: an examination of nitrifier denitrification. Appl Environ Microbiol 49:1134–1141

Rakshit S, Matocha CJ, Coyne MS (2008) Nitrite reduction by siderite. Soil Sci Soc Am J 72:1070–1077. https://doi.org/10.2136/sssaj2007.0296

Rapson TD, Dacres H (2014) Analytical techniques for measuring nitrous oxide. TrAC Trends Anal Chem 54:65–74

Ravishankara AR, Daniel JS, Portmann RW (2009) Nitrous oxide ($N_2O$): the dominant ozone-depleting substance emitted in the 21st century. Science 326(5949):123–125

Rehr B, Klemme J-H (1988) Competition for nitrate between denitrifying *Pseudomonas stutzeri* and nitrate ammonifying enterobacteria. FEMS Microbiol Ecol 62:51–58

Remde A, Conrad R (1990) Production of nitric oxide in *Nitrosomonas europaea* by reduction of nitrite. Arch Microbiol 15:187–191

Robertson GP, Huston MA, Evans FC, Tiedje JM (1988) Spatial variability in a successional plant community: patterns of nitrogen availability. Ecology 69:1517–1524

Schlesinger WA (2013) An estimate of the global sink for nitrous oxide in soils. Glob Chang Biol 19:2929–2931. https://doi.org/10.1111/gcb.12239

Schmidt I, van Spanning RJM, Jetten JSM (2004) Denitrification and ammonia oxidation by *Nitrosomonas europaea* wild-type and NirK- and NorB-deficient mutants. Microbiology 150:4107–4114

Sexstone AJ, Revsbech NP, Parkin TB, Tiedje JM (1985) Direct measurement of oxygen profiles and denitrification rates in soil aggregates. Soil Sci Soc Am J 49:645–651

Shaw LJ, Nicol GW, Smith G et al (2006) *Nitrosospira* spp. can produce nitrous oxide via a nitrifier denitrification pathway. Environ Microbiol 8:214–222

Shen JP, Zhang LM, Zhu YG, Zhang JB, He JZ (2008) Abundance and composition of ammonia-oxidizing bacteria and ammonia-oxidizing archaea communities of an alkaline sandy loam. Environ Microbiol 10:1601–1611

Shoun H, Kim DH, Uchiyama H, Sugiyama J (1992) Denitrification by fungi. FEMS Microbiol Lett 94:277–281

Shutan MA, Wang J, Xiaoyuan Y (2019) Is nitrous oxide reduction primarily regulated by the fungi-bacteria abundance ratio in fertilized soil? Pedosphere 29:569–576. https://doi.org/10.1016/S1002-0160(19)60830-1

Smil V (2001) Enriching the Earth: Fritz Haber, Carl Bosch, and the transformation of world food production. MIT Press, Cambridge

Smith MS (1982) Dissimilatory reduction of $NO_2^-$ to $NH_4^+$ and $N_2O$ by a soil *Citrobacter* sp. Appl Environ Microbiol 43:854–860

Smith MS (1983) Nitrous oxide production by *Escherichia coli* is correlated with nitrate reductase activity. Appl Environ Microbiol 45:1545–1547
Smith MS, Zimmerman K (1981) Nitrous oxide production by nondenitrifying soil nitrate reducers. Soil Sci Soc Am J 45:865–871
Soderlund R, Svensson BH (1976) The global nitrogen cycle. In: Stevenson RH, Soderlund R (eds) Nitrogen, phosphorus, and sulphur-global cycles. SCOPE Report 7. Ecological Bulletin, Stockholm, pp 61–81
Stoll S (2007) Isotopomers and isotopologues: the history behind the confusion. Chem Educ 12:1–3
Sutka RL, Ostrom NE, Ostrom PH, Gandhi H, Breznak JH (2003) Nitrogen isotopomer site preference of $N_2O$ produced by *Nitrosomonas europaea* and *Metylococcus capsulatus* bath. Rapid Commun Mass Spectrom 17:738–745
Syakila A, Kroeze C (2011) The global nitrous oxide budget revisited. GHG Measure Manage 1:17–26
Taylor AE, Vajrala N, Giguere AT, Gitelman AI, Arp DJ, Myrold DD, Sayavedra-Soto L, PJ (2013) Use of aliphatic *n*-alkynes to discriminate soil nitrification activities of ammonia-oxidizing thaumarchaea and bacteria. Appl Environ Microbiol 79: 21 https://doi.org/10.1128/AEM.01928-13
Tiedje JM (1988) Ecology of denitrification and dissimilatory nitrate reduction to ammonium. In: Zehnder AJB (ed) Biology of anaerobic organisms. Wiley, New York, pp 179–243
Tomich TP, Brondt SB, Dahlgren RA, Scow MK (eds) (2016) The California nitrogen assessment: challenges and solutions for people, agriculture, and environment. University of California Press, Berkeley
Vedenina IY, Zavarzin GA (1977) Biological removal of nitrous oxide under oxidizing conditions. Mikrobiologiya 46:898–903
Vedenina IY, Miller YM, Kapustin OA, Zavarzin GA (1980) Oxidation of nitrous oxide during decomposition of hydrogen peroxide by catalase. Mikrobiologiya 49:5–8
Ward B (2012) The global nitrogen cycle. In: Knoll AH, Canfield DE, Konhauser KO (eds) Fundamentals of geobiology. Blackwell Publishing, London, pp 36–48
Wunderlin P, Mohn J, Joss A, Emmenegger L, Siegrist H (2012) Mechanisms of $N_2O$ production in biological wastewater treatment under nitrifying and denitrifying conditions. Water Res 46:1027–1037. https://doi.org/10.1016/j.watres.2011.11.080
Yoshida T, Alexander M (1970) Nitrous oxide formation by *Nitrosomonas europaea* and heterotrophic microorganisms. Soil Sci Soc Am Proc 34:880–882
Zaehle S (2013) Terrestrial nitrogen-carbon cycle interactions at the global scale. Philos Trans R Soc Britain 368:20120125
Zumft WG (1997) Cell biology and molecular basis of denitrification. Microbiol Mol Biol Rev 61:533–616

# Soil Ecosystem Services and the Greenhouse Gas Emission and Cycling

**Khaidem Jackson, Yumnam Bijilaxmi Devi, Thounaojam Thorny Chanu, and Thounaojam Thomas Meetei**

## Introduction

Ecosystems are complex systems that encompass intricate relationships between living organisms and their habitats. They are defined by the vegetative diversity in a given space as well as the interactions between biotic, abiotic, climatic, and soil conditions. Ecosystems provide essential goods and services that satisfy human needs and enrich social and economic experiences. The Millennium Ecosystem Assessment (2005) classified ecosystems into four major types: (i) provisioning services, (ii) regulating services, (iii) cultural services, and (iv) supporting services (Fig. 1). To sustain the basic needs of the ever-expanding global population, it is critical to measure the natural capital and ecosystem services they provide (Dominati et al. 2010). Natural refers to the stock of natural assets, such as soil, forests, and water, that yield a flow of goods and services. Soils are a prime supplier of important ecosystem services (Haygarth and Ritz 2009), but their value is often

K. Jackson
Department of Soil Science and Agricultural Chemistry, Lovely Professional University, Jalandhar, Punjab, India

Y. Devi
College of Horticulture and Forestry, Natural Resource Management, Rani Lakshmi Bai Central Agricultural University, Jhansi, Uttar Pradesh, India

T. T. Chanu
Department of Botany, School of Life Science, Assam Don Bosco University, Sonapur, Assam, India

T. T. Meetei (✉)
Department of Soil Science and Agricultural Chemistry, Lovely Professional University, Jalandhar, Punjab, India

College of Agriculture, Department of Soil Science, Rani Lakhsmi Bai Central Agricultural University, Jhansi, Uttar Pradesh, India



S. R. Mohanty, B. Kollah (eds.), *Greenhouse Gas Regulating Microorganisms in Soil Ecosystems*, https://doi.org/10.1007/978-3-031-70569-4_11

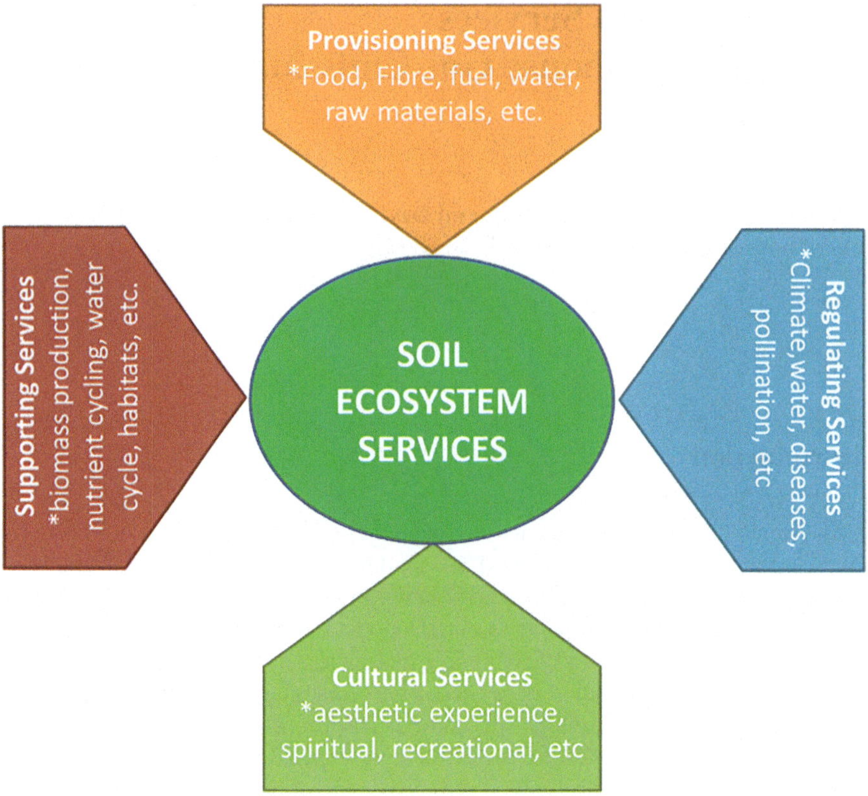


**Fig. 1** Soil ecosystem services

overlooked or undervalued. This is partly due to the fact that most human activities occur above ground and partly due to the limitations of existing frameworks for valuing natural capital and ecosystem services (De Groot et al. 2002; MEA 2005). Biodiversity is essential for many ecosystem services. Soil contains at least one-third of the living organisms on Earth, but only about 1% of these organisms have been identified (Breure et al. 2012). Soils provide a hefty supply of food resources, accounting for almost 99% of the food materials consumed by humans (Pimentel 2006). Soils also act as natural filters and aqueducts, providing filtered groundwater for consumption. Essential nutrients for plants are supplied from the lattices of the soil structure (Daily et al. 1997). The microbial population residing in the soil has various activities, including biomass decomposition. Soil microbial flora and fauna are also very important to the pharmaceutical sector, as they provide many medicines, probiotics, and antibiotics (D'Costa et al. 2006; Minton 2003; Turbé et al. 2010). Greenhouse gas evolution has a significant impact on global climate. Soil-based ecosystem services are clearly essential for the normal functioning of the biosphere. Soils can be distinguished as a natural capital because they provide various services that are beneficial to humans. Soils are also an important resource for achieving sustainable agriculture systems. Soil is a non-renewable resource that provides inexhaustible benefits to life on Earth. It takes millions of years to convert consolidated rocks and mineral constituents into unconsolidated materials that can be called soil. Supportive services are one of the most important soil ecosystem services (SES), as they provide the foundation for all other SES. However, due to rapid population growth, other soil ES, such as provisioning and regulatory services, are becoming increasingly vital (Haines-Young and Potschin 2010). Provisioning services of soil ES include the products that supply nutrients and energy from the ecosystem. To feed the growing population, significant work has been done under provisioning ES. However, different farming communities have focused on crop production while neglecting soil health. For example, the Green Revolution was initiated in India in 1960 to increase food production and feed the growing population (Eliazer Nelson et al. 2019). However, the excessive use of inorganic fertilizers and other chemicals to feed high-yielding varieties (HYV) crops disrupted soil health and led to deterioration. One of the most important soil health indicators affected during the Green Revolution is soil organic carbon (SOC). Due to intensive agriculture practices during the Green Revolution, only a minimum amount of crop residues are returned to the soils, which affects the SOC content. When there is a minimum amount of organic matter returned to the soil, it drastically affects the physical stability, resulting in low microbial activity (Bot and Benites 2005). This affects the nutrient cycling processes, which are very important as far as soil health. Different chemicals used during crop production also pollute the soil. As a result, soil health deterioration drastically decreases the soil's ability to retain these pollutants. Therefore, during every irrigation or rainfall, some amount of pollutants reaches freshwater bodies, causing eutrophication and water pollution (Lal 2020). Regulatory services of soil ES include the benefits that soil provides to regulate natural processes, such as climate regulation, water purification, and pest control. Soil plays a vital role in regulating the global climate by storing carbon and releasing it slowly into the atmosphere. Soil also plays a role in water purification by filtering contaminants from rainwater and runoff. Additionally, soil helps to control pests and diseases by providing habitat for beneficial organisms (Bernhardt-Römermann et al. 2011). Cultural services of soil ES include the non-material benefits that soil provides, such as recreation, aesthetics, and education. Soil provides opportunities for recreation, such as gardening and hiking. Soil also contributes to the aesthetic value of landscapes. Additionally, soil is an important educational resource, as it can teach us about the natural world and our place in it.

## *Green House Gases (GHG)'s Evolution Systems in Agriculture*

Emission of gaseous substances such as nitrous oxide ($N_2O$), carbon dioxide ($CO_2$), and methane ($CH_4$) from agricultural soils contributes to global warming and atmospheric pollution. Earthworms are considered ecosystem engineers because they cause bioturbation, which enhances the nutrient content of soil. However, during the process, they induce emission of $N_2O$ from the soil (De Vries et al. 2013). Nutrient management practices involve the production of greenhouse gases (GHG's), especially $N_2O$, which is a potent GHG and considered to be an antithesis for ecosystem services (Galloway et al. 2004). Among soil-based ecosystem services, certain biogeochemical cycles and GHG emissions seem to be directly affected by various nutrient cycling processes and carbon sequestration (Ballantyne IV et al. 2008; Allen and Gillooly 2009). Organic matter decomposition and mineralization are two frequently occurring ecosystem functions that are associated with the nutrient cycling ES (de Bello et al. 2010). Altering the soil ES leads to certain abnormalities in the soil functions, which ultimately leads to the development of higher quantities of GHG's in the atmosphere. For example, human alteration of the nitrogen cycle has enhanced the health and well-being of society, but anything in excess is toxic. Excess alteration to the nitrogen cycle has led to increased production of several GHGs, including $CO_2$, $N_2O$, and $CH_4$. Nitrogen deposition from the atmosphere can benefit and enhance plant growth. It can also remarkably influence carbon uptake (Liu and Greaver 2009; Quinn Thomas et al. 2010). Carbon sequestration is extensively involved in GHG emissions and climate change (Ghaley et al. 2014). The addition of nitrogen to soil can reduce $CO_2$ release into the atmosphere and retain carbon stocks in the soil (Janssens and Luyssaert 2009). However, this does not mean that other GHGs are not produced from the system. Nitrogen addition can compound the production of $CH_4$ and $N_2O$, which are major GHGs. Another study by Forster et al. 2007, also found that $N_2O$ concentration increases rapidly as a result of nitrogen incorporation into the soil by stimulating nitrification and denitrification processes. The availability of nitrogen in the soil system can vivify $N_2O$ production, with $CH_4$ production being a concomitant effect of another competing process (Compton et al. 2011). The increase of these gases in the biosphere is known to pose significant negative impacts on the climate. Greenhouse gases, such as carbon dioxide ($CO_2$), methane ($CH_4$), and nitrous oxide ($N_2O$), play a vital role in regulating the Earth's climate. They trap and absorb infrared radiation emitted from the Earth's surface, warming the planet and making life possible (Bian 2020). However, human activities are increasing the concentration of greenhouse gases in the atmosphere at an unprecedented rate, causing the Earth's climate to change rapidly. This is leading to global warming and other serious environmental problems, such as more extreme weather events, rising sea levels, and mass extinctions (Heller and Keoleian 2015).

To reduce greenhouse gas emissions and mitigate the effects of climate change, we need to transition to renewable energy sources, improve energy efficiency, reduce deforestation, and change our agricultural practices. We must also invest in new technologies to capture and store carbon dioxide. Taking action now is essential to protect our planet and ensure a sustainable future for all (Lal 2010).

Soil is an important source and sink of the three major greenhouse gases, namely carbon dioxide ($CO_2$), methane ($CH_4$), and nitrous oxide ($N_2O$). $CO_2$ is the most significant greenhouse gas, accounting for about 80% of human-caused emissions. During the eighteenth century, the annual $CO_2$ emissions globally were estimated to be around 0.028 Pg C that gradually increased with time, and now it has reached around 35 PgC (Wang et al. 2017a, b).

Land use change (LUC) is a major contributor of $CO_2$ emissions. In the mid-1800s, LUC contributed more $CO_2$ emissions than fossil fuel combustion. However, after 1947, fossil fuel combustion surpassed LUC emissions due to rapid urbanization and industrialization.

In 2020, fossil fuel burning contributed 34.81 Pg of $CO_2$ emissions, while LUC contributed 3.21 Pg. LUC emissions are primarily caused by the conversion of natural ecosystems to agricultural land and other uses (Wang et al. 2017a, b) (Fig. 2). Changes in land environs to agricultural land in certain developing countries have been a major cause for these emissions. Moreover, in developed countries, forest lands have been converted to cropping lands, grazing grounds, and pastures in the early twentieth century. Currently, the focus has been made on conversion to real estates in sub-urban developmental projects. When forests and other natural ecosystems are cleared for agriculture or other purposes, the soil carbon pool is reduced. The carbon pools can be reduced by 30–50% over 100 years in temperate climatic zones and 50–75% in just 20–50 years in the tropical region (Lal 2008a). Temperature plays an important role in the decomposition of organic materials. If the temperature increases at a fixed rate, the rate of decomposition may be higher, and the soil carbon pools may be deeply affected. The only way to remove emitted $CO_2$ from the atmosphere is to sequester it in the terrestrial ecosystem as soil organic carbon (SOC). Soil organic carbon (SOC) pools are nearly four times the size of biotic

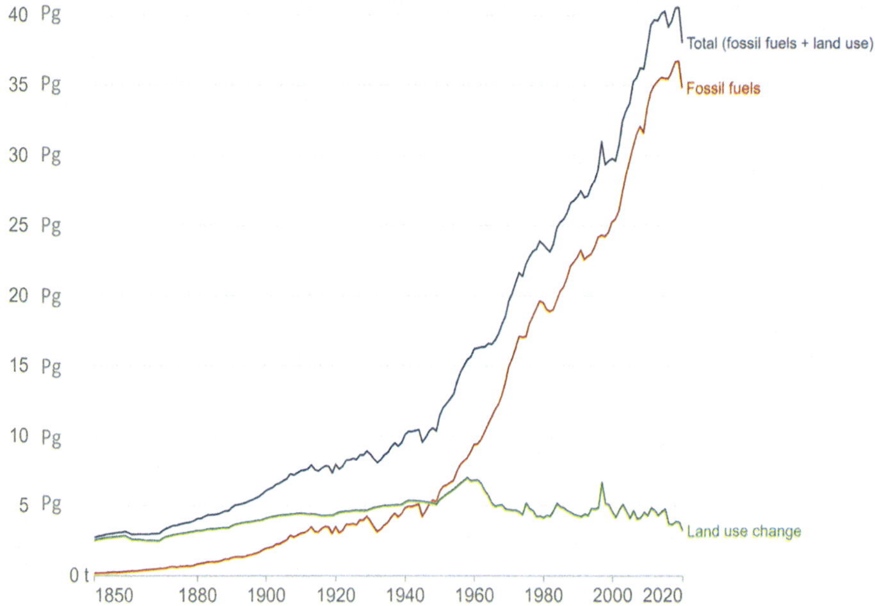


**Fig. 2** Global $CO_2$ emissions from fossil fuels and land use change (Source: Ritchie and Roser 2019)

pools and three times the size of the atmospheric carbon pool. At a depth of 2 m, SOC pools contain approximately 2.5 × $10^{12}$ tonnes of carbon. Terrestrial carbon pools contain 57% of the total physical carbon pools and four times the carbon content of the atmosphere. The terrestrial C pool amounts to 3.2 × $10^{12}$ tonnes which can be classified into 2.5 × $10^{12}$ tonnes in soil, 6.6 × $10^{10}$ tonnes in litter, and 6.2 × $10^{11}$ tonnes in live organic materials (Lal 2003). Any alteration in the soil ecosystem can lead to a cascade of changes in soil physics, biology, chemistry, and biochemistry (Troeh et al. 2004). These changes can induce oxidation-reduction reactions and mineralization of organic matter in the soil ecosystem. Subsequent physical and chemical weathering processes can release greenhouse gases such as $CO_2$, $N_2O$, and $CH_4$ (Lal 2003). Between 1959 and 2010, half of the $CO_2$ emitted from human activities was absorbed by carbon sinks in the ocean and soil (Raupach 2013). Carbon sequestration is the process of capturing and storing atmospheric $CO_2$ in soil. Plants sequester carbon in soil through photosynthesis, where they absorb $CO_2$ and $H_2O$ and convert them into carbohydrates and oxygen. The carbon in plant parts or whole plants is transferred to the soil as soil organic carbon (SOC) after death and decay. Therefore, soil is an important carbon sink that stores the most carbon in the terrestrial ecosystem (Lal 2018). Soil carbon storage is an essential soil ecosystem service that provides both provisioning and regulatory services. Soil organic carbon (SOC) is a food source for soil organisms, which enhances microbial diversity and stimulates nutrient transformations. Crops assimilate essential nutrients from the soil and convert them into organic constituents. After crop death and decay, these nutrients must return to the soil through nutrient cycling. SOC is essential for nutrient cycling and transformation as it supports the growth and reproduction of soil microbes. Another important function of SOC is to improve soil aggregate stability (Chaplot and Cooper 2015). The soil primary particles (sand, silt, and clay) formed after pedogenesis have to be rearranged/aggregated for a good soil health. Only after the formation soil aggregates through these particles, soil is considered as a 'natural habitat' which possess the basic components of life i.e. solid (mineral matter and organic matter) and pores (liquid and gas). The most important cementing agents responsible for soil particle aggregation are organic matter and humus (Bongiovanni and Lobartini 2006). Therefore, with higher SOC content in soil, the physical, chemical, and biological properties of soil increase turning overall soil health in a good shape.

Agriculture practices can contribute up to 49% of the total global $CH_4$ emissions (FAO 2003). Residue retention in paddy fields is a primary source of available C, which promotes $CH_4$ formation under anaerobic conditions (Zou et al. 2005). Stacking of paddy straws can further enhance this process by creating an anaerobic environment inside the stacks. This suggests that a decrease in the $CH_4$ oxidation can promote more emission of $CH_4$ into the atmosphere. Flooded rice cultivation follows the same condition of anaerobic decomposition and thus releases $CH_4$. Livestock production and meat consumption are major contributors to $CH_4$ emissions. As the demand for meat increases, the adverse effects of $CH_4$ release are likely to skyrocket in the near future (FAO 2003). Ruminant animals have a specialized stomach that can ferment the plant materials they consume. During digestion,

$CH_4$ is released as a by-product. Livestock manure also produces $CH_4$ emissions as microbes decompose the organic matter. Methane emissions from livestock account for 3–18% of total global $CH_4$ emissions (Pitesky et al. 2009). Methane ($CH_4$) is an important greenhouse gas because it traps heat energy in the atmosphere (Ledley et al. 1999). $CH_4$ concentrations have increased by 2.5 times since the pre-industrial era. A recent study estimated that total annual global $CH_4$ emissions are around 570 million tonnes (Mt) (Dangal et al. 2017). Anthropogenic activities contribute around 60% of these emissions, while natural sources contribute the remaining 40%. Wetlands are the largest source of methane ($CH_4$) emissions, followed by agriculture and waste mismanagement, which contribute 194, 145, and 68 million tonnes (Mt) of $CH_4$ emissions per year, respectively (Le Mer and Roger 2001; see Fig. 3). The impact of different greenhouse gases (GHGs) on the greenhouse effect varies depending on how long they remain in the atmosphere (Shine et al. 2005). $CH_4$ is the second most abundant GHG after carbon dioxide ($CO_2$), but it has a much shorter residence time (around 12 years) than $CO_2$ (which can remain in the atmosphere for centuries). This means that the overall impact of $CH_4$ on the greenhouse effect is lower than that of $CO_2$. However, $CH_4$ is a much more potent GHG than $CO_2$, absorbing 25 times more heat energy per molecule (MacCarty et al. 2008).

To maintain or reduce methane ($CH_4$) emissions, the regulatory services of soil ecosystems (ES) are essential. $CH_4$ is the end product of continuous anaerobic respiration. When organic matter decomposes in the absence of oxygen for a long period, $CH_4$ is produced. Therefore, soil is the ultimate source and sink of atmospheric $CH_4$, making agriculture and wetlands the largest contributors of methane to the atmosphere. The ability of soil to emit and consume $CH_4$ depends on land use and ecosystem type. Soils of different natural forests can consume a certain amount of $CH_4$ from the atmosphere at a rate of a few to a fraction of mg $m^{-2}$ $d^{-1}$. In contrast, agricultural and wetland soils emit $CH_4$ at a rate of several to a few hundred mg $m^{-2}d^{-1}$.

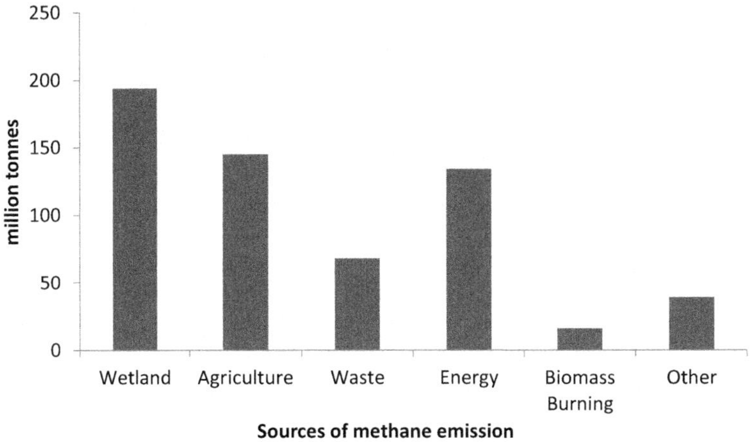


**Fig. 3** Global methane emission from different sources

Nitrogen (N) is an essential component of soil that is required by various organisms for protein synthesis. N is mainly available to soil through atmospheric $N_2$ fixation by some microorganisms and electrical discharge in the atmosphere. Over 90% of total soil N is in organic form, which must be mineralized to be used by organisms. The most available forms of inorganic N for organism assimilation from soil are $NO_3^-$ and $NH_4^+$. Mineralized N can be lost from soil through leaching, volatilization, denitrification, fixation, and immobilization (Hofman and Van Cleemput 2004). During denitrification, N is emitted as nitrogen dioxide ($NO_2$), nitrous oxide ($N_2O$), and dinitrogen ($N_2$). $N_2O$ is a greenhouse gas that traps heat energy and contributes to global warming (Kweku et al. 2018). Gaseous loss of N from soil is an essential part of the N cycle that helps to prevent N pollution. It is a natural mechanism to eliminate soil and water pollution, as anything in excess can be harmful. However, several anthropogenic activities, such as the overuse of inorganic fertilizers, have been shown to stimulate increased emissions of nitrous oxide ($N_2O$) from the soil (Tao et al. 2018). $N_2O$, also known as laughing gas, is a potent greenhouse gas that is 300 times more effective at trapping heat than $CO_2$ (Palm et al. 2014). Although $N_2O$ emissions are relatively low compared to $CO_2$ emissions, they account for about 6% of total annual GHG emissions. Despite its significant impact on global climate change, $N_2O$ has often been neglected in climate policy making. In the past four decades, $N_2O$ emissions have increased by 30%, primarily due to the expansion of agriculture and the increased use of inorganic N fertilizers.

Nitrification and denitrification are the major processes driving nitrous oxide ($N_2O$) emissions into the atmosphere (Thamdrup 2012). Nitrification occurs under aerobic conditions, while denitrification occurs under anaerobic conditions. Heterotrophic denitrification is considered to be a major source of $N_2O$ emissions (Wang et al. 2017a, b). Nitrification can also contribute to $N_2O$ emissions when it occurs in soils with less than 40% water-filled pore space (Dalal et al. 2003). The frequency and intensity of these emissions are closely linked to soil structure, which is influenced by bulk density, soil carbon content, and soil aggregation. Tillage practices and organic and inorganic inputs can all affect soil structure. Organic residue management and other agricultural practices can alter soil nitrate ($NO_3^-$) availability. This can affect $N_2O$ emissions by changing the decomposability of organic matter with high carbon (C) substrates (Firestone and Davidson 1989).

When soil $NO_3$- concentrations and labile C fractions are high, the conversion of $N_2O$ to atmospheric nitrogen is reduced (Hutchinson and Davidson 1993; Weier et al. 1993; Senbayram et al. 2012).

Under anaerobic conditions with high water saturation, soil organic matter decomposes readily, creating a suitable environment for denitrification. This promotes $N_2O$ production in the system (Dalal et al. 2003). Nitrous oxide ($N_2O$) emission is a natural process involved in the nitrogen cycle. However, high rates of $N_2O$ emission can be associated with higher application of inorganic nitrogen fertilizers to the soil. The higher the uptake of this nitrogen by crops, the higher the emission of $N_2O$ will be. Globally, 50% of applied nitrogen fertilizer remains in the plant system, 5% remains in the soil ecosystem, 25% is emitted as $N_2O$, and the remainder enters aquatic ecosystems (Galloway et al. 2004). The production of chemical nitrogen fertilizers is also an energy-intensive process that produces additional greenhouse gases.

## Mitigation of Climate Change Through Soil Ecosystem Services

The services provided by the soil ecosystem must be used in a sustainable manner. The most important greenhouse gases discussed above are essential for the proper functioning of the soil ecosystem in one way or another. However, certain anthropogenic activities have caused them to become non-essential. To maintain the regulatory and provisioning services of soil ecosystems, we need good research, policymaking, and revised practices in the farming community. As discussed earlier, soil organic carbon (SOC) is essential for soil aggregate stability, microbial biodiversity, nutrient transformation, buffering actions, nutrient retention, and more (Lal 2016). However, if inappropriate land use management practices are used in the field, SOC content can lead to the emission of other greenhouse gases ($CH_4$ and $N_2O$), which are more potent than $CO_2$.

The first and most important thing to consider for reducing GHG emissions from soil is to improve soil physical properties (Nawaz et al. 2013. Soil health sustainability is confirmed when a medium for organism habitat (mineral matter), nutrient source (organic matter), good aeration (air), and optimum moisture (water) are present.

However, the rapid growth of the population is inducing an increase in demand, which is leading to increased conversion of land use systems from natural to agricultural and other construction purposes, thereby disturbing the natural habitat. Research and policymaking are still underway to reduce the anthropogenic impact of GHG emissions globally. In the agricultural sector, various measures can be considered or revised to reduce GHG emissions. Producing food to feed the world's population is always a priority, as the world faces hunger and malnutrition (Ritchie and Roser 2020). However, maintaining soil and environmental sustainability is also important for future generations, as we cannot compromise their future for our own. Therefore, now is a crucial time to enhance the regulatory and provisioning services of soil ecosystems through advanced research and policymaking to reduce GHG emissions from soil.

The soil ecosystem services (ES) and its regulatory functions are essential for good agricultural productivity. The ability of the soil to retain certain nutrients alongside carbon (C) and nitrogen (N) can predict how susceptible it is to degradation. Healthy soils should possess this ability plus the ability to regulate greenhouse gases (GHGs) (Balbi et al. 2015).

Management practices that involve the production of manure from animal wastes can help reduce emissions. The application of leguminous crops in the field can enhance N-cycling processes, which can improve N use efficiency. The N-fixing bacteria present in the root nodules of legumes can halve $CO_2$ emissions (Drinkwater and Snapp 2007). Additionally, this process can reduce the use of inorganic N, which in turn reduces $N_2O$ emissions into the atmosphere. Sequestering carbon in the soil can increase the storage of carbon pools and reduce $CO_2$ emissions (Lal 2008a, b). One way to accomplish this is to plant perennial plants in areas with low soil organic carbon (SOC) content (Poulton et al. 2003). This process could potentially sequester a large amount of carbon globally, but it remains theoretical due to some practical constraints (Smith et al. 2008). Agricultural practices often involve clearing land, which can produce greenhouse gases (GHGs) that are harmful to the global climate (Searchinger et al. 2008). Slowing down the decomposition process can increase SOC content and enhance carbon pools in the soil. The subsoil has a lower concentration of organic carbon than the topsoil, but it has the potential to store more carbon with enhanced stabilization (Jenkinson and Coleman 2008). If we better understood the mechanisms involved in this process, it could play a pivotal role in carbon sequestration (Fontaine et al. 2007). 4-milli initiative. The 4-milli initiative launched by France on December 1, 2015, aims to play a crucial role in ensuring food security and combating climate change. Leaders from 195 countries met in Paris to negotiate an accord for combating climate change (UNFCCC 2015). Under the accord, participating governments pledged to reduce carbon emissions and footprints to keep the global temperature below 2 °C of pre-industrial levels. Updated records and plans on cutting $CO_2$ emissions are tentatively reported to the public every 5 years, starting in 2020.

The agreement has historic significance for recognizing the harmful effects of greenhouse gases and planning to reduce their emissions from land use practices, while promoting conservation and sustainable management to enhance carbon stocks. Captured carbon from the atmosphere can be stored in biomass, soils, or the deep sea for an infinite amount of time.

## Conclusion

The increasing global population cannot be stopped, but the greenhouse gas (GHG) emissions caused by population growth that are impacting climate change must be controlled to ensure environmental sustainability. To achieve this, we must use the regulatory services provided by soil ecosystems without compromising the provisioning services of food production. This will require good research, policymaking, and revised land use management practices in the farming community. One of the most important nutrient cycles that contribute to soil health and environmental sustainability is the carbon cycle. Soil is an important carbon sink, and it can store large quantities of carbon from the atmosphere through carbon sequestration. More carbon reserved in the soil improves soil physical, chemical, and biological properties, resulting in soil health and sustainability. However, if the sequestered carbon in the soil is not maintained properly, it can be released to the atmosphere as $CO_2$. Additionally, mismanagement of carbon can also lead to the production of other, more potent GHGs ($CH_4$ and $N_2O$) that can impact climate change. Therefore, increasing carbon sequestration and implementing land use management practices that conserve more carbon in the soil are the best ways to control GHG emissions from agricultural soil. By taking these steps, we can ensure that soil ecosystems continue to provide the essential services that we rely on while also mitigating the effects of climate change.

## References

Allen AP, Gillooly JF (2009) Towards an integration of ecological stoichiometry and the metabolic theory of ecology to better understand nutrient cycling. Ecol Lett 12(5):369–384

Balbi S, del Prado A, Gallejones P, Geevan CP, Pardo G, Pérez-Miñana E, Manrique R, Hernandez-Santiago C, Villa F (2015) Modeling trade-offs among ecosystem services in agricultural production systems. Environ Model Softw 72:314–326

Ballantyne F IV, Menge DN, Ostling A, Hosseini P (2008) Nutrient recycling affects autotroph and ecosystem stoichiometry. Am Nat 171(4):511–523

Bernhardt-Römermann M, Gray A, Vanbergen AJ, Bergès L, Bohner A, Brooker RW, De Bruyn L, De Cinti B, Dirnböck T, Grandin U, Hester AJ (2011) Functional traits and local environment predict vegetation responses to disturbance: a pan-European multi-site experiment. J Ecol 99(3):777–787

Bian Q (2020) Waste heat: the dominating root cause of current global warming. Environ Syst Res 9(1):1–1

Bongiovanni MD, Lobartini JC (2006) Particulate organic matter, carbohydrate, humic acid contents in soil macro-and microaggregates as affected by cultivation. Geoderma 136(3-4):660–665

Bot A, Benites J (2005) The importance of soil organic matter: key to drought-resistant soil and sustained food production. Food & Agriculture Org

Breure AM, De Deyn GB, Dominati E, Eglin T, Hedlund K, Van Orshoven J, Posthuma L (2012) Ecosystem services: a useful concept for soil policy making! Curr Opin Environ Sustain 4(5):578–585

Chaplot V, Cooper M (2015) Soil aggregate stability to predict organic carbon outputs from soils. Geoderma 243:205–213

Compton JE, Harrison JA, Dennis RL, Greaver TL, Hill BH, Jordan SJ, Walker H, Campbell HV (2011) Ecosystem services altered by human changes in the nitrogen cycle: a new perspective for US decision making. Ecol Lett 14(8):804–815

D'Costa VM, McGrann KM, Hughes DW, Wright GD (2006) Sampling the antibiotic resistome. Science 311(5759):374–377

Daily GC, Matson PA, Vitousek PM (1997) Ecosystem services supplied by soil. In: Nature's services: societal dependence on natural ecosystems, pp 113–132

Dalal RC, Wang W, Robertson GP, Parton WJ (2003) Nitrous oxide emission from Australian agricultural lands and mitigation options: a review. Soil Research 41(2):165–195

Dangal SR, Tian H, Zhang B, Pan S, Lu C, Yang J (2017) Methane emission from global livestock sector during 1890–2014: magnitude, trends and spatiotemporal patterns. Glob Chang Biol 23(10):4147–4161

de Bello F, Lavorel S, Díaz S, Harrington R, Cornelissen JH, Bardgett RD, Berg MP, Cipriotti P, Feld CK, Hering D, Martins da Silva P (2010) Towards an assessment of multiple ecosystem processes and services via functional traits. Biodivers Conserv 19(10):2873–2893

De Groot RS, Wilson MA, Boumans RM (2002) The dynamics and value of ecosystem services: integrating economic and ecological perspectives. A typology for the classification, description and valuation of ecosystem functions, goods and services. Ecol Econ 41:393–408

De Vries FT, Thébault E, Liiri M, Birkhofer K, Tsiafouli MA, Bjørnlund L, BrachtJørgensen H, Brady MV, Christensen S, De Ruiter PC, d'Hertefeldt T (2013) Soil food web properties explain ecosystem services across European land use systems. Proc Natl Acad Sci 110(35):14296–14301

Dominati E, Patterson M, Mackay A (2010) A framework for classifying and quantifying the natural capital and ecosystem services of soils. Ecol Econ 69(9):1858–1868

Drinkwater LE, Snapp S (2007) Nutrients in agroecosystems: rethinking the management paradigm. Adv Agron 92:163–186

Eliazer Nelson AR, Ravichandran K, Antony U (2019) The impact of the Green Revolution on indigenous crops of India. J Ethnic Foods 6(1):8

FAO (2003) World agriculture: towards 2015/2030.Interim report. FAO, Rome

Firestone MK, Davidson EA (1989) Microbiological basis of NO and N 2 O production and consumption in soil. In: Exchange of trace gases between terrestrial ecosystems and the atmosphere, vol 47. Willey, pp 7–21

Fontaine S, Barot S, Barré P, Bdioui N, Mary B, Rumpel C (2007) Stability of organic carbon in deep soil layers controlled by fresh carbon supply. Nature 450(7167):277–280

Forster P, Ramaswamy V, Artaxo P, Berntsen T, Betts R, Fahey DW, Haywood J, Lean J, Lowe DC, Myhre G, Nganga J, Prinn R, Raga G, Schulz M, Van Dorland R (2007) Chapter 2: Changes in atmospheric constituents and in radiative forcing. IPCC fourth assessment report WG 1. IPCC. Cambridge, Cambridge University Press.

Galloway JN, Dentener FJ, Capone DG, Boyer EW, Howarth RW, Seitzinger SP, Asner GP, Cleveland CC, Green PA, Holland EA, Karl DM (2004) Nitrogen cycles: past, present, and future. Biogeochemistry 70(2):153–226

Ghaley BB, Porter JR, Sandhu HS (2014) Soil-based ecosystem services: a synthesis of nutrient cycling and carbon sequestration assessment methods. Int J Biodivers Sci Ecosyst Serv Manag 10(3):177–186

Haines-Young R, Potschin M (2010) The links between biodiversity, ecosystem services and human well-being. Ecosyst Ecol New Synth 1:110–139

Haygarth PM, Ritz K (2009) The future of soils and land use in the UK: soil systems for the provision of land-based ecosystem services. Land Use Policy 26:S187–S197

Heller MC, Keoleian GA (2015) Greenhouse gas emission estimates of US dietary choices and food loss. J Ind Ecol 19(3):391–401

Hofman G, Van Cleemput O (2004) Soil and plant nitrogen. International Fertilizer Industry Association

Hutchinson GL, Davidson EA (1993) Processes for production and consumption of gaseous nitrogen oxides in soil. In: Agricultural ecosystem effects on trace gases and global climate change, vol 55, pp 79–93

Janssens IA, Luyssaert S (2009) Carbon cycle: Nitrogen's carbon bonus. Nat Geosci 2(5):318–319

Jenkinson DS, Coleman K (2008) The turnover of organic carbon in subsoils. Part 2. Modelling carbon turnover. Eur J Soil Sci 59(2):400–413

Kweku DW, Bismark O, Maxwell A, Desmond KA, Danso KB, Oti-Mensah EA, Quachie AT, Adormaa BB (2018) Greenhouse effect: greenhouse gases and their impact on global warming. J Sci Res Rep 17(6):1–9

Lal R (2003) Global potential of soil carbon sequestration to mitigate the greenhouse effect. Crit Rev Plant Sci 22(2):151–184

Lal R (2008a) Sequestration of atmospheric $CO_2$ in global carbon pools. Energy Environ Sci 1(1):86–100

Lal R (2008b) Soil carbon stocks under present and future climate with specific reference to European ecoregions. Nutr Cycl Agroecosyst 81(2):113–127

Lal R (2010) Managing soils and ecosystems for mitigating anthropogenic carbon emissions and advancing global food security. Bioscience 60(9):708–721

Lal R (2016) Soil health and carbon management. Food Energy Secur 5(4):212–222

Lal R (2018) Digging deeper: a holistic perspective of factors affecting soil organic carbon sequestration in agroecosystems. Glob Chang Biol 24(8):3285–3301
Lal R (2020) Soil quality and sustainability. In: Methods for assessment of soil degradation, vol 26. CRC Press, pp 17–30
Le Mer J, Roger P (2001) Production, oxidation, emission and consumption of methane by soils: a review. Eur J Soil Biol 37(1):25–50
Ledley TS, Sundquist ET, Schwartz SE, Hall DK, Fellows JD, Killeen TL (1999) Climate change and greenhouse gases. EOS Trans Am Geophys Union 80(39):453–458
Liu L, Greaver TL (2009) A review of nitrogen enrichment effects on three biogenic GHGs: the $CO_2$ sink may be largely offset by stimulated $N_2O$ and $CH_4$ emission. Ecol Lett 12(10):1103–1117
MacCarty N, Ogle D, Still D, Bond T, Roden C (2008) A laboratory comparison of the global warming impact of five major types of biomass cooking stoves. Energy Sustain Dev 12(2):56–65
Millennium ecosystem assessment ME (2005) Ecosystems and human well-being. Island Press, Washington, DC
Minton NP (2003) Clostridia in cancer therapy. Nat Rev Microbiol 1(3):237–242
Nawaz MF, Bourrie G, Trolard F (2013) Soil compaction impact and modelling. A review. Agron Sustain Dev 33(2):291–309
Palm C, Blanco-Canqui H, DeClerck F, Gatere L, Grace P (2014) Conservation agriculture and ecosystem services: an overview. Agric Ecosyst Environ 187:87–105
Pimentel D (2006) Soil erosion: a food and environmental threat. Environ Dev Sustain 8(1):119–137
Pitesky ME, Stackhouse KR, Mitloehner FM (2009) Clearing the air: Livestock's contribution to climate change. Adv Agron 103:1–40
Poulton PR, Pye E, Hargreaves PR, Jenkinson DS (2003) Accumulation of carbon and nitrogen by old arable land reverting to woodland. Glob Chang Biol 9(6):942–955
Quinn Thomas R, Canham CD, Weathers KC, Goodale CL (2010) Increased tree carbon storage in response to nitrogen deposition in the US. Nat Geosci 3(1):13–17
Raupach MR (2013) Ecosystem services and the global carbon cycle. In: Ecosystem services and carbon sequestration in the biosphere. Springer, Dordrecht, pp 155–181
Ritchie H, Roser M (2019) $CO_2$ emissions. Retrieved from Our World in Data: https://ourworldindata.org/$CO_2$-emissions
Ritchie H, Roser M (2020) Environmental impacts of food production. Our world in data
Searchinger T, Heimlich R, Houghton RA, Dong F, Elobeid A, Fabiosa J, Tokgoz S, Hayes D, Yu TH (2008) Use of US croplands for biofuels increases greenhouse gases through emissions from land-use change. Science 319(5867):1238–1240
Senbayram M, Chen R, Budai A, Bakken L, Dittert K (2012) $N_2O$ emission and the $N_2O/(N_2O+N_2)$ product ratio of denitrification as controlled by available carbon substrates and nitrate concentrations. Agric Ecosyst Environ 147:4–12
Shine KP, Fuglestvedt JS, Hailemariam K, Stuber N (2005) Alternatives to the global warming potential for comparing climate impacts of emissions of greenhouse gases. Clim Chang 68(3):281–302
Smith P, Martino D, Cai Z, Gwary D, Janzen H, Kumar P, McCarl B, Ogle S, O'Mara F, Rice C, Scholes B (2008) Greenhouse gas mitigation in agriculture. Philos Trans R Soc B: Biol Sci 363(1492):789–813
Tao R, Wakelin SA, Liang Y, Hu B, Chu G (2018) Nitrous oxide emission and denitrifier communities in drip-irrigated calcareous soil as affected by chemical and organic fertilizers. Sci Total Environ 612:739–749
Thamdrup B (2012) New pathways and processes in the global nitrogen cycle. Annu Rev Ecol Evol Syst 43:407–428
Troeh FR, Hobbs JA, Donahue RL (2004) Soil and water conservation for productivity and environmental protection. Prentice-Hall, Inc
Turbé A, De Toni A, Benito P, Lavelle P, Lavelle P, Camacho NR, Van Der Putten WH, Labouze E, Mudgal S (2010) Soil biodiversity: functions, threats and tools for policy makers. Bio Intelligence Service, IRD, and NIOO, Report for European Commission (DG Environment).

UNFCCC (2015) Adoption of the Paris agreement. COP 25th session Paris
Wang H, Marshall CW, Cheng M, Xu H, Li H, Yang X, Zheng T (2017a) Changes in land use driven by urbanization impact nitrogen cycling and the microbial community composition in soils. Sci Rep 7(1):1–2
Wang R, Goll D, Balkanski Y, Hauglustaine D, Boucher O, Ciais P, Janssens I, Penuelas J, Guenet B, Sardans J, Bopp L (2017b) Global forest carbon uptake due to nitrogen and phosphorus deposition from 1850 to 2100. Glob Chang Biol 23(11):4854–4872
Weier KL, Doran JW, Power JF, Walters DT (1993) Denitrification and the dinitrogen/nitrous oxide ratio as affected by soil water, available carbon, and nitrate. Soil Sci Soc Am J 57(1):66–72
Zou J, Huang Y, Jiang J, Zheng X, Sass RL (2005) A 3-year field measurement of methane and nitrous oxide emissions from rice paddies in China: effects of water regime, crop residue, and fertilizer application. Glob Biogeochem Cycles 19(2):GB2021

# Diversity of Greenhouse Gas Producing and Mitigating Microbes in Soil

**Thounaojam Thomas Meetei, Nongmaithem Shitaljit Singh, Yumnam Bijilaxmi Devi, and Kangujam Bokado**

## Introduction

A layer of gases known as greenhouse gases (GHG) surrounds our planet and is essential for human survival, as it keeps our planet warm through a process known as the greenhouse effect. The greenhouse effect is the process by which solar energy radiated from the Earth's surface is trapped by GHG in the atmosphere, keeping the planet warm. Without GHG, our planet's average temperature would be much colder, making it unfit for our survival. However, due to activities related to the rapid increase in the global human population, more GHG are being emitted, making the blanket of gases surrounding the Earth much thicker and escalating the atmospheric greenhouse effect (Sharma et al. 2021). Given its effect on global temperature, global warming is now a cause for concern. With a global population of 3.3 billion (and still growing) and its related activities, our climate is more vulnerable to change. Anthropogenic activities such as industrialization, urbanization, and agriculture have increased global temperature by 1 °C since the pre-industrial era and are likely to increase by 1.5 °C by 2050 (Simoni 2022). The significant relationship between increased population and global atmospheric temperature is associated with the increased emission of greenhouse gases (GHG) into the atmosphere,

T. T. Meetei (✉) · N. S. Singh
College of Agriculture, Department of Soil Science, Rani Lakshmi Bai Central Agricultural University, Jhansi, Uttar Pradesh, India

Department of Soil Science and Agricultural Chemistry, Lovely Professional University, Jalandhar, Punjab, India

Y. B. Devi
College of Horticulture and Forestry, Natural Resource Management, Rani Lakshmi Bai Central Agricultural University, Jhansi, India

K. Bokado
Department of Agronomy, Lovely Professional University, Jalandhar, Punjab, India



S. R. Mohanty, B. Kollah (eds.), *Greenhouse Gas Regulating Microorganisms in Soil Ecosystems*, https://doi.org/10.1007/978-3-031-70569-4_12

which exacerbates the greenhouse effect, making the atmospheric temperature so high that it results in global climate change.

The primary producers of GHG in the atmosphere, induced by the rapid increase in human population, are rapid industrialization, which includes transportation, heat and electricity production, manufacturing and construction, and industrial processes (Mebarki et al. 2020). In the past five years, the world has experienced various natural disasters related to climate change and global warming, affecting millions of people and causing economic crises.

The major greenhouse gases (GHG) associated with the greenhouse effect include carbon dioxide ($CO_2$), methane ($CH_4$), nitrous oxide ($N_2O$), and fluorinated gases. $CO_2$ is the most common GHG, emitted primarily through the extraction and combustion of fossil fuels (Li et al. 2021). It is also naturally produced during respiration and other oxidation processes. $CO_2$ is the largest contributor to the greenhouse effect in the Earth's atmosphere, but it is less potent than other GHG (Ritchie et al. 2020). Even though $CO_2$ is not a particularly potent GHG, it can have a strong greenhouse effect and contribute to global warming when its concentration exceeds certain critical thresholds. Since the eighteenth century (the beginning of the industrial era), human activities have increased the concentration of $CO_2$ in the atmosphere by 50%, which is equivalent to 150% of its value in 1750 (Skrable et al. 2022). This increase in atmospheric $CO_2$ concentrations has led to an increase in the average global temperature by 1.15 °C in 2022 (Allabakash and Lim 2022).

Consequently, the years 2015–2022 are likely to be the warmest years on record. The second most abundant GHG after $CO_2$ is methane ($CH_4$), which accounts for about 16–25% of global warming (Rosentreter et al. 2021). Although $CH_4$ is less abundant than $CO_2$, it has a 25 times greater potential to cause global warming (Zhao et al. 2023). At least 30% of global warming associated with human activity is driven by $CH_4$ emissions. $CH_4$ can have both a direct and indirect impact on the greenhouse effect that causes global warming. The direct effect is its ability to trap radiated sunlight from the Earth's surface, while the indirect effect includes the production of $CO_2$ and water vapor, which are produced when emitted $CH_4$ is oxidized in the atmosphere. Global $CH_4$ emissions in 2021 were 15% higher than in 1984–2006 (Hamidu et al. 2022). Various industrial activities emit $CH_4$, and it is also the end result of the anaerobic breakdown of organic materials or any substances that contain carbon (Harindintwali et al. 2022). Therefore, any human-induced activities that promote anaerobic conditions for a long period of time can be a factor responsible for its emission. Examples include hydromorphic soils, landfills, swamps, marshes, paddy cultivation, cattle, earthworms, and some other ruminant animals (Ezenwa et al. 2020).

In addition to $CO_2$ and $CH_4$, nitrous oxide ($N_2O$), commonly known as laughing gas, is a human-related GHG released from the terrestrial ecosystem and a hazard to the current global warming. Anthropogenic activities, such as fuel combustion, industrial processes, unscientific waste management, and agriculture, emit significant levels of $N_2O$ into the atmosphere (Wang et al. 2021). Naturally, $N_2O$ is also released into the atmosphere through the Earth's nitrogen cycle, an important biogeochemical cycle for maintaining the concentration of nitrogen (N) in both soil and the atmosphere (Aryal et al. 2022). However, human activities have disrupted the Earth's nitrogen cycle, polluting both soil and the atmosphere.

Since the start of the Green Revolution, excessive amounts of inorganic fertilizers have been used to feed high-yield crops. Additionally, due to the increase in large-scale farming with the use of both organic and inorganic fertilizers, N input to the soil has greatly intensified, disrupting the natural N cycle. Additionally, faulty land use management practices or mismanagement of agricultural practices can lead to the release of more N from the soil to the atmosphere in the form of $N_2O$. Once released, $N_2O$ is difficult to convert back to other forms of less toxic substances because of its ability to stay in the atmosphere for an average of 114 years (Mehmood et al. 2022). Therefore, the release of $N_2O$ from the soil poses a serious threat, as it has a 300-fold greater potential to cause global warming than $CO_2$ (Hinger et al. 2019). Fluorinated gases are produced by various industries and are used as substitutes for ozone-depleting substances, such as refrigerants, fire retardants, and aerosol propellants. There are four main categories of fluorinated gases: hydrofluorocarbons (HFCs), perfluorocarbons (PFCs), sulfur hexafluoride (SF6), and nitrogen trifluoride (NF3) (Dunse et al. 2019). Fluorinated gases have a very high global warming potential relative to other greenhouse gases, and they can remain in the atmosphere for thousands of years. Therefore, even a small amount of fluorinated gases can lead to significant global warming.

## GHG Emission from Agriculture Soil

GHG can come from both natural and anthropogenic sources. Natural sources include organic matter decomposition, respiration, volcanic eruptions, and many biogeochemical cycles taking place on earth (Walling and Vaneeckhaute 2020). On the other hand, the factors associated with anthropogenic activities are related to the global increase in human population that leads to uncontrolled discharge of GHG. It predominantly includes the energy-related activities, i.e., burning of fossil fuels for electricity and transportation (Lamb et al. 2021). Another important source of GHG emitted related to human intervention is from agriculture. The contribution of GHG through agriculture is also a mostly discussed topic in the recent era for its emission (Tubiello et al. 2021). There is a direct and indirect impact of agriculture on the current global climate change as a result of the rise in food demand for the rapidly rising worldwide human population. The direct effect can include the expansion of agriculture holding where land use changes takes place. The conversion of natural forest and grassland to agriculture land disrupts the carbon (C) cycle, affecting the C sink to the soil (Yu and Song 2023). The highest C reservoir is soil, which holds about 1500 Pg of carbon. Therefore, during the expansion of agricultural holdings with an aim to increase food production, different natural landforms are destroyed, encouraging the emission of GHG mainly from the terrestrial ecosystem. Furthermore, deforestation for land use change leads to less C sequestration, making a larger source of C emission from the soil in the form of $CO_2$, exacerbating the present global warming. Indirectly, due to mismanagement of the agricultural practices, especially highly potent GHG like methane ($CH_4$) and nitrous oxides ($N_2O$) are emitted. Therefore, the major GHG released from agricultural soils are primarily $CO_2$, $CH_4$, and $N_2O$.

Soil is timing with different organisms (both micro and macro) that make soil a living component. All the biogeochemical cycles taking place in nature are directly or indirectly related to the activities of these organisms (Basu et al. 2021). Principally, soil microorganisms play an important role in organic matter decomposition, nutrient transformation, and nutrient cycling processes. During these processes, different by-products are released that could be beneficial or toxic to the environment. Carbon (C) is the most critical factor for the growth of microorganisms that depends on the organic matter present in soil. The required C for the microbial growth is assimilated through soil organic matter decomposition, which can be classified into two types, viz., aerobic and anaerobic (Meena et al. 2021). Depending upon the types of decomposition, the end product also differs. $CO_2$ is almost invariably the main greenhouse gas emitted during the aerobic breakdown of organic matter, whereas $CH_4$ is released during the anaerobic process (Lin et al. 2019). Different groups of microorganisms are associated during the decomposition processes through the secretion of specific enzymes that catalyzed the reaction. Another GHG emitted mostly from agriculture is $N_2O$, and its emission is also indirectly related to the organic matter decomposition. The emission of $N_2O$ from soil is a part of the nitrogen (N) cycle that is essential for maintaining the N balance in soil and atmosphere (Aryal et al. 2022). However, due to excess substrate availability and faulty land use management practices, more $N_2O$ is emitted from the soil, causing the atmosphere to become more concentrated. $N_2O$, also known as laughing gas, is a highly potent GHG, which has more potential (300 times) than $CO_2$ in imparting a greenhouse effect, resulting in more global warming. The process of denitrification involves the release of $N_2O$ from the soil (Aryal et al. 2022). The organism that causes the loss of nitrogen during denitrification is classified as a facultative anaerobe. This particular set of microbes used the nitrate ($NO_3^-$) form of nitrogen when anaerobic conditions were present, which led to the generation of $N_2O$, a powerful greenhouse gas (Aryal et al. 2022). Therefore, it is evident from the fact that the $CO_2$ and $CH_4$ emitted from the soil are mainly due to the decomposition of organic matter mediated by soil microorganisms, either through aerobic or anaerobic. As a result, when we talk about the GHG emission from the terrestrial ecosystem, it is crucial to consider the role of soil microbes in GHG emission. The microbial-mediated emission of these greenhouse gases increases when ecosystems are disturbed. Agriculture, which is the practice of cultivating soil, growing plants, and raising livestock, is one such activity (Laborde et al. 2021). Agriculture alone could produce about 30% of the world's GHG emissions in the modern period (Panchasara et al. 2021). Naturally, the main drivers of GHG from soil include microbial activities, floral and faunal respiration, and aerobic and anaerobic decompositions of organic matter (Oertel et al. 2016). However, due to human intervention and activities altering soil moisture, structure, temperature, cropping patterns, fertilizer, and other poorly managed agricultural practices, the emission rate increases. Increased population and rapid expansion of agricultural holdings and practices have disrupted the physical, chemical, and biological characteristics of soil. Utilization of chemical fertilizers, insecticides, and animal manure has increased, primarily following the green revolution, stimulating the emission brought on by increased substrate availability for nutrient transformation or decomposition. Furthermore, the problem is made worse by intensive agricultural methods using traditional tillage operations.

## Greenhouse Gas Producing Microbes in Soil

The likelihood that any GHG will trap heat energy and cause global warming depends on three important criteria: gas concentration in the atmosphere, residence time in the atmosphere, and global warming potential (GWP). GWP is the potential of a gas to trap heat relative to carbon dioxide ($CO_2$) over a specific time period. Microorganisms emit three major GHG that contribute to the greenhouse effect: $CO_2$, methane ($CH_4$), and nitrous oxide ($N_2O$) (Gupta et al. 2021). $CO_2$ is the most common GHG in the atmosphere, accounting for about 75% of human-induced GHG emissions. It has a long residence time, meaning it can stay in the atmosphere for hundreds of years (Yoro and Daramola 2020). $CH_4$ has a higher GWP than $CO_2$ (25 times), but a shorter residence time (12 years). It accounts for about 16% of human-induced GHG emissions. $N_2O$ is the most potent GHG, with a GWP 300 times greater than $CO_2$ and a residence time of about 120 years. It accounts for about 6% of human-induced GHG emissions.

## Emission of Carbon Dioxide ($CO_2$)

Carbon dioxide ($CO_2$) is the most important greenhouse gas (GHG) because it absorbs and radiates heat, increasing the global temperature. Although $CO_2$ is not as potent as other GHG such as nitrous oxide ($N_2O$) and methane ($CH_4$), it is the largest contributor to present-day global warming because it is so common and accounts for about 60% of the total greenhouse effect (Yoro and Daramola 2020). The main cause of $CO_2$ emissions into the atmosphere is human activity, such as the exploitation and burning of fossil fuels and the production of various industrial by-products. Different aerobic organisms (macro and micro) residing under different ecosystems respire to produce energy that involves the exchange of gases, especially $O_2$ and $CO_2$. During this process, the movement of $O_2$ from the outside environment to the cells within the tissue and the removal of $CO_2$ from the cell to the outside environment take place. Numerous macro- and microorganisms that respire in the soil and release $CO_2$ into the environment are a major source of greenhouse gases (GHG). The respiration rate increases with increasing temperature that releases more $CO_2$ that could impact the present global warming (Sushko et al. 2019). GHG emitted due to the ever-emerging human-induced activities are expected to increase the atmospheric temperature by more than 1.5–2 °C. An increase in atmospheric temperature could also lead to an increase in soil temperature that promotes microbial activities and increased $CO_2$ emission through respiration and organic matter decomposition. The amount of C emitted in the form of $CO_2$ depends on the quantity and types of organic residue, climatic condition, primarily temperature, and moisture (Ray et al. 2020). Different microorganisms present in soil decomposed organic material returned to the soil facilitated by temperature and moisture, emitting a significant amount of $CO_2$ that could impact the present global climate change (Sushko et al. 2019). Due to the huge amount of organic matter (OM) present in the soil and its potential to be sensitive to temperature change, permafrost zone soil may emit significant amounts of gases as a result of rising temperatures (Wu et al. 2019). Organic material has been accumulating for thousands of years beneath permafrost due to the region's low temperature. Because of the repercussions of rising atmospheric temperature, even a small increase in global temperature causes increased degradation of the reserved OM by microbes impacting the global climate change.

## Emission of Nitrous Oxide ($N_2O$)

Nitrous oxide ($N_2O$) is one of the potent GHG which are mainly emitted from soil, organic manure applications, residue burning, the ocean, and industries. However, more than two-thirds of its emission to the atmosphere is from the microbial-mediated processes of nitrification and denitrification, whereas one-third from the ocean (Wrage-Mönnig et al. 2018). $N_2O$ can remain in the atmosphere for around 120 years without converting into another form. On average, 60% of the emitted $N_2O$ to the atmosphere is from the soil ecosystem alone due to the application of inorganic nitrogenous fertilizers and organic manures (Bhattacharyya et al. 2012; Kumar et al. 2019). And this will greatly rise as a result of human-induced activities linked to agricultural intensification for the expanding population, which are expected to continue to rise in the near future and have a considerable impact on global warming. It has been reported that for every 1000 kg of nitrogenous fertilizers applied to the soil, 10–15 kg of N will be lost in the form of $N_2O$ (Wrage et al. 2001). Therefore, it is now important to understand the key pathway of $N_2O$ emission from soil mediated by different microbes present in soil. Literally the key pathway includes nitrification, denitrification, nitrifier denitrification, anaerobic ammonia oxidation (anammox) and nitrification coupled denitrification. All these processes are the part of N cycle which is facilitated by microorganisms. Recently, algae and plants are also considered another source of $N_2O$ emission which was initially missed out creating certain uncertainties in the global budget.

Nitrification is a microbial-mediated, two-step natural process that oxidizes ammonia ($NH_3$) to nitrite ($NO_2^-$) and eventually to nitrate ($NO_3^-$) by some specific oxidizing bacteria and archaea. For a long time, it was believed that the nitrification process is mediated only by oxidizing bacteria, but now the role of oxidizing archaea is also explicated (Soler-Jofra et al. 2021). Archeaea are the distinct group of organisms that can survive in very harsh conditions such as high temperature, high salt concentration, and under oxygen stress conditions. Most studied microorganisms responsible for nitrification are bacteria, and they belong to Chemolithoautotropic microorganisms that use the nitrification process as an energy source for carbon dioxide ($CO_2$) fixing. The most common genera of bacteria responsible for the first steps of nitrification are *Nitrosomonas* and *Nitrosococcus,* whereas *Nitrobacter* and *Nitrospina* are responsible for the second steps. Heterotrophic nitrifiers are organisms that utilized organic C as a source of energy, and C is also known for its potential for the nitrification process; however, it is poorly known, and its contribution to soil is still under investigation (Martikainen 2022). Compared to bacteria, fungi are more likely to engage in heterotrophic nitrification. The process of nitrification is the most crucial stage of the N cycle, which changes the reduced form of N to the oxidized form, subsequently producing substrate for denitrification and the anaerobic ammonia oxidation (anammox) process that is linked to the production of $N_2O$. Additionally, during the incomplete oxidation of an intermediate product of nitrification (hydroxylamine, $NH_2OH$), $N_2O$ is released as a by-product (Liu et al. 2019).

Denitrification is another microbial-mediated process that undergoes the reduction of $NO_2^-$and $NO_3^-$ formed during the nitrification process to $N_2$ as a response to lowering the partial pressure of oxygen ($O_2$) in soil (Yang et al. 2020). From terrestrial to aquatic ecosystems, temperate to tropical regions, freshwater to marine water, surface water to aquifers, and wastewater to organic matter treatment, denitrification is a ubiquitous phenomenon in nature. It is a significant natural occurrence that helps to lessen the hazardous effects of $NO_2^-$ and $NO_3^-$ that could damage soil and freshwater bodies (Busico et al. 2020). However, excess denitrification can lead to loss of soil fertility (N is an essential nutrient required by plants) and emit $N_2O$, a GHG that elevates global warming. Denitrification process is solely a microbial-mediated process that a large range of microorganisms are responsible. Important microorganisms that are involved in the denitrification pathway include heterotrophic denitrifications, nitrifier denitrification, and chemo-denitrification. Most denitrifiers are heterotrophs and facultative aerobes that are able to utilize $NO_3^-$ in exchange for $O_2$ as an electron acceptor when the $O_2$ concentration decreases (Yang et al. 2020). Heterotrophic denitrifiers are responsible for the sequential reduction of the substratea $NO_3^-$ to $NO_2^-$, nitric oxide (NO), $N_2O$, and dinitrogen gas ($N_2$). The major denitrifiers in various settings have been identified as belonging to the genera *Pseudomonas, Ralstonia, Alcaligenes, Paracoccus, Rhodobacter, Rubrivivax, Thauera, Burkholderia, Bacillus*, Thiobacillus, Propionibacterium, and Streptomyces, while *Shewanella baltica* and Marinobacter species were the predominant marine denitrifiers.

Chemo-denitrification is a non-biological, non-enzymatic chemical decomposition of intermediates produced during the nitrification process that are facilitated by the presence of organic (amine) and inorganic ($Fe^{2+}$) compounds (Chen et al. 2020) produced by heterotrophic $Fe^{3+}$reducing microorganisms. The majority of studies neglected chemodenitrification in favor of focusing primarily on nitrification and denitrification for $N_2O$ emission. However, chemodenitrification, which contributes between 31 and 75 percent of all $N_2O$ emissions in agricultural soil, is an important factor (Wang et al. 2019). This reaction occurs most frequently in soil that is inherently acidic and rich in organic materials. In submerged soil, due to the rapid chemical reaction between $NO_2^-$ and $Fe^{2+}$ and the buildup of $NO_2^-$ during microbial nitrate reduction, chemodenitrification and the resulting abiotic $N_2O$ emission are more likely to occur (Otte et al. 2019). The main product of chemo-denitrification is nitric oxide (NO); however, significant productions of $N_2O$ are also reported by the nitrate respire *Serratiamarcescens* (Anderson and Levine 1986). The most common chemodenitrification taking place in nature is the conversion of nitrous acid ($HNO_2$) to NO that is facilitated by low soil pH. Any pH value of soil greater than five could lead to an insignificant chemodinitrification process (Skiba 2008).

Nitrifier denitrification is a pathway carried out by autotropic ammonia and methane-oxidizing bacteria under $O_2$ stress conditions that leads to the incomplete nitrification process (stopped at $NO_2^-$formation) followed by the reduction of $NO_2^-$ via NO and $N_2O$ to $N_2$ (Thakur and Medhi 2019). In this pathway, not all the $NO_2^-$ produced are reduced into $N_2$ but also to $N_2O$, which is a very potent GHG. The pathway can contribute $N_2O$ to the atmosphere as much as it can contribute by heterotrophic denitrification in the terrestrial and aquatic ecosystems. This pathway should not be confused with the coupled nitrification-dentrification process that occurred both under aerobic and anaerobic conditions carried out by ammonia oxidizer and heterotrophic denitrifiers, respectively. This pathway is also sometimes called aerobic denitrification, facilitated under $O_2$ stress conditions with high $NO_2^-$concentrations. The ammonia-oxidizing bacteria *Nitrosomonas*, *Nitrosospira* spp.,and the methane oxidizer *Methylococcus capsulatus* are known for the nitrifier denitrification process.

## Emission of Methane ($CH_4$)

$CH_4$ is the second most important GHG, having 300 times more GWP than the $CO_2$ with an average residence time of 120 years (Mehmood et al. 2022). Emissions of $CH_4$ into the atmosphere can come from both manmade and natural sources (Etiope and Klusman 2002). Natural emissions include decomposition of plant materials in wetlands, seepage of gas from undergrounds, volcanoes, wildfires, and termites released, contributing 40% of the total emission, whereas anthropogenic activities such as agriculture, landfilling, mining, combustion of fossil fuels, and certain industrial processes, contributing 60% of the total emission. Methanogenesis, which is one of the final stages of anaerobic breakdown of organic molecules in an ecosystem mediated by strictly anaerobic microorganisms, produces $CH_4$ (Liu et al. 2022). The commonly studied methanogenic microorganisms belong to methanogenic archaea or simply methanogens. The methanogenesis process takes place in strictly anaerobic conditions with the availability of substrate for the process to proceed. Therefore, depending upon the availability and reduction of the substrate for $CH_4$ production, methanogenesis can be of different types. Acetoclastic methanogenesis, methylotrophic methanogenesis, and hydrogenotrophic methanogenesis (Chen et al. 2022). Acetoclastic methanogenesis refers to the process of methanogenesis in which methanogens use acetate as a substrate, as opposed to methylotrophicmethanogemesis, which uses methanol or methylamines. When $H_2$ and $CO_2$ are the primary substrates utilized by methanogens, it is called hydrogenotrophic methanogens. Different microbiomes are responsible for the production of substrate that is responsible for methanogenesis under strict anaerobic conditions. Contrary to popular belief, mounting research indicates that methane is paradoxically produced by some microbes in oxygen-saturated aquatic and terrestrial habitats, including cyanobacteria, algae, fungi, purple non-sulfur bacteria, and cryptogamic cover (Liu et al. 2022). Therefore, prolonged anaerobic condition due to any human-induced activities could lead to the emission of $CH_4$ into the atmosphere that impacts the recent global climate change.

Ruminant livestock also contributes a significant amount of $CH_4$ to the environment, a source from the livestock sector, which accounts for about 81% of GHG emitted from this sector (Tapio et al. 2017). The production is a result of the rumen microbial methanogenesis. Rumen in the fermentation chamber of ruminant, in which different microbiomes exist to decompose the cellulose in grasses and animal feeds. There are a ton of anaerobic bacteria, fungus, ciliate protozoa, and archaea inside the rumen. The anaerobic bacteria contribute the largest microbial biomass, followed by protozoa (30–40%), fungi (8–12%), and archaea (0.3–4%) (Tapio et al. 2017; Liang et al. 2020). *Fibrobactersuccinogenes*, *Ruminococcusalbus, and Ruminococcusflavefaciens* are the most common cellulose degraders found in rumen, although many other cellulose degraders are found that are highly efficient in hemicellulose degradation, such as members from the genera *Prevotella, Butyrivibrio, and Pseudobutyrivibrio* (Sha et al. 2020). Ciliated protozoa are responsible for the digestion overall fiber intake that contributes about 30–40%. In high-grain diets, the Entodinium genus is the most prevalent protozoan (Nagaraja and Titgemeyer 2007). Despite being less common than bacteria, fungi are among the greatest organisms for breaking down carbohydrates. The anaerobic fungi found in rumen fall into the class *Neocallimastigomycetes*, which includes six genera, namely Anaeromyces, Caecomyces, Cyllamyces, Neocallimastix, Orpinomyces, and Piromyces with 21 known species (Joshi et al. 2022). Rumen archaea are strictly anaerobic and responsible for the production of $CH_4$ by utilizing the substrate produced by other microorganisms. It has been reported that around 90% of rumen methanogens come from the following genera: Methanobrevibacter (63.2%), Methanomicrobium (7.7%), Methanosphaera (9.8%), Thermoplasma (7.4%), and Methanobacterium (1.2%) (Matthews et al. 2019). These microbiomes produced $CO_2$ and $H_2$, a substrate utilized by methanogenic archaea to convert it into $CH_4$.

Similarly, termites are thought to be responsible for 1–3% of the $CH_4$ emissions worldwide that are responsible for the current global climate change (Nauer et al. 2018). Frequently in tropical and subtropical regions, termites consumed large quantities of plant materials. During the decomposition of the lignocellulose part of the plant materials in the termite guts, the symbiotic methanogen archaea produced $CH_4$ through the process of methanogenesis. Likewise, some species of earthworm can also produce $CH_4$ which is mediated by high microbial diversity in the earthworm gut (Yasmin et al. 2022). The low concentration of molecular $O_2$ and higher quantity of lignocellulose material from the ingestion of plant residue leads to the production of certain GHG that include $CH_4$.

## Diversity of Greenhouse Gas Mitigating Microbes in Soil

The significant role of microorganisms in emitting GHG from the terrestrial and marine ecosystems in global climate change has been elaborated in the sections above. However, microorganisms not only emit the most potent GHG, but fortunately, they also consume them when conditions favor. Microbes are earth-original and most adaptable inhabitants that can evolve very quickly. In our earth's ecosystems, only 10% of the total microbial diversity is explored, while remaining 90% is unexplored and is residing and evolving with the change in climate since the earth's creation. Therefore, a variety of microbiomes may be able to utilize the powerful GHG in the environment to mitigate the current global climate change (Hu et al. 2017). Earth is a closed system, and so all the nutrient elements and energy have to recycle from one tropic level to another, and microbes ensure that all are recycled on our planet. Cyanobacteria, over a period of two billion years, changed our atmosphere through photosynthesis; they turn $CO_2$ back into $O_2$ and sugar. Microbiomes are so diverse that even if some species die out due to climate change, others might survive and take their place, allowing nature's recycling to continue. Another way microbiomes cope with climate change is to build healthy soil. Many soil microbiomes secrete some chemicals that can cement the soil particles, maintaining soil aggregate stability that can resist soil erosion and hold more water and nutrients for plant growth. Therefore, understanding the microbial biodiversity that can absorb and reduce the concentration from the atmosphere is important to mitigate the present global climate change.

Plants, cyanobacteria, and different other autotrophic microorganisms harness energy from the sun to fix or transform $CO_2$ into C-containing plant biomass, thereby emitting $O_2$. This process maintained $CO_2$ balanced in the atmosphere, thereby maintaining atmospheric temperature. Phytoplankton, an autotrophic microscopic microorganism, is thought to be responsible for the maximum emission of $O_2$ (50–80%) on earth through photosynthesis (Naselli-Flores and Padisák 2022). They are extremely diverse in nature and include photosynthesizing bacteria to algae, and its important groups include non-motile, fast-growing diatoms, motile flagellates, dinoflagellates, cyanobacteria, green algae, and chalk-coated coccolithophores. Due do activities related to humans, such as deforestation, mining of reserved C (coal, fossil fuels, etc.), and its combustion for energy and transportation, excessive use of chemical fertilizers and pesticides, faulty land management practices, and intensive agricultural practices, break the balance of the C cycle, thereby breaking the $CO_2$ source and sink ratio and making the atmospheric temperature to increased. However, the overall impact of temperature increase on phytoplankton is favorable since it will lead to higher growth rates of phytoplankton with higher $CO_2$ absorption rates.

The primary contributor of GHG emission to the atmosphere belongs to the energy sector, followed by transportation, manufacturing, and agriculture. The energy sector is the largest contributor of GHG (25% of the total emission), mainly due to the utilization of fossil fuels for generating energy, thereby emitting a significant quantity that has an impact on the global climate change (Zheng et al. 2019). Therefore, to reduce the emission of GHG from this sector, it is essential to curb the consumption of fossil fuels. Nevertheless, the desire for energy is a vital one that all people must have on a daily basis from morning till night. With the rate at which humanity is expanding worldwide, the need for energy will likewise rise. Finding a replacement for fossil fuels that has a C-neutral or C-negative impact on the atmosphere is therefore crucial at this time. C-neutral fuels are fuels that do not produce net GHG, whereas C-negative are those that remove more C from the atmosphere than emissions. Fuels derived from plants and microbes are part of the climate solution because they don't rely on fossil C sources like coal and petroleum products. The most appropriate examples of C-neutral or C-negative fuels are biofuels that are produced within a short period of time by using microorganisms on organic materials (Dey and Bhaskarwar 2021). Biofuels are considered C-neutral because the C emitted during combustion equals the C sequestered, i.e., they neither contribute nor emit C to the atmosphere. Additionally, biofuels are renewable, environmentally friendly, sustainable, and biodegradable. The most common types of biofuels used as an alternative to fossil fuels are biomethane, bioethanol, and biodiesel (Azeez and Al-Zuhairi 2020). The organisms responsible for the production of biofuels from the organic waste generated from agriculture are presented in Table 1. GHG emissions are reduced when biofuels are used instead of fossil fuels. Utilizing biomethane results in a reduction in GHG emissions (Buchspies et al. 2020), whereas bioethanol could reduce the GHG emission by as much as 86% (Iram et al. 2022). When biodiesel is used in place of fossil fuels carbon monoxide (CO) emissions

**Table 1** The diversity of biofuels producing microbes

| Sl No | Microbes | Biofuels |
|---|---|---|
| 1 | *Serratia* | Biodiesel |
| 2 | *Ralstoniaeutropha* | Biodiesel |
| 3 | *Ralstoniaeutropha* H16 | Bioethanol |
| 4 | *Methanothermobacter* | Biogas |
| 5 | *Clostridium tyrobutyricum* | 1-Butanol |
| 6 | *Caldicellulosiruptorsaccharolyticus* | Hydrogen |
| 7 | *Saccharomyces cerevisiae* | Bioethanol |
| 8 | *Zymomonasmobilis* | Bioethanol |
| 9 | *Streptococcus fragilis* | Bioethanol |
| 10 | *Kluyveromycesfragilis* | Bioethanol |
| 11 | *Clostridium thermocellum* | Bioethanol |
| 12 | *Clostridium acetobutylicum* | Acetone |

decreased by 10%, and hydrocarbon emissions and particulate exhausts decreased by 20% and 10%, respectively (Loganathan et al. 2020).

Methane removal (converted to less harmful products) from the environment is another important practice to consider. It is possible to use both biological and non-biological ways to remove $CH_4$ from the environment. However, non-biological contributes more than the biological because biological methods are confined only to the soil. The non-biological method of methane removal takes place in the lowest layers of earth's atmosphere (troposphere and stratosphere) through an oxidation reaction. The oxidation of $CH_4$ in the troposphere layer is the largest $CH_4$ sink from the environment, contributing around 506 Mt. of $CH_4$ per year (Broucek 2014), followed by the stratosphere removing 40 Mt. from the total $CH_4$ burden (Elghandour et al. 2019). The oxidation of $CH_4$ by the hydroxyl radical (OH) both in the troposphere and stratosphere is the major mechanism responsible for conversion into carbon monoxide (CO), nitrogen dioxide ($NO_2$), and hydroperoxide ($HO_2$), which can then be removed from the atmosphere. The biological method for $CH_4$ removal from the environment is done through soil methanotrophic bacteria that can oxidize $CH_4$ by a process known as methanotrophy, which is a high affinity oxidation process (Jiang et al. 2010; Salem et al. 2021). This method is only confined to the soil since the methanotroph cannot break down $CH_4$ from the atmosphere. Methanotrophs are gram-negative bacteria that require $CH_4$ as a source of C and energy for their survival. They can grow and utilize $CH_4$ both under aerobic and anaerobic conditions. The methanotroph occurs mostly in soil where $CH_4$ is available, such as natural wetlands, landfills, paddy fields, and mud. $CH_4$ produced mostly in these sites gets trapped between the soil particles. These trapped $CH_4$ are actually oxidized by methanotrophic microorganisms to harness energy under aerobic and anaerobic conditions to ultimately generate carbon dioxide ($CO_2$) and water. Methanotrophic microorganisms can consume around 30 Mt. of $CH_4$ per year that can reduce the impact of $CH_4$, especially, on terrestrial ecosystem to the present global climate change (Salem et al. 2021).

Another potent GHG that has 300 times more potential in trapping heat energy than $CO_2$ on a per-molecule basis is $N_2O$. It is also a potent ozone depleting substance in the stratosphere. Approximately 65% of the $N_2O$ emitted to the atmosphere is expected to be contributed from the soil ecosystem (Mengmeng et al. 2021). The emission of $N_2O$ from the soil ecosystem is mainly from the application of nitrogenous fertilizers (62%), manures, (31%) and biomass burning (7%) (Hu et al. 2015). Application of excess nitrogenous fertilizers coupled with faulty land use practices leads to the maximum production of $N_2O$. It is estimated that for every 100 kg of nitrogenous applied to soil, around 1–5 kg of nitrogen will be lost in the form of $N_2O$, and there is an exponential increase in emission relative to the nitrogenous fertilizer application (Shcherbak et al. 2014). Considering the increasing trend of the global human population and its food security, it is expected that agriculture soil will contribute up to 59% of total $N_2O$ emissions in 2030 (Domeignoz-Horta et al. 2018). The emission of $N_2O$ from the soil ecosystem discussed in the previous section is solely mediated by soil microorganisms and is an intermediate product of the denitrification process. Different enzymes secreted by microbes are responsible for the denitrification process. However, the emission of $N_2O$ during the process is due to more activity of the first three denitrification enzymes—nitrate reductase, nitrite reductase, and nitric oxide reductase—than nitrous oxide reductase (Brown et al. 2000). Nitrous oxide reductase is an important enzyme responsible for the reduction of $N_2O$ to dinitrogen ($N_2$). However, some abiotic variables, especially pH, $O_2$ level, nitrate load, and C availability in soil, defined the ability of denitrifying bacteria to reduce $N_2O$ to $N_2$. Very few mitigation methods involving microbes are reported till now. But few inoculums, such as *Azoarcus, Niastella, Burkholderia* spp., *Azospirillum, Herbaspirillum,* and *Bradyrhizobium diazoefficiens*, along with fertilizers or manures, reduce $N_2O$ from the soil, as reported by Usyskin-Tonne et al. (2019). (Usyskin-Tonne et al. 2019). Therefore, the conversion/removal of environment $N_2O$ to the most stable form ($N_2$) primarily depends on maintaining abiotic factors and including inoculums responsible for reducing $N_2O$ or secreting more nitrous oxide reductase enzyme. The principal factor responsible for $N_2O$ emission is soil $O_2$ level that decreases with an increase in $O_2$ level. Another important factor is soil pH, which has a negative relation with the emission of $N_2O$. Nitrate load is another important factor that brings the denitrification process emitting more $N_2O$ from soil. $NO_3$ load in soil is directly related to the nutrient application in soil, either in organic or inorganic form. C availability in soil is also another abiotic factor that brings the reactions towards $N_2O$ emission. With an increase in C released from the C sources, the denitrification process increases, thereby increasing $N_2O$ emissions. Therefore, by keeping both abiotic and biotic factors responsible for $N_2O$ emission checked, the impact of $N_2O$ can be reduced.

## Conclusion

Climate change associated with the progressive increase of GHG in the atmosphere due to human-induced activities is well documented. Emissions of these GHG are directly related to the activities of different organisms present in soil. Naturally, the GHG released to the atmosphere are a part of the natural biogeochemical cycle that checks the balance between the terrestrial ecosystem and the atmosphere. However, with human intervention, the biogeochemical cycles are disrupting, breaking the balance, thereby making more emission than absorption. Soil is teeming with different organisms and is by far the most biologically diverse part of earth. Most of the important organisms are yet to explore. Exploring organisms that can utilize the potent GHG is a possible option to mitigate the present climate change because different groups of organisms can use these gases for their metabolic pathways. Therefore, more study on the organisms, especially organisms absorbing or utilizing these GHG, is vital to maintaining the equilibrium between the terrestrial ecosystem and the atmosphere. When there is more absorption than the emission, less will be the impact of the recent climate change.

## References

Allabakash S, Lim S (2022) Anthropogenic influence of temperature changes across East Asia using CMIP6 simulations. Sci Rep 12(1):1–14

Anderson IC, Levine JS (1986) Relative rates of nitric oxide and nitrous oxide production by nitrifiers, denitrifiers, and nitrate respirers. Appl Environ Microbiol 51(5):938–945

Aryal B, Gurung R, Camargo AF, Fongaro G, Treichel H, Mainali B et al (2022) Nitrous oxide emission in altered nitrogen cycle and implications for climate change. Environ Pollut:120272

Azeez RA, Al-Zuhairi FK (2020) Biofuels (Bioethanol, Biodiesel, and Biogas) from lignocellulosic biomass: a review. J Univ Babylon Eng Sci:202–215

Basu S, Kumar G, Chhabra S, Prasad R (2021) Role of soil microbes in biogeochemical cycle for enhancing soil fertility. In: New and future developments in microbial biotechnology and bioengineering. Elsevier, pp 149–157

Bhattacharyya P, Roy KS, Neogi S, Adhya TK, Rao KS, Manna MC (2012) Effects of rice straw and nitrogen fertilization on greenhouse gas emissions and carbon storage in tropical flooded soil planted with rice. Soil Tillage Res 124:119–130

Broucek J (2014) Production of methane emissions from ruminant husbandry: a review. J Environ Prot 5(15):1482

Brown K, Tegoni M, Prudêncio M, Pereira AS, Besson S, Moura JJ et al (2000) A novel type of catalytic copper cluster in nitrous oxide reductase. Nat Struct Biol 7(3):191–195

Buchspies B, Kaltschmitt M, Junginger M (2020) Straw utilization for biofuel production: a consequential assessment of greenhouse gas emissions from bioethanol and biomethane provision with a focus on the time dependency of emissions. GCB Bioenergy 12(10):789–805

Busico G, Kazakis N, Colombani N, Khosravi K, Voudouris K, Mastrocicco M (2020) The importance of incorporating denitrification in the assessment of groundwater vulnerability. Appl Sci 10(7):2328

Chen D, Yuan X, Zhao W, Luo X, Li F, Liu T (2020) Chemodenitrification by Fe (II) and nitrite: pH effect, mineralization and kinetic modeling. Chem Geol 541:119586

Chen Y, Wu N, Liu C, Mi T, Li J, He X et al (2022) Methanogenesis pathways of methanogens and their responses to substrates and temperature in sediments from the South Yellow Sea. Sci Total Environ 815:152645

Dey N, Bhaskarwar AN (2021) Bioconversion of food waste into ethanol: a review. In: Sustainable bioconversion of waste to value added products (pp 45–58)

Domeignoz-Horta LA, Philippot L, Peyrard C, Bru D, Breuil MC, Bizouard F et al (2018) Peaks of in situ $N_2O$ emissions are influenced by $N_2O$-producing and reducing microbial communities across arable soils. Glob Chang Biol 24(1):360–370

Dunse BL, Derek N, Fraser PJ, Krummel PB, Steele LP (2019) Australian and global HFC, PFC, sulfur hexafluoride nitrogen trifluoride and sulfuryl fluoride emissions, Report prepared for Australian Government Department of the Environment and Energy

Elghandour MM, Adegbeye MJ, Barbabosa-Pilego A, Perez NR, Hernández SR, Zaragoza-Bastida A, Salem AZ (2019) Equine contribution in methane emission and its mitigation strategies. J Equine Vet Sci 72:56–63

Etiope G, Klusman RW (2002) Geologic emissions of methane to the atmosphere. Chemosphere 49(8):777–789

Ezenwa VO, Civitello DJ, Barton BT, Becker DJ, Brenn-White M, Classen AT et al (2020) Infectious diseases, livestock, and climate: a vicious cycle? Trends Ecol Evol 35(11):959–962

Gupta K, Kumar R, Baruah KK, Hazarika S, Karmakar S, Bordoloi N (2021) Greenhouse gas emission from rice fields: a review from Indian context. Environ Sci Pollut Res 28(24):30551–30572

Hamidu I, Afotey B, Ayatul-Lahi Z (2022) Design and development of a low-cost sensor iot computing device for greenhouse gas momitor from selected industry locations. Scalab Comput 23(4):363–376

Harindintwali JD, Xiang L, Wang F, Chang SX, Zhao Z, Mei Z et al (2022) Syntrophy of bacteria and archaea in the anaerobic catabolism of hydrocarbon contaminants. Crit Rev Environ Sci Technol:1–27

Hinger I, Pelikan C, Mußmann M (2019, January) Role of the ubiquitous bacterial family Woeseiaceae for $N_2O$ production in marine sediments. In: Geophysical Research Abstracts (vol 21)

Hu HW, Chen D, He JZ (2015) Microbial regulation of terrestrial nitrous oxide formation: understanding the biological pathways for prediction of emission rates. FEMS Microbiol Rev 39(5):729–749

Hu HW, Trivedi P, He JZ, Singh BK (2017) Microbial nitrous oxide emissions in dryland ecosystems: mechanisms, microbiome and mitigation. Environ Microbiol 19(12):4808–4828

Iram A, Cekmecelioglu D, Demirci A (2022) Integrating 1G with 2G bioethanol production by using Distillers' Dried Grains with Solubles (DDGS) as the feedstock for lignocellulolytic enzyme production. Fermentation 8(12):705

Jiang H, Chen Y, Jiang P, Zhang C, Smith TJ, Murrell JC, Xing XH (2010) Methanotrophs: multifunctional bacteria with promising applications in environmental bioengineering. Biochem Eng J 49(3):277–288

Joshi A, Young D, Huang L, Mosberger L, Munk B, Vinzelj J et al (2022) Effect of growth media on the diversity of Neocallimastigomycetes from non-Rumen habitats. Microorganisms 10(10):1972

Kumar R, Kumar R, Prakash O (2019) Chapter-5 the impact of chemical fertilizers on our environment and ecosystem. Chief Ed 35:69

Laborde D, Mamun A, Martin W, Piñeiro V, Vos R (2021) Agricultural subsidies and global greenhouse gas emissions. Nat Commun 12(1):2601

Lamb WF, Wiedmann T, Pongratz J, Andrew R, Crippa M, Olivier JG et al (2021) A review of trends and drivers of greenhouse gas emissions by sector from 1990 to 2018. Environ Res Lett 16(7):073005

Li Y, Shang J, Zhang C, Zhang W, Niu L, Wang L, Zhang H (2021) The role of freshwater eutrophication in greenhouse gas emissions: a review. Sci Total Environ 768:144582

Liang J, Nabi M, Zhang P, Zhang G, Cai Y, Wang Q et al (2020) Promising biological conversion of lignocellulosic biomass to renewable energy with rumen microorganisms: a comprehensive review. Renew Sust Energ Rev 134:110335

Lin L, Xu F, Ge X, Li Y (2019) Biological treatment of organic materials for energy and nutrients production—Anaerobic digestion and composting. In: Advances in bioenergy, vol 4. Elsevier, pp 121–181

Liu S, Schloter M, Hu R, Vereecken H, Brüggemann N (2019) Hydroxylamine contributes more to abiotic $N_2O$ production in soils than nitrite. Front Environ Sci 7:47

Liu LY, Xie GJ, Ding J, Liu BF, Xing DF, Ren NQ, Wang Q (2022) Microbial methane emissions from the non-methanogenesis processes: a critical review. Sci Total Environ 806:151362

Loganathan M, Thanigaivelan V, Madhavan VM, Anbarasu A, Velmurugan A (2020) The synergetic effect between hydrogen addition and EGR on cashew nut shell liquid biofuel-diesel operated engine. Fuel 266:117004

Martikainen PJ (2022) Heterotrophic nitrification–an eternal mystery in the nitrogen cycle. Soil Biol Biochem 168:108611

Matthews C, Crispie F, Lewis E, Reid M, O'Toole PW, Cotter PD (2019) The rumen microbiome: a crucial consideration when optimising milk and meat production and nitrogen utilisation efficiency. Gut Microbes 10(2):115–132

Mebarki B, Draoui B, Allaoua B, Draoui A (2020) Road transportation industry facing the energy and climate challenges. In: Renewable energy-resources, challenges and applications. IntechOpen

Meena AL, Karwal M, Kj R, Narwal E (2021) Aerobic composting versus Anaerobic composting: comparison and differences. Food Sci Rep 2:23–26

Mehmood T, Hassan MA, Li X, Ashraf A, Rehman S, Bilal M et al (2022) Mechanism behind sources and sinks of major anthropogenic Greenhouse gases. In: Climate change alleviation for sustainable progression. CRC Press, pp 114–150

Mengmeng JI, Hao TIAN, Xiaogang WU, Ji LI, Ying ZHU, Guojun WU et al (2021) Enhanced $N_2O$ emission rate in field soil undergoing conventional intensive fertilization is attributed to the shifts of denitrifying guilds. Pedosphere 31(1):145–156

Nagaraja TG, Titgemeyer EC (2007) Ruminal acidosis in beef cattle: the current microbiological and nutritional outlook. J Dairy Sci 90:E17–E38

Naselli-Flores L, Padisák J (2022) Ecosystem services provided by marine and freshwater phytoplankton. Hydrobiologia:1–16

Nauer PA, Hutley LB, Arndt SK (2018) Termite mounds mitigate half of termite methane emissions. Proc Natl Acad Sci 115(52):13306–13311

Oertel C, Matschullat J, Zurba K, Zimmermann F, Erasmi S (2016) Greenhouse gas emissions from soils—A review. Geochemistry 76(3):327–352

Otte JM, Blackwell N, Ruser R, Kappler A, Kleindienst S, Schmidt C (2019) $N_2O$ formation by nitrite-induced (chemo) denitrification in coastal marine sediment. Sci Rep 9(1):10691

Panchasara H, Samrat NH, Islam N (2021) Greenhouse gas emissions trends and mitigation measures in australian agriculture sector—a review. Agriculture 11(2):85

Ray RL, Griffin RW, Fares A, Elhassan A, Awal R, Woldesenbet S, Risch E (2020) Soil $CO_2$ emission in response to organic amendments, temperature, and rainfall. Sci Rep 10(1):5849

Ritchie H, Roser M, Rosado P (2020) $CO_2$ and greenhouse gas emissions. Our world in data

Rosentreter JA, Borges AV, Deemer BR, Holgerson MA, Liu S, Song C et al (2021) Half of global methane emissions come from highly variable aquatic ecosystem sources. Nat Geosci 14(4):225–230

Salem R, ElDyasti A, Audette GF (2021) Biomedical applications of biomolecules isolated from methanotrophic bacteria in wastewater treatment systems. Biomol Ther 11(8):1217

Sha Y, Hu J, Shi B, Dingkao R, Wang J, Li S et al (2020) Characteristics and functions of the rumen microbial community of cattle-yak at different ages. BioMed Res Int 2020

Sharma GD, Shah MI, Shahzad U, Jain M, Chopra R (2021) Exploring the nexus between agriculture and greenhouse gas emissions in BIMSTEC region: The role of renewable energy and human capital as moderators. J Environ Manag 297:113316

Shcherbak I, Millar N, Robertson GP (2014) Globalmetaanalysis of the nonlinear response of soil nitrous oxide ($N_2O$) emissions to fertilizer nitrogen. Proc Natl Acad Sci 111(25):9199–9204

Simoni M (2022) Towards more sustainable cement production: a no-combustion approach for the decarbonisation of CaCO3 (Doctoral dissertation, University of Sheffield)

Skiba U (2008) Denitrification. Ecological Processes Centre for Hydrology and Ecology, Penicuik

Skrable K, Chabot G, French C (2022) World atmospheric $CO_2$, its 14C specific activity, non-fossil component, anthropogenic fossil component, and emissions (1750–2018). Health Phys 122(2):291–305

Soler-Jofra A, Pérez J, van Loosdrecht MC (2021) Hydroxylamine and the nitrogen cycle: a review. Water Res 190:116723

Sushko S, Ananyeva N, Ivashchenko K, Vasenev V, Kudeyarov V (2019) Soil CO 2 emission, microbial biomass, and microbial respiration of woody and grassy areas in Moscow (Russia). J Soils Sediments 19:3217–3225

Tapio I, Snelling TJ, Strozzi F, Wallace RJ (2017) The ruminalmicrobiome associated with methane emissions from ruminant livestock. J Anim Sci Biotechnol 8(1):1–11

Thakur IS, Medhi K (2019) Nitrification and denitrification processes for mitigation of nitrous oxide from waste water treatment plants for biovalorization: challenges and opportunities. Bioresour Technol 282:502–513

Tubiello FN, Rosenzweig C, Conchedda G, Karl K, Gütschow J, Xueyao P et al (2021) Greenhouse gas emissions from food systems: building the evidence base. Environ Res Lett 16(6):065007

Usyskin-Tonne A, Hadar Y, Minz D (2019) Altering $N_2O$ emissions by manipulating wheat root bacterial community. Sci Rep 9(1):1–9

Walling E, Vaneeckhaute C (2020) Greenhouse gas emissions from inorganic and organic fertilizer production and use: A review of emission factors and their variability. J Environ Manag 276:111211
Wang M, Hu R, Ruser R, Schmidt C, Kappler A (2019) Role of chemodenitrification for $N_2O$ emissions from nitrate reduction in rice paddy soils. ACS Earth Space Chem 4(1):122–132
Wang C, Amon B, Schulz K, Mehdi B (2021) Factors that influence nitrous oxide emissions from agricultural soils as well as their representation in simulation models: a review. Agronomy 11(4):770
Wrage N, Velthof GL, Van Beusichem ML, Oenema O (2001) Role of nitrifierdenitrification in the production of nitrous oxide. Soil Biol Biochem 33(12–13):1723–1732
Wrage-Mönnig N, Horn MA, Well R, Müller C, Velthof G, Oenema O (2018) The role of nitrifierdenitrification in the production of nitrous oxide revisited. Soil Biol Biochem 123:A3–A16
Wu X, Zang S, Ma D, Ren J, Chen Q, Dong X (2019) Emissions of $CO_2$, $CH_4$, and $N_2O$ fluxes from forest soil in permafrost region of Daxing'an Mountains, northeast China. Int J Environ Res Public Health 16(16):2999
Yang J, Feng L, Pi S, Cui D, Ma F, Zhao HP, Li A (2020) A critical review of aerobic denitrification: insights into the intracellular electron transfer. Sci Total Environ 731:139080
Yasmin N, Jamuda M, Panda AK, Samal K, Nayak JK (2022) Emission of greenhouse gases (GHGs) during composting and vermicomposting: measurement, mitigation, and perspectives. Energy Nexus:100092
Yoro KO, Daramola MO (2020) $CO_2$ emission sources, greenhouse gases, and the global warming effect. In: Advances in carbon capture. Woodhead Publishing, pp 3–28
Yu H, Song W (2023) Research progress on the impact of land use change on soil carbon sequestration. Land 12(1):213
Zhao X, Chen T, Wang Y, Li K, Zhan R, Lin H (2023) Effect of electric field and water on Pd-based catalyst for complete oxidation of methane: study on promotion and deactivation. Chem Eng J 457:141126
Zheng X, Streimikiene D, Balezentis T, Mardani A, Cavallaro F, Liao H (2019) A review of greenhouse gas emission profiles, dynamics, and climate change mitigation efforts across the key climate change players. J Clean Prod 234:1113–1133

# Strategies to Improve N Use Efficiency to Minimize $N_2O$ Emission from Agricultural Soil

**Y. K. Jhala, H. K. Patel, H. N. Shelat, R. V. Vyas, and D. G. Panpatte**

## Introduction

Agriculture plays a crucial role in ensuring food security and sustainable development. Plants consume carbon dioxide and nitrogen from the air and soil, respectively, and redistribute them among various pools, such as above- and below-ground living biomass, dead residues, and soil organic matter. Within the agriculture sector, carbon dioxide, methane, and nitrous oxide are released as part of biogeochemical cycles through plant respiration, the decomposition of plant and animal biomass, and combustion. Human activities related to land use, such as cropland management, livestock rearing, crop cultivation, manuring, and biomass burning, are the main sources of greenhouse gas emissions in the agriculture sector. Among these gases, nitrous oxide ($N_2O$) is particularly important due to its high global warming potential, which is 298 times that of carbon dioxide. It has an atmospheric life span of 114 years and contributes approximately 6.24% to the overall global radiative forcing. The global atmospheric concentration of nitrous oxide has been increasing, with 30% of it being emitted by anthropogenic activities, mainly related to agriculture. Approximately 60% of nitrous oxide of anthropogenic origin comes from the use of nitrogen fertilizer and manure in crop cultivation. This is because about half of the applied nitrogenous fertilizers are lost from the ecosystem through accumulation in the soil, leaching, and emissions to the atmosphere. Typically, plants absorb nitrogen in the form of nitrate ($NO_3^-$) or ammonium ($NH_4^+$) ions. Ammonium ions are adsorbed onto the soil and undergo a nitrification process, while nitrate can

Y. K. Jhala (✉) · H. K. Patel · H. N. Shelat · R. V. Vyas
Department of Agricultural Microbiology, B. A. College of Agriculture, Anand Agricultural University, Anand, Gujarat, India

D. G. Panpatte
Department of Plant Pathology, College of Agriculture, Dongarshelki Tanda, Udgir, Latur, Maharashtra, India

 
S. R. Mohanty, B. Kollah (eds.), *Greenhouse Gas Regulating Microorganisms in Soil Ecosystems*, https://doi.org/10.1007/978-3-031-70569-4_13

easily be lost from the system through leaching. The application of nitrogenous fertilizers results in the leaching of nitrate into groundwater and the emission of nitrous oxide into the atmosphere, which poses a significant threat to livelihoods (Timilsena et al. 2014). The degradation of nitrous oxide in the stratosphere through photocatalysis leads to the formation of NO, which contributes to the destruction of ozone molecules (Crutzen 1981). Reducing nitrous oxide emissions from fertilized soil is challenging due to the complex processes involved, which can vary depending on climatic conditions and management practices. Furthermore, other forms of nitrogen, such as nitrate, ammonia, and nitric oxide, contribute to N2O emissions once they reach terrestrial and aquatic ecosystems. Therefore, a thorough understanding of the complex microbial processes involved in nitrous oxide emission from crop cultivation systems is necessary for designing effective mitigation strategies.

## How Nitrous Oxide Gets Emitted from Soil?

Nitrous oxide is produced through microbial transformations of nitrogen in soils and water, specifically through nitrification and denitrification processes (Butterbach-Bahl et al. 2013). Autotrophic aerobic nitrification, carried out by ammonia-oxidizing bacteria and nitrite-oxidizing bacteria, and anaerobic denitrification, mediated by denitrifying bacteria, are the main microbial processes responsible for nitrous oxide formation. During the nitrification process, chemoautotrophic bacteria rapidly oxidize ammonium to nitrate when there is an abundant supply of ammonical nitrogen from fertilizers, manures, and crop residues. A small amount of nitrogen is released as $N_2O$ during this process (Butterbach-Bahl et al. 2013). On the other hand, denitrification occurs in wet soils with sufficient carbon and nitrogen, where nitrate ions are reduced through microbial respiration (Zhu et al. 2019). As a result of denitrification, soil oxygen levels decrease, and microbes utilize nitrate as a terminal electron acceptor. Although most of the nitrate is reduced to $N_2$, $N_2O$ is produced as an intermediate in these processes. Soil microbes play a crucial role in nitrification and denitrification processes, accounting for up to 70% of global $N_2O$ emissions (Syakila and Kroeze 2011; Braker and Conrad 2011). Microbes possess a diverse metabolic potential, leading to various processes that either produce or consume nitrous oxide. Additionally, certain abiotic processes, such as the chemical decomposition of nitrite (chemidenitrification) and the oxidation of hydroxylamine ($NH_2OH$), contribute to nitrous oxide formation (Bremner 1997). Chemidenitrification typically occurs in acidic or neutral soil, where $NO_2$ volatilization and fixation occur in soil organic matter (Bremner et al. 1980; Bremner 1997). Hydroxylamine is an intermediate compound in the oxidation of $NH_4^+$ to $NO_3^-$ and can produce N2O (Bremner 1997).

Nitrous oxide emission from agricultural soils is governed by following processes (Fig. 1):

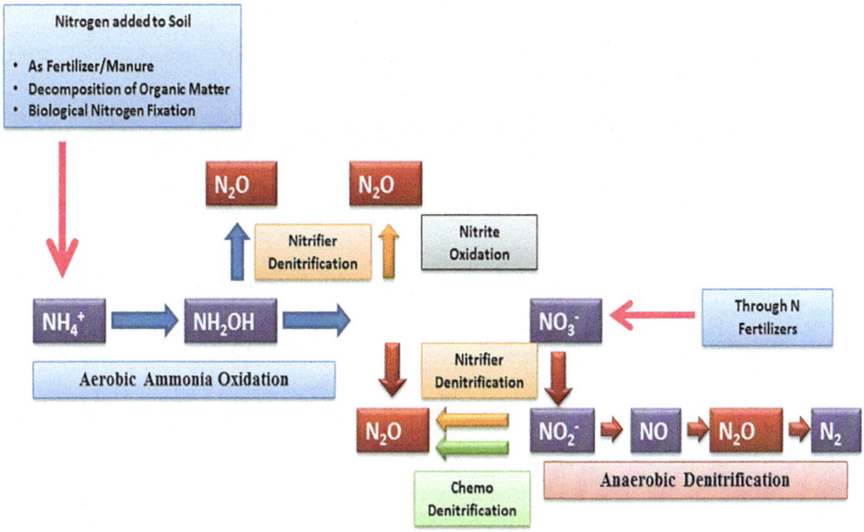


**Fig. 1** Processes that produce $N_2O$ in soil

- Chemical decomposition of hydroxylamine during autotrophic and heterotrophic nitrification
- Chemodenitrification of soil nitrite and abiotic decomposition of ammonium nitrate in the presence of light, humidity, and reacting surfaces
- Nitrifier-denitrification within the same nitrifying microorganism
- Coupled nitrification–denitrification by distinct microorganisms (production of nitrate by nitrite oxidizers, which is immediately denitrified in situ by denitrifiers)
- Denitrification conducted by organisms capable of using nitrogen oxides as alternative electron acceptors under O2-limiting environmental conditions
- Co-denitrification of organic N compounds with NO
- Nitrate ammonification or dissimilatory nitrate reduction to ammonium

## Why Controlling Nitrous Oxide Emission Is a Tough Task?

None of the biological systems are entirely efficient, resulting in the leakage of nitrogen into the atmosphere in the form of nitrous oxide ($N_2O$). Nevertheless, the proportion of nitrogen cycling through agricultural systems that escapes into the atmosphere as $N_2O$ can be minimized through effective nutrient management (Ribaudo et al. 2011; Snyder et al. 2009). Nitrogen use efficiency worldwide rarely exceeds 50% and is even lower in developing countries. The need to feed a global population of 7 billion has increased the demand for synthetic N fertilizers in agriculture, posing a greater threat of increasing N2O emissions (Fig. 2).

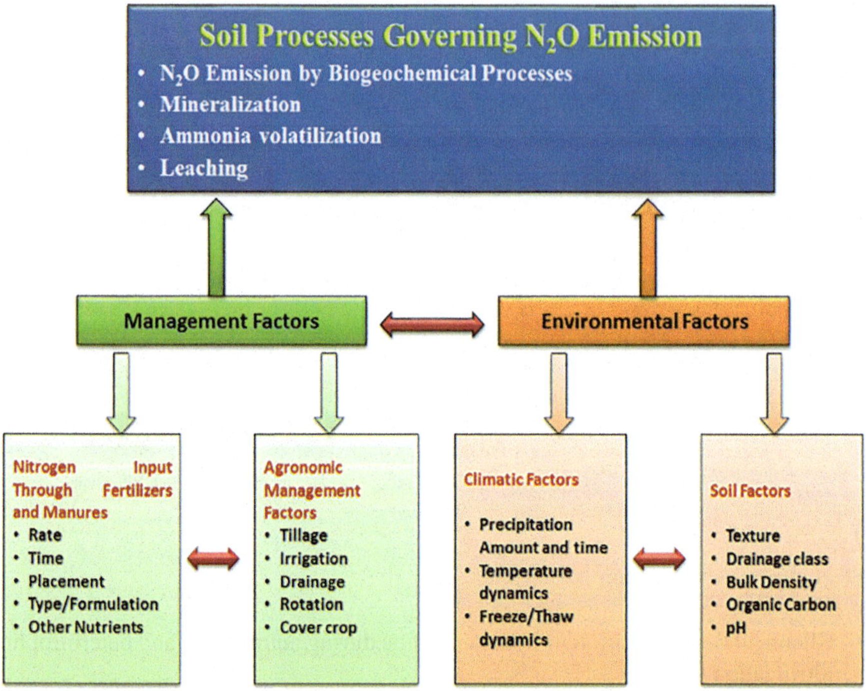


**Fig. 2** Factors influencing $N_2O$ emission

The primary challenge in reducing nitrogen losses from agroecosystems lies in optimizing crop N-use efficiency. In this regard, the design of strategies for mitigating nitrous oxide emissions faces numerous challenges due to the unique physical and biological factors associated with this gas (Galloway et al. 2004), such as:

- Practices in nitrogen management that result in increased accumulation of fertilizer N in crop biomass lead to the highest $N_2O$ emissions (Fujinuma et al. 2011; Gagnon et al. 2011). This is due to the fact that most of the nitrogen losses occur in the form of ammonia ($NH_3$) volatilization and/or nitrate ($NO_3^-$) leaching, which account for approximately 10% and 30% of applied nitrogen, respectively, compared to N2O losses of less than 3% (De Klein et al. 2006). Therefore, increasing nitrogen use efficiency by limiting $NH_3$ and/or $NO_3^-$ losses may result in an increase in the amount of available nitrogen for uptake and nitrous oxide production. Consequently, while improving nitrogen use efficiency, it is crucial to consider the key factors and soil processes that contribute to nitrous oxide formation; otherwise, increasing nitrogen use efficiency may only be a theoretical concept.
- The most effective approach to reducing nitrous oxide emissions is to decrease the rate of nitrogenous fertilizer application. However, limiting the rate of N fertilization is directly associated with yield losses and can impact farmers'

profits. Reducing the N fertilizer rate is only feasible when the current rate exceeds the economically optimal rate for a given cropping system (Millar et al. 2010). Fine-tuning N fertilizer application based on economic yield and considering weather parameters is essential. Therefore, improving N fertilizer application strategies by selecting better sources, adjusting the timing of application, and optimizing placement appears to be an appealing option. Despite the high operational costs associated with altering N fertilizer application strategies, it is important to note that adopting these improved strategies would result in an overall decrease in nitrogen fertilizer rates, offsetting the operational costs while reducing environmental risks.

- Four soil processes govern nitrous oxide production: nitrification, nitrifier-denitrification, chemo-denitrification, and denitrification. When we apply N to the soil, various processes contribute to the production of $N_2O$, and various management and environmental factors stimulate these processes. The occurrence of these soil processes can vary under a wide range of soil conditions, making it difficult to optimize specific soil conditions to reduce $N_2O$ production.
- Reducing nitrous oxide emissions through smaller and more frequent applications of nitrogenous fertilizer, as well as utilizing carbon-rich manures instead of nitrogen-rich ones, can help lower inorganic nitrogen levels and limit emissions. Implementing strategies such as conducting tillage under cool conditions may also be effective in reducing nitrous oxide emissions, but the feasibility of these approaches on a large scale depends on factors such as soil type, nutrient distribution, and local rainfall and irrigation patterns.

## Measures to Reduce Nitrous Oxide Emission from Crop Cultivation

The primary method known to decrease nitrous oxide concentrations in the atmosphere is by reducing its emission, as natural sinks for nitrous oxide are not readily available. Effective control of nitrous oxide emissions can be achieved by managing inorganic nitrogen concentrations in soils, particularly in conditions of high temperature and moisture where denitrification or nitrification processes occur rapidly.

### *Management of Nitrogen Fertilizer Application*

#### Application Rates

The unregulated use of nitrogenous fertilizer presents a significant challenge in achieving high nitrogen use efficiency (Ju et al. 2009; Peiling and Fucang 2012; Zhang et al. 2021). It is imperative to adjust agronomic practices to enhance nitrogen use efficiency while maintaining optimal fertilizer application rates to ensure crop yield (Luo et al. 2018). The appropriate level of fertilizer application is contingent upon soil properties and agroclimatic factors (Schumann 2000). A reduction in nitrogen application rates may lead to delayed leaf senescence, ultimately resulting in increased yield. Excessive application of nitrogenous fertilizers beyond crop requirements can lead to elevated nitrous oxide emissions (Schils et al. 2008). Typically, the recommended doses of nitrogenous fertilizers for crops are determined based on the economic optimum rate, beyond which additional fertilizer application may not yield economic benefits. These recommended rates are generally below the biological optimum requirement necessary for achieving maximum yield. It is suggested that determining the recommended fertilizer dose should account for the biological optimum as well as the potential nitrous oxide emissions and nitrate leaching per unit of grain production (Hoben et al. 2010). A uniform reduction in fertilizer application can effectively mitigate nitrous oxide emissions (MacLeod et al. 2010) while also reducing cultivation costs, thereby benefiting farmers and the overall agroecosystem. When applying nitrogen to the soil, it is essential to consider the crop's nitrogen requirements, existing soil nitrogen levels, and soil conditions. Recommendations also advise that nitrogen from synthetic fertilizers and manure should only be applied when soil nitrogen supply is insufficient to meet crop needs. The emission reductions achieved through this approach are anticipated to be modest.

The majority of crops primarily derive their nitrogen supply from the mineralization of soil organic matter, which is further supplemented by the application of inorganic fertilizers. Typically, the distribution and availability of nitrogen throughout a field are not taken into account when planning fertilizer applications. Site-specific nitrogen management aims to precisely meet the nutrient requirements of crops. This can be achieved through techniques such as grid soil sampling, assessment of residual soil nitrate nitrogen values, utilization of nitrogen availability maps, and creation of nitrogen fertilizer recommendation maps. Monitoring the nitrogen status of crops can be accomplished through methods like canopy reflectance of light and leaf color chart analysis. Modern advancements in remote sensing technology enable high-resolution monitoring of crop growth and development. Satellite-based monitoring of canopy reflectance allows for accurate determination of nitrogen concentrations in crop tissues, facilitating the generation of fertilizer recommendations tailored to specific areas within a field. Factors such as poor soil conditions, including compaction, low soil organic matter levels, or limited soil depth, can restrict nitrogen uptake by crops. In such cases, excessive nitrogen application may lead to increased nitrogen losses without a corresponding enhancement in crop nitrogen recovery. Remote sensing and sensor-based monitoring systems enable the detection of variations in soil nitrogen supply across fields, thereby supporting decision-making processes to adjust nitrogen application rates accordingly. Precision nitrogen management strategies have the potential to reduce $N_2O$ emissions. The leaf color chart represents a cost-effective tool for real-time or crop-need-based nitrogen management. This method offers a simple and economical approach to monitoring nitrogen deficiency by assessing leaf nitrogen content based on chlorophyll levels at different growth stages, serving as an alternative to more sophisticated tools like chlorophyll meters or SPAD meters.

## Application Time

The timing of nitrogen fertilizer application is a critical factor influencing nitrogen use efficiency, with various factors such as crop growth stage, water availability, and environmental conditions affecting this efficiency. Applying nitrogenous fertilizers during key growth stages can lead to higher nitrogen accumulation in crops, reducing inorganic nitrogen accumulation in soil and minimizing nitrous oxide emissions without compromising yield. Adjusting application timing based on soil and environmental conditions, such as avoiding warm and moist conditions to reduce $N_2O$ emissions, can be beneficial. Nitrous oxide emissions are influenced by microbial processes, soil moisture, and temperature post-fertilizer application. Berry et al. (2010) suggested that applying fertilizers when soil temperatures are low may decrease $N_2O$ emissions, but this approach must be carefully managed to prevent increased $NO_3^-$ leaching with significant rainfall. While targeting application during dry soil conditions can reduce emissions, it may limit crop nitrogen uptake. Farmers face challenges in determining optimal fertilizer application timing based on environmental factors like temperature and moisture, highlighting the importance of field drainage systems and accurate weather forecasting to aid in decision-making. Proper timing of nitrogen fertilizer application has the potential to significantly reduce nitrous oxide emissions, although the response may vary depending on nitrogen uptake. Studies have shown that under anaerobic conditions, nitrous oxide emissions occur due to incomplete denitrification, which can be exacerbated by factors such as rising water tables and waterlogging caused by natural or anthropogenic compaction and heavy rainfall.

## Application Method

A significant variability exists in soil fertility under field conditions, which greatly influences crop yield and nutrient uptake. This issue can be addressed through the targeted application of nitrogen fertilizers based on specific site conditions. Tailoring nitrogen fertilization to specific sites has the potential to improve yield and quality and ultimately increase nitrogen use efficiency. Yong et al. (2018) conducted a study testing various nitrogen fertilizer concentrations applied between rows of maize-soybean relay intercropped fields at different distances (15 cm, 30 cm, and 45 cm), finding that the 15 cm and 30 cm treatments resulted in better nitrogen use efficiency and yield. Therefore, optimal results were achieved with lower nitrogen application at 15–30 cm from the fertilizer application location to the maize row.

In a separate study, it was noted that the productivity of lowland rice is heavily influenced by the selection of varieties and their capacity to utilize nutrients effectively. Insufficient nitrogen fertilizer application may occur, particularly when nitrogen is immobilized after fertilizing ratoon crops in unburned sites (Rossetto et al. 2010).

Drip fertigation, a method that combines water and fertilizer application in precise quantities, has been recognized as a beneficial technique for crop production (Neelam and Rajput 2011; Ayyadurai and Manickasundaram 2014). Research indicates that utilizing drip fertigation with mulching can lead to improved crop yields (Dai and Dong 2014). Studies have shown that this method enhances nutrient and water use efficiency, resulting in increased crop yield per unit of nutrient and water applied (Neelam and Rajput 2011; Bar-Yosef 1999). Additionally, drip fertigation allows for targeted application of fertilizers to the crop's root zone, leading to enhanced fertilizer use efficiency without compromising crop yield (Ayyadurai and Manickasundaram 2014; Yan et al. 2015; Ma et al. 2016). Furthermore, employing drip fertigation with high plant populations has been found to conserve nitrogen and sustain crop yields in arid climates (Dong et al. 2010, 2012; Mao 2013; Anas et al. 2020).

The most effective method for reducing nitrogen loss through urea fertilizer application is deep point placement, with the exception of soils with high percolation rates. By placing nitrogenous fertilizers at a depth of 5–10 cm in anaerobic soil layers, the concentration of nitrogen in flood water and the surface oxidized layer can be limited, resulting in decreased nitrogen losses. Alternatively, foliar application of urea or ammonium nitrate solutions is a less vulnerable method to surface runoff, microbial immobilization, volatilization, and denitrification prior to plant absorption. The rationale for utilizing foliar fertilizers includes situations where: (i) soil conditions restrict the availability of soil-applied nutrients; (ii) high rates of nutrient loss from soil applications are anticipated; (iii) plant growth stage, internal plant nutrient demand, and environmental conditions collectively impede the delivery of nutrients to critical plant organs (Fernández et al. 2013).

## Types of Fertilizers

Stabilized fertilizers contain nitrogen stabilizers that prolong the time required for nitrogen to become available in the soil. These fertilizers, which include nitrification and urease inhibitors, directly inhibit microbial processes. Nitrification inhibitors consist of compounds that hinder the initial enzyme involved in nitrification, specifically ammonia monooxygenase, thereby impeding the conversion of $NH_4^+$ to $NH_2^-$ by nitrifying bacteria (Trenkel 2010; Akiyama et al. 2009; Ruser and Schulz 2015). By slowing down the nitrification process, plants have more time to absorb $NO_3^-$, leading to increased nitrogen-use efficiency. Consequently, the nitrate concentration in the soil decreases, reducing nitrate leaching and nitrous oxide emissions through denitrification (Trenkel 2010). Therefore, it can be inferred that nitrification inhibitors indirectly decrease nitrous oxide emissions by inhibiting the denitrification process, a major source of nitrous oxide emissions in agricultural ecosystems (Ruser and Schulz 2015). Nitrification inhibitors can be used in conjunction with both chemical and organic fertilizers. Three commonly utilized nitrification inhibitors are N-(n-butyl) thiophosphorictriamide (NBPT), dicyandiamide (DCD), and 3,4-dimethylpyrazole phosphate (DMPP). Among these, NBPT is the most prevalent due to its similar solubility and diffusivity as urea, making it highly effective. However, NBPT's efficiency is limited in tropical conditions as it degrades rapidly under high temperatures and humidity (Trenkel 2010; Martins et al. 2017). Urease inhibitors inhibit the hydrolytic activity of urease, thereby slowing down urea hydrolysis and reducing $NH_3$ emissions into the atmosphere (Halvorson et al. 2014). Dicyandiamide (DCD) and 3,4-dimethylpyrazole phosphate (DMPP) are the most commonly available nitrification inhibitors. Despite efforts to reduce nitrogen losses from autumn-applied slurries using chemicals like DCD, their cost-effectiveness in terms of enhancing yields and reducing nitrate leaching losses has been questioned by the farming community (Chambers et al. 2006). The persistence of nitrification inhibitors in soils is crucial for their effectiveness in reducing $N_2O$ emissions. Research indicates that the majority of N2O emissions occur within the first 4–6 weeks after the application of manufactured fertilizer N and organic manures (Thorman et al. 2007). Therefore, nitrification inhibitors should be applied either concurrently with or shortly after fertilizer/manure application and should remain active in the soil for at least two months to effectively reduce $N_2O$ emissions. The extent of emission reductions achieved through nitrification inhibitor application is expected to be substantial, although the effectiveness may vary depending on the adoption of the practice.

One alternative approach is the utilization of controlled-release fertilizers, which offer a consistent release of nutrients (Trenkel 2010). These fertilizers are designed with a coating that allows for a gradual release of nutrients into the soil, thereby enhancing crop yields and reducing environmental contamination. The process of creating controlled-release fertilizers involves coating fertilizer molecules with substances such as sulfur, polymers, or superabsorbent materials for water retention (Timilsena et al. 2014; Trenkel 2010; Azeem et al. 2014; Li et al. 2016a; Naz and Sulaiman 2016; Yamamoto et al. 2016). For instance, nitrogen release in coated fertilizers is facilitated through a semi-permeable coating that enables diffusion (Halvorson et al. 2014). This diffusion process involves water penetrating the coating, condensing on the solid fertilizer core, and gradually dissolving the nutrients. As a result of increased osmotic pressure within the granules, nitrogen is released slowly based on concentration gradients (Azeem et al. 2014). The majority of research in this area has focused on coated urea, which demonstrates enhanced nitrogen release with rising temperatures and aligns nutrient release closely with plant nitrogen requirements. By modifying coating properties, the release of nutrients can be controlled. Controlled-release urea, for example, reduces nitrogen availability for nitrification and denitrification processes in the soil, thereby limiting nitrous oxide emissions (Halvorson et al. 2014; Azeem et al. 2014; Naz and Sulaiman 2016). The use of controlled and slow-release fertilizers ensures a steady supply of nutrients throughout the growing season, promoting high nitrogen use efficiency. Environmental advantages of these fertilizers include reductions in nitrate leaching, ammonia volatilization, and nitrous oxide emissions from microbial activities. However, it is important to note that controlled and slow-release fertilizers are costly, and there is a lack of concrete research findings to confirm their effectiveness in mitigating nitrous oxide emissions.

## Organic Amendments

Organic manures such as slurries, farmyard manure, and poultry manures are generated within agricultural settings and subsequently applied to farmland. Additional sources of manures encompass treated sewage sludges, composts, digestates, and industrial residues like paper crumble and by-products from the food industry. The utilization of organic manures stands out as a highly sustainable practice for managing agricultural waste effectively, supplying essential nutrients to crops, and preserving soil fertility. The judicious application of the nitrogen provided by organic manures, coupled with adjustments in manufactured fertilizer rates, plays a pivotal role in reducing diffuse nitrogen pollution from agricultural systems (Chambers et al. 2006). Challenges associated with manures include variations in nutrient composition, difficulties in precise application, and the potential for nitrogen loss post-application, which can impede farmers' ability to accurately quantify the nitrogen supplied through organic manure applications. Strategies aimed at nitrogen supply through manures to mitigate nitrous oxide emissions should consider the nutrient content of manure and the reduction of nutrient losses post-land spreading. While animal manure and plant residue applications have historically been the conventional means of fertilizing agricultural lands, the prevalent use of chemical fertilizers has surged in recent times due to heightened food demand. Nevertheless, owing to the adverse environmental repercussions and relatively higher costs associated with chemical fertilizers compared to organic manures, as well as their limited availability in various regions globally, the adoption of organic manures is steadily on the rise (Thangarajan et al. 2013). The adverse environmental impacts of organic manures, such as nutrient leaching and greenhouse gas emissions, are notably less severe in comparison to those of chemical fertilizers.

Animal manures, such as those derived from livestock (such as cattle, pigs, or sheep) and poultry, are widely utilized organic amendments that are experiencing a notable increase in production levels due to the global rise in meat production. These manures are rich in essential nutrients like nitrogen (N), phosphorus (P), and potassium (K), which play a crucial role in enhancing soil fertility and overall health (Thangarajan et al. 2013). Composts, which can be produced from crop residues or manure, serve as a valuable carbon (C) source and contribute to improving soil structure and preventing erosion. However, most composts have relatively low nutrient content, with nitrogen levels typically ranging from 1% to 2% (in contrast to urea, which contains 46% nitrogen), necessitating additional supplementation with other fertilizers or higher application rates. Furthermore, the composting process often results in significant losses of carbon and nitrogen through greenhouse gas (GHG) emissions (Dambreville et al. 2006; Thangarajan et al. 2013).

Plant residues, encompassing components such as stalks, stems, leaves, and seed pods, can serve as a source of nitrogen for the soil. These residues are predominantly derived from cereals or legumes (e.g., red clover, winter vetch, and ryegrass). Factors such as the method of residue application (e.g., surface application or soil incorporation), soil moisture levels, and residue quality can influence GHG emissions, potentially leading to increased denitrification post-incorporation. Residues with high nitrogen content and low carbon-to-nitrogen ratios, such as legumes, may release nitrogen more rapidly, resulting in elevated nitrous oxide ($N_2O$) emissions. Conversely, higher carbon-to-nitrogen ratios could mitigate $N_2O$ emissions by retaining nitrogen. The application of organic amendments directly impacts $N_2O$ emissions through nitrogen transformation processes, with factors like soil pH alteration and carbon-to-nitrogen ratios playing significant roles. While a high carbon-to-nitrogen ratio promotes soil organic matter accumulation, a low ratio may elevate $N_2O$ emissions from soils (Thangarajan et al. 2013; Li et al. 2016b; Shan and Yan 2013; Huang et al. 2004).

One potential strategy to mitigate $N_2O$ emissions resulting from organic amendments involves combining them with slow-release or controlled-release fertilizers. When paired with nitrification inhibitors, the application of liquid manures like cattle slurry or pig manure can lead to substantial reductions in $N_2O$ emissions (Ruser and Schulz 2015; Merino et al. 2006; Misselbrook et al. 2014; Guardia et al. 2017). Organic manures have the potential to be environmentally friendly if soil conditions, such as moisture levels and pH, are carefully assessed and best management practices, such as no-tillage or split application, are implemented. Additionally, optimizing organic manure rates, timing of application, and aligning with crop nitrogen requirements, or utilizing a combination of organic manure and stabilized fertilizers, are viable approaches to minimize $N_2O$ emissions post-organic manure application (Thangarajan et al. 2013).

Livestock slurries derived from cattle and pig waste, as well as poultry manure, exhibit elevated levels of easily accessible nitrogen in comparison to farmyard manure derived from straw, which contains predominantly organically bound nitrogen with limited readily available nitrogen content. Synchronizing the application of slurries and poultry manure with the peak period of crop nitrogen uptake can diminish the presence of inorganic nitrogen in the soil, consequently mitigating nitrous oxide emissions.

## *Agronomic Approaches*

### Introduction of Efficient Varieties

The amount of nitrogen absorbed by plants varies depending on the plant species and cultivars, influenced by both plant morphological characteristics and environmental factors. Enhancing nitrogen use efficiency involves not only the effective uptake of applied fertilizers but also the efficient utilization of nitrogen within the plant. Plants exhibiting increased uptake of applied nitrogen have the potential to decrease nitrous oxide emissions. Additionally, cultivars with higher nitrogen accumulation tend to have lower levels of residual nitrogen post-harvest, leading to reduced nitrous oxide emissions. While the immediate reduction in nitrous oxide emissions may be modest, the long-term impact is expected to be significant.

Plant traits that impact the nitrogen use efficiency of crops include:

1. The efficiency in the uptake of nitrogenous compounds and its regulation (Schmidt et al. 2002).
2. Root morphological characteristics such as distribution, quantity, maturity, and exudate quality play a significant role in influencing nitrogen mineralization, transformation, and transportation (Ladha et al. 2003).
3. Biological nitrogen fixation in the rhizospheric region (Shrestha and Ladha 1996).
4. Physiological nitrogen requirements for plant growth, the capacity to translocate, distribute, and mobilize absorbed nitrogen within different plant organs, leaf nitrogen import-export mechanisms, leaf senescence patterns, and the photosynthetic efficiency of plants (Ladha et al. 2003).

## Cover Crops

Cover crops can help decrease nitrous oxide emissions by absorbing excess nitrogen from crop production. Non-legume cover crops can be planted to consume nitrogen during periods when the soil is not cultivated, reducing the availability of nitrogen that can be converted to nitrous oxide after crop harvest. When these cover crops are harvested, nitrogen is slowly released into the soil, reducing the need for fertilizer. Alternatively, leguminous cover crops can release nitrogen slowly for cash crops, reducing the need for inorganic fertilizers. There has been a long-standing debate about whether legume residues from nitrogen-fixing crops contribute to nitrous oxide emissions (Baggs 2008; Shelp et al. 2000). Some studies have shown that nitrous oxide emissions from leguminous crop residues are similar to those from fertilizer nitrogen additions (Ghosh et al. 2002).

## Conservation Agriculture (CA)

Improving soil health, especially the nutrient supply capacity, can be achieved through conservation agriculture methods (Fageria 2002). Intensive tillage can lead to increased soil organic carbon loss due to higher oxidation rates, making the soil more vulnerable to erosion (Soon et al. 2001). To reduce nutrient loss, zero and minimal tillage techniques are being increasingly used along with planting crops in permanent beds to enhance soil health in the long term (Burgess et al. 2002). Implementing conservation agriculture practices for crop cultivation can enhance soil physical, chemical and biological health, leading to higher and sustained soil carbon levels (Ambus and Jensen 2001), improved aggregate stability (Calviño et al. 2003), and changes in macro-porosity (Burgess et al. 2002). Effective management of nitrogen mineralization from crop residue degradation and the use of nitrogenous fertilizers to meet crop requirements can enhance nitrogen use efficiency in conservation agriculture (Calviño et al. 2003).

## Residue Management

Crop residues are the remains of crops left in the field after harvesting (Malhi et al. 2010). They play a crucial role in plant growth by influencing the availability of nutrients (Mohanty and Mishra 2014). These residues are important sources and sinks for carbon and nitrogen cycles (Dinnes et al. 2002). Providing nitrogen slowly to the crops over a long period, crop residues release inorganic nitrogen into the soil during later stages of plant growth when nitrogen demand is high (Pankhurst et al. 2002). Cereal crop residues typically supply 40–100 kg N/ha/season and contribute to the buildup of organic carbon in the soil, improving soil nitrogen levels (Burgess et al. 2002). Legume residues, with their high nitrogen content and lower C/N ratios compared to cereals, are effective sources of nitrogen (Dinnes et al. 2002). Leguminous residues can mineralize quickly, making half of the nitrogen available to plants within two months of incorporation (Ram et al. 2011).

## Green Manuring

Leguminous plants are frequently employed for green manuring owing to their nitrogen-fixing potential. The annual nitrogen accumulation rate of legumes is believed to be between 20 and 300 kg/ha (Thind et al. 2012). Fast-growing short-duration leguminous crops are usually favored due to their nitrogen-fixing ability, capacity to accumulate larger biomass, and reduced need for cultural practices, making them suitable for intensive cropping systems (Sharma et al. 2011).

## Proper Crop Rotations

The significance of implementing appropriate crop sequences lies in the improvement of nitrogen utilization efficiency in crops. The use of optimal crop sequences ensures the efficient utilization of essential agricultural resources, including mineral nutrients and soil moisture by crops, thereby contributing to the long-term stability of the production system (Thind et al. 2012). The practice of rotating legume and cereal crops, which has been in existence for a long time, is beneficial as legumes have the ability to fix atmospheric nitrogen, reducing the nitrogen requirements of subsequent cereal crops (Helmers et al. 2001). The low C:N ratios of legume residues play a crucial role in enhancing soil physical, chemical, and biological health, ultimately increasing nitrogen availability (Gan et al. 2003; Varvel and Wilhelm 2003).

### Genetic Improvement

The enhancement of nitrogen use efficiency through a transgenic approach is centered around genes associated with nitrate assimilation and metabolism. The primary nitrate assimilation pathway, which encompasses nitrate transporters, nitrate reductase (NR), nitrite reductase (NiR), plastidic glutamine synthetase (GS2), and ferredoxin glutamate synthase (Fd-GOGAT), is a primary focus for improving nitrogen utilization. Furthermore, secondary nitrogen assimilation and remobilization pathways, including cytosolic glutamine synthetase (GS1), reduced nicotinamide adenine dinucleotide glutamate synthase (NADH-GOGAT), glutamate dehydrogenase (GDH), and various aminotransferases, present numerous possibilities for intervention across uptake, assimilation, translocation, and remobilization stages.

The initial step in nitrate assimilation involves the uptake of nitrogen by plants, which can be enhanced to improve nitrogen use efficiency. Plant nitrate uptake mechanisms consist of two components: a low-affinity transport system controlled by the NRT1 gene family and a high-affinity transport system regulated by the NRT2 gene family. While previous studies have focused on overexpressing high-affinity nitrate/nitrite transporters in tobacco and Arabidopsis to enhance nitrate uptake and assimilation, the expected improvement in nitrogen use efficiency was not observed (Pathak et al. 2008). Researchers have highlighted nitrogen remobilization as a crucial factor in enhancing plant nitrogen use efficiency (Masclaux-Daubresse et al. 2010). Overexpression of AS genes involved in amino acid remobilization and translocation has been shown to increase seed and total protein content. By genetically engineering rice to introduce barley alanine aminotransferase cDNA under the control of a tissue-specific OsAnt1 promoter, plants exhibited enhanced biomass and grain yield, as well as alterations in key metabolites and nitrate content, indicating improved nitrogen use efficiency (Shrawat et al. 2008). Recent studies have implicated the Arabidopsis NRT1 transporter NRT1.7 in nitrate remobilization from source to sink tissues through mutant analysis (Fan et al. 2007).

## Conclusion

The emission of nitrous oxide from agroecosystems is primarily driven by nitrification and denitrification processes, which are influenced by both biotic and abiotic factors. The availability of inorganic nitrogen in the soil plays a crucial role in nitrous oxide emissions, and this availability is directly affected by the application of nitrogenous fertilizers. While nitrogen inputs are necessary for achieving desirable crop yields, the full utilization of applied nitrogen is often not possible, leading to the formation of nitrous oxide and contributing to the greenhouse effect. To effectively mitigate nitrous oxide emissions from agricultural soils, it is important to have a comprehensive understanding of the emission pathways through different nitrogenous fertilizer sources, as well as the influence of soil biology and physico-chemical properties. Additionally, finding the optimal management practices that maximize plant yield while minimizing $N_2O$ emissions, considering nitrogenous fertilizer management, is crucial. The information on how different N-fertilizers induce $N_2O$ emissions is valuable for greenhouse gas inventories and contributes to the accuracy of final values in these studies. It is also important to consider the overall nitrogen budget to estimate the potential benefits of any designed mitigation strategy. Furthermore, the correlation between $N_2O$ emissions and the annual contribution of different periods, such as sowing, after harvest, and rainy season, as well as N turnover, should be taken into account for a comprehensive understanding of the emission patterns.

Furthermore, the vital processes of nitrification and denitrification are primarily driven by microorganisms, making it imperative to concentrate research efforts on these aspects in order to mitigate $N_2O$ emissions. By adopting targeted strategies, not only can greenhouse gas (GHG) reduction be achieved, but also the efficiency and profitability of agricultural enterprises can be improved. In the long run, the implementation of precision farming practices that account for the temporal and spatial heterogeneity of soil, along with the introduction of new high-yielding crop varieties requiring less fertilizer, should be considered. The successful execution of these strategies at the farm level will ultimately yield favorable outcomes, necessitating collaborative endeavors between the government and the public for the overall betterment of mankind.

## References

Akiyama H, Yan X, Yagi K (2009) Evaluation of effectiveness of enhanced-efficiency fertilizers as mitigation options for N2O and NO emissions from agricultural soils: meta-analysis. Glob Chang Biol 16(6):1837–1846. https://doi.org/10.1111/j.1365-2486.2009.02031.x

Ambus P, Jensen ES (2001) Crop residue management strategies to reduce N-losses—interaction with crop N supply. Commun Soil Sci Plant Anal 32(7–8):981–996. https://doi.org/10.1081/css-100104100

Anas M, Liao F, Verma KK, Sarwar MA, Mahmood A, Chen ZL, Li Q, Zeng XP, Liu Y, Li YR (2020) Fate of nitrogen in agriculture and environment: agronomic, eco-physiological and molecular approaches to improve nitrogen use efficiency. Biol Res 53:47. https://doi.org/10.1186/s40659-020-00312-4

Ayyadurai P, Manickasundaram P (2014) Growth, nutrient uptake and seed cotton yield as influenced by foliar nutrition and drip fertigation in cotton hybrid. Int J Agric Sci 10:276–279

Azeem B, KuShaari K, Man ZB, Basit A, Thanh TH (2014) Review on materials & methods to produce controlled release coated urea fertilizer. J Control Release 181:11–21. https://doi.org/10.1016/j.jconrel.2014.02.020

Baggs EM (2008) A review of stable isotope techniques for N2O source partitioning in soils: recent progress, remaining challenges and future considerations. Rapid Commun Mass Spectrom 22(11):1664–1672. https://doi.org/10.1002/rcm.3456

Bar-Yosef B (1999) Advances in fertigation. Adv Agron:1–77. https://doi.org/10.1016/s0065-2113(08)60910-4

Berry PM, Kindred DR, Olesen JE, Jorgensen LN, Paveley ND (2010) Quantifying the effect of interactions between disease control, nitrogen supply and land use change on the greenhouse gas emissions associated with wheat production. Plant Pathol 59:753–763. https://doi.org/10.1111/j.1365-3059.2010.02276.x

Braker G, Conrad R (2011) Diversity, structure, and size of N2O-producing microbial communities in soils—what matters for their functioning? Adv Appl Microbiol:33–70. https://doi.org/10.1016/b978-0-12-387046-9.00002-5

Bremner JM (1997) Sources of nitrous oxide in soils. Nutr Cycl Agroecosys, Dordrecht 49(1–3):7–16

Bremner J, Blackmer A, Waring S (1980) Formation of nitrous oxide and dinitrogen by chemical decomposition of hydroxylamine in soils. Soil Biol Biochem 12(3):263–269. https://doi.org/10.1016/0038-0717(80)90072-3

Burgess MS, Mehuys GR, Madramootoo CA (2002) Nitrogen dynamics of decomposing corn residue components under three tillage systems. Soil Sci Soc Am J 66(4):1350–1358. https://doi.org/10.2136/sssaj2002.1350

Butterbach-Bahl K, Baggs EM, Dannenmann M, Kiese R, Zechmeister-Boltenstern S (2013) Nitrous oxide emissions from soils: how well do we understand the processes and their controls? Philos Trans R Soc B 368(1621):20130122. https://doi.org/10.1098/rstb.2013.0122

Calviño PA, Andrade FH, Sadras VO (2003) Maize yield as affected by water availability, soil depth, and crop management. Agron J 95(2):275–281. https://doi.org/10.2134/agronj2003.0275

Chambers B, Smith K, Pain B (2006) Strategies to encourage better use of nitrogen in animal manures. Soil Use Manag 16:157–166. https://doi.org/10.1111/j.1475-2743.2000.tb00220.x

Crutzen PJ (1981) Atmospheric chemical processes of the oxides of nitrogen, including nitrous oxide. In: Delwiche CC (ed) Denitrification, nitrification and atmospheric nitrous oxide. John Wiley and Sons, Chichester, pp 17–44

Dai J, Dong H (2014) Intensive cotton farming technologies in China: achievements, challenges and countermeasures. Field Crop Res 155:99–110. https://doi.org/10.1016/j.fcr.2013.09.017

Dambreville C, Hallet S, Nguyen C, Morvan T, Germon J, Philippot L (2006) Structure and activity of the denitrifying community in a maize-cropped field fertilized with composted pig manure or ammonium nitrate. FEMS Microbiol Ecol 56(1):119–131. https://doi.org/10.1111/j.1574-6941.2006.00064.x

De Klein C, Novoa RSA, Ogle S, Smith K, Rochette P, Wirth TC, Mcconkey BG, Mosier A, Rypdal K (2006) N2O Emissions from Managed Soils, and CO2 Emissions from Lime and Urea Application. 2006 IPCC Guidelines for National Greenhouse Gas Inventories

Dinnes DL, Karlen DL, Jaynes DB, Kaspar TC, Hatfield JL, Colvin TS, Cambardella CA (2002) Nitrogen management strategies to reduce nitrate leaching in tile-drained midwestern soils. Agron J 94(1):153–171. https://doi.org/10.2134/agronj2002.0153

Dong H, Kong X, Li W, Tang W, Zhang D (2010) Effects of plant density and nitrogen and potassium fertilization on cotton yield and uptake of major nutrients in two fields with varying fertility. Field Crop Res 119(1):106–113. https://doi.org/10.1016/j.fcr.2010.06.019

Dong H, Li W, Eneji AE, Zhang D (2012) Nitrogen rate and plant density effects on yield and late-season leaf senescence of cotton raised on a saline field. Field Crop Res 126:137–144. https://doi.org/10.1016/j.fcr.2011.10.005

Fageria NK (2002) Soil quality vs. environmentally-based agricultural management practices. Commun Soil Sci Plant Anal 33(13–14):2301–2329. https://doi.org/10.1081/css-120005764

Fan X, Jia L, Li Y, Smith SJ, Miller AJ, Shen Q (2007) Comparing nitrate storage and remobilization in two rice cultivars that differ in their nitrogen use efficiency. J Exp Bot 58(7):1729–1740. https://doi.org/10.1093/jxb/erm033

Fernández V, Sotiropoulos T, Brown PH (2013) Foliar fertilisation: principles and practices. International Fertilizer Industry Association (IFA), Paris

Fujinuma R, Venterea RT, Rosen C (2011) Broadcast urea reduces N 2 O but increases NO emissions compared with conventional and shallow-applied anhydrous ammonia in a coarse-textured soil. J Environ Qual 40(6):1806–1815. https://doi.org/10.2134/jeq2011.0240

Gagnon B, Ziadi N, Rochette P, Chantigny MH, Angers DA (2011) Fertilizer source influenced nitrous oxide emissions from a clay soil under corn. Soil Sci Soc Am J 75(2):595–604. https://doi.org/10.2136/sssaj2010.0212

Galloway JN, Dentener FJ, Capone DG, Boyer EW, Howarth RW, Seitzinger SP, Asner GP, Cleveland CC, Green PA, Holland EA, Karl DM, Michaels AF, Porter JH, Townsend AR, Vosmarty CJ (2004) Nitrogen cycles: Past, present, and future. Biogeochemistry 70(2):153–226. https://doi.org/10.1007/s10533-004-0370-0

Gan YT, Miller PR, McConkey BG, Zentner RP, Stevenson FC, McDonald CL (2003) Influence of diverse cropping sequences on durum wheat yield and protein in the semiarid northern Great Plains. Agron J 95(2):245–252. https://doi.org/10.2134/agronj2003.0245

Ghosh S, Majumdar D, Jain M (2002) Nitrous oxide emissions from kharif and rabi legumes grown on an alluvial soil. Biol Fertil Soils 35(6):473–478. https://doi.org/10.1007/s00374-002-0498-9

Guardia G, Cangani MT, Sanz-Cobena A, Junior JL, Vallejo A (2017) Management of pig manure to mitigate NO and yield-scaled N2O emissions in an irrigated Mediterranean crop. Agric Ecosyst Environ 238:55–66. https://doi.org/10.1016/j.agee.2016.09.022

Halvorson AD, Snyder CS, Blaylock AD, Del Grosso SJ (2014) Enhanced-efficiency nitrogen fertilizers: potential role in nitrous oxide emission mitigation. Agron J 106(2):715–722. https://doi.org/10.2134/agronj2013.0081

Helmers GA, Yamoah CF, Varvel GE (2001) Separating the impacts of crop diversification and rotations on risk. Agron J 93(6):1337–1340. https://doi.org/10.2134/agronj2001.1337

Hoben JP, Gehl RJ, Millar N, Grace PR, Robertson GP (2010) Nonlinear nitrous oxide (N2O) response to nitrogen fertilizer in on-farm corn crops of the US Midwest. Glob Chang Biol 17(2):1140–1152. https://doi.org/10.1111/j.1365-2486.2010.02349.x

Huang Y, Zou J, Zheng X, Wang Y, Xu X (2004) Nitrous oxide emissions as influenced by amendment of plant residues with different C:N ratios. Soil Biol Biochem 36(6):973–981. https://doi.org/10.1016/j.soilbio.2004.02.009

Ju XT, Xing GX, Chen XP, Zhang SL, Zhang LJ, Liu XJ, Cui ZL, Yin B, Christie P, Zhu ZL, Zhang FS (2009) Reducing environmental risk by improving N management in intensive Chinese agricultural systems. Proc Natl Acad Sci U S A 106:9., (3041–3046). https://doi.org/10.1073/pnas.0813417106

Ladha J, Dawe D, Pathak H, Padre A, Yadav R, Singh B, Singh Y, Singh Y, Singh P, Kundu A, Sakal R, Ram N, Regmi A, Gami S, Bhandari A, Amin R, Yadav C, Bhattarai E, Das S et al (2003) How extensive are yield declines in long-term rice–wheat experiments in Asia? Field Crop Res 81(2–3):159–180. https://doi.org/10.1016/s0378-4290(02)00219-8

Li X, Li Q, Xu X, Su Y, Yue Q, Gao B (2016a) Characterization, swelling and slow-release properties of a new controlled release fertilizer based on wheat straw cellulose hydrogel. J Taiwan Inst Chem Eng 60:564–572. https://doi.org/10.1016/j.jtice.2015.10.027

Li X, Sørensen P, Olesen JE, Petersen SO (2016b) Evidence for denitrification as main source of N 2 O emission from residue-amended soil. Soil Biol Biochem 92:153–160. https://doi.org/10.1016/j.soilbio.2015.10.008

Luo Z, Liu H, Li W, Zhao Q, Dai J, Tian L, Dong H (2018) Effects of reduced nitrogen rate on cotton yield and nitrogen use efficiency as mediated by application mode or plant density. Field Crop Res 218:150–157. https://doi.org/10.1016/j.fcr.2018.01.003

Ma HY, Shi WQ, Liu YA, Xian AM, Wang JG (2016) Effect of different irrigation and fertilization methods on yield and water and fertilizer use efficiency of pineapple. Chin J Trop Crops 37:1882–1888

MacLeod M, Moran D, Eory V, Rees R, Barnes A, Topp CF, Ball B, Hoad S, Wall E, McVittie A, Pajot G, Matthews R, Smith P, Moxey A (2010) Developing greenhouse gas marginal abatement cost curves for agricultural emissions from crops and soils in the UK. Agric Syst 103(4):198–209. https://doi.org/10.1016/j.agsy.2010.01.002

Malhi SS, Nyborg M, Goddard T, Puurveen D (2010) Long-term tillage, straw and N rate effects on quantity and quality of organic C and N in a gray Luvisol soil. Nutr Cycl Agroecosyst 90(1):1–20. https://doi.org/10.1007/s10705-010-9399-8

Mao S (2013) Cotton farming in China Shanghai. Shanghai Scientific and Technical Press, Shanghai, p 2013

Martins M, Sant'Anna S, Zaman M, Santos R, Monteiro R, Alves B, Jantalia C, Boddey R, Urquiaga S (2017) Strategies for the use of urease and nitrification inhibitors with urea: impact on N 2 O and NH 3 emissions, fertilizer- 15 N recovery and maize yield in a tropical soil. Agric Ecosyst Environ 247:54–62. https://doi.org/10.1016/j.agee.2017.06.021

Masclaux-Daubresse C, Daniel-Vedele F, Dechorgnat J, Chardon F, Gaufichon L, Suzuki A (2010) Nitrogen uptake, assimilation and remobilization in plants: challenges for sustainable and productive agriculture. Ann Bot 105(7):1141–1157. https://doi.org/10.1093/aob/mcq028

Merino P, Menéndez S, Pinto M, González-Murua C, Estavillo J (2006) 3, 4-Dimethylpyrazole phosphate reduces nitrous oxide emissions from grassland after slurry application. Soil Use Manag 21(1):53–57. https://doi.org/10.1111/j.1475-2743.2005.tb00106.x

Millar N, Robertson GP, Grace PR, Gehl RJ, Hoben JP (2010) Nitrogen fertilizer management for nitrous oxide (N2O) mitigation in intensive corn (maize) production: an emissions reduction protocol for US Midwest agriculture. Mitig Adapt Strateg Glob Chang 15(2):185–204. https://doi.org/10.1007/s11027-010-9212-7

Misselbrook TH, Cardenas LM, Camp V, Thorman RE, Williams JR, Rollett AJ, Chambers BJ (2014) An assessment of nitrification inhibitors to reduce nitrous oxide emissions from UK agriculture. Environ Res Lett 9(11):115006. https://doi.org/10.1088/1748-9326/9/11/115006

Mohanty A, Mishra KN (2014) Influence of conservation agriculture production system on available soil nitrogen, phosphorus, potassium and maize equivalent yield on a fluventichaplustepts in the north central plateau zone of Odisha. Trends Biosci J 7:3962–3967

Naz MY, Sulaiman SA (2016) Slow release coating remedy for nitrogen loss from conventional urea: a review. J Control Release 225:109–120. https://doi.org/10.1016/j.jconrel.2016.01.037

Neelam P, Rajput T (2011) Simulation and modeling of water movement in potato (*Solanumtuberosum*) under subsurface drip system. Indian J Agric Sci 2011(81):25–32

Pankhurst C, McDonald H, Hawke B, Kirkby C (2002) Effect of tillage and stubble management on chemical and microbiological properties and the development of suppression towards cereal root disease in soils from two sites in NSW, Australia. Soil Biol Biochem 34(6):833–840. https://doi.org/10.1016/s0038-0717(02)00014-7

Pathak RR, Ahmad A, Lochab S, Raghuram N (2008) Molecular physiology of plant N-use efficiency and biotechnological options for its enhancement. Curr Sci 94(11):1394–1403

Peiling L, Fucang Z (2012) Effects of water and nitrogen coupling on cotton nitrogen absorption and utilization under alternate root partition drip[J]. Transactions of the Chinese Society of Agricultural Engineering (Transactions of the CSAE) 28(25):112–116

Ram H, Singh Y, Saini KS, Kler DS, Timsina J, Humphreys EJ (2011) Agronomic and economic evaluation of permanent raised beds, no tillage and straw mulching for an irrigated maize-wheat system in Northwest India. Exp Agric 48(1):21–38. https://doi.org/10.1017/s0014479711000809

Ribaudo M, Delgado J, Hansen L, Livingston M, Mosheim R, Williamson J (2011) Nitrogen in agricultural systems: implications for conservation policy. USDA, Washington, DC

Rossetto R, Dias F, Landell M, Cantarella H, Tavares S, Vitti A, Perecin D (2010) N and K fertilisation of sugarcane ratoons harvested without burning. Proc Int Soc Sugar Cane Technol 27:1–8

Ruser R, Schulz R (2015) The effect of nitrification inhibitors on the nitrous oxide (N 2 O) release from agricultural soils—a review. J Plant Nutr Soil Sci 178(2):171–188. https://doi.org/10.1002/jpln.201400251

Schils RL, Van Groenigen JW, Velthof GL, Kuikman PJ (2008) Nitrous oxide emissions from multiple combined applications of fertiliser and cattle slurry to grassland. Plant Soil 310(1–2):89–101. https://doi.org/10.1007/s11104-008-9632-2

Schmidt JP, DeJoia AJ, Ferguson RB, Taylor RK, Young RK, Havlin JL (2002) Corn yield response to nitrogen at multiple in-field locations. Agron J 94(4):798–806. https://doi.org/10.2134/agronj2002.7980

Schumann AW (2000) Prospects for improving nitrogen fertiliser use efficiency with a new soil test and ammonia volatilization model. Pro Ann Congr S Afr Sugar Technol Assoc 73:70–78

Shan J, Yan X (2013) Effects of crop residue returning on nitrous oxide emissions in agricultural soils. Atmos Environ 71:170–175. https://doi.org/10.1016/j.atmosenv.2013.02.009

Sharma P, Abrol V, Sharma R (2011) Impact of tillage and mulch management on economics, energy requirement and crop performance in maize–wheat rotation in rainfedsubhumidinceptisols, India. Eur J Agron 34(1):46–51. https://doi.org/10.1016/j.eja.2010.10.003

Shelp ML, Beauchamp EG, Thurtell GW (2000) Nitrous oxide emissions from soil amended with glucose, alfalfa, or corn residues. Commun Soil Sci Plant Anal 31(7–8):877–892. https://doi.org/10.1080/00103620009370484

Shrawat AK, Carroll RT, DePauw M, Taylor GJ, Good AG (2008) Genetic engineering of improved nitrogen use efficiency in rice by the tissue-specific expression ofalanine aminotransferase. Plant Biotechnol J 6(7):722–732. https://doi.org/10.1111/j.1467-7652.2008.00351.x

Shrestha RK, Ladha JK (1996) Genotypic variation in promotion of rice Dinitrogen fixation as determined by nitrogen-15 dilution. Soil Sci Soc Am J 60(6):1815–1821. https://doi.org/10.2136/sssaj1996.03615995006000060029x

Snyder C, Bruulsema T, Jensen T, Fixen P (2009) Review of greenhouse gas emissions from crop production systems and fertilizer management effects. Agric Ecosyst Environ 133(3–4):247–266. https://doi.org/10.1016/j.agee.2009.04.021

Soon YK, Clayton GW, Rice WA (2001) Tillage and previous crop effects on dynamics of nitrogen in a wheat–soil system. Agron J 93(4):842–849. https://doi.org/10.2134/agronj2001.934842x

Syakila A, Kroeze C (2011) The global nitrous oxide budget revisited. GHG Measure Manage 1(1):17–26. https://doi.org/10.3763/ghgmm.2010.0007

Thangarajan R, Bolan NS, Tian G, Naidu R, Kunhikrishnan A (2013) Role of organic amendment application on greenhouse gas emission from soil. Sci Total Environ 465:72–96. https://doi.org/10.1016/j.scitotenv.2013.01.031

Thind HS, Kumar A, Gupta RK, Kaul A, Vashistha M (2012) Fixed-time adjustable dose site-specific fertilizer nitrogen management in transplanted irrigated rice (Oryza sativa L.) in South Asia. Field Crop Res 126:63–69. https://doi.org/10.1016/j.fcr.2011.09.007

Thorman R, Chadwick D, Harrison R, Boyles L, Matthews R (2007) The effect on N2O emissions of storage conditions and rapid incorporation of pig and cattle farmyard manure into tillage land. Biosyst Eng 97(4):501–511. https://doi.org/10.1016/j.biosystemseng.2007.03.039

Timilsena YP, Adhikari R, Casey P, Muster T, Gill H, Adhikari B (2014) Enhanced efficiency fertilisers: a review of formulation and nutrient release patterns. J Sci Food Agric 95(6):1131–1142. https://doi.org/10.1002/jsfa.6812

Trenkel ME (2010) Slow-and controlled-release and stabilized fertilizers: an option for enhancing nutrient use efficiency in agriculture. International Fertilizer Industry Association (IFA), Paris. https://doi.org/10.1017/CBO9781107415324.004

Varvel GE, Wilhelm WW (2003) Soybean nitrogen contribution to corn and sorghum in western corn belt rotations. Agron J 95(5):1220–1225. https://doi.org/10.2134/agronj2003.1220

Yamamoto CF, Pereira EI, Mattoso LH, Matsunaka T, Ribeiro C (2016) Slow release fertilizers based on urea/urea–formaldehyde polymer nanocomposites. Chem Eng J 287:390–397. https://doi.org/10.1016/j.cej.2015.11.023

Yan KF, Peng FT, Qi YJ, Dang ZQ, Zhang JH, Jiang XM (2015) Effects of different irrigation and fertilizer methods on space variation of nitrate nitrogen and nitrogen fertilization absorption and utilization of plants. Water Saving Irrigat 9:33–38

Yong TW, Ping C, Qian D, Qing D, Feng Y, Wang XC, Liu WG, Yang WY (2018) Optimized nitrogen application methods to improve nitrogen use efficiency and nodule nitrogen fixation in a maize-soybean relay intercropping system. J Integr Agric 17:664–676

Zhang H, Zhao Q, Wang Z, Wang L, Li X, Fan Z, Zhang Y et al (2021) Effects of nitrogen fertilizer on photosynthetic characteristics, biomass, and yield of wheat under different shading conditions. Agronomy 11(10):1989. MDPI AG. https://doi.org/10.3390/agronomy11101989

Zhu G, Song X, Ju X, Zhang J, Müller C, Sylvester-Bradley R, Thorman RE, Bingham I, Rees RM (2019) Gross N transformation rates and related N2O emissions in Chinese and UK agricultural soils. Sci Total Environ 666:176–186. https://doi.org/10.1016/j.scitotenv.2019.02.241

# Impact of Fertilizers on Greenhouse Gas Mitigating Microbes in Agroecosystem

**V. Manasa, Bandeppa, R. Gobinath, Kiran Kumar, S. Vijay kumar, P. C. Latha, and Amol Phule**

## Introduction

As the population grows and per capita consumption patterns change, farmers alter food, feed, livestock, and fiber production energy use, land-use composition, and social equity. All these changes require more addition of chemical inputs, especially nitrogen. Nitrogen (N) significantly influences the diversity and functioning of various terrestrial, freshwater, and marine ecosystems. While nitrogenous fertilizers are essential for achieving increased crop yields, their excessive application can lead to environmental damage (Chen et al. 2004). Synthetic nitrogen fertilization stands out as a major contributor to human-induced emissions of $N_2O$ from agricultural soils (Bouwman 1996). The rapid increase in the use of synthetic N fertilizers coupled with low nitrogen use efficiency hampered the sustainable development of agriculture and led to diversified environmental problems (Liu et al. 2013).

The entire process involved in the production of synthetic nitrogen fertilizers, encompassing fossil fuel mining, transportation, ammonia synthesis, and conversion to various nitrogen fertilizer products, serves as a noteworthy source of greenhouse gas emissions. Consequently, the utilization of synthetic nitrogen fertilizers emerges as a key driver of greenhouse gas emissions in the agricultural sector.

Among the greenhouse gases (GHG), methane and $N_2O$ are huge contributors to global warming, with a potential (GWP) ~23 times and 360 times higher than that of $CO_2$, respectively (Gebert et al. 2004). Specifically, GWP is an index that provides the relative impact a specific gas could have on global climate over a defined time period.Because $CH_4$ has a short life (~8 years) relative to other GHGs such as

V. Manasa · Bandeppa (✉) · R. Gobinath · S. Vijay kumar · P. C. Latha · A. Phule
ICAR-Indian Institute of Rice Research, Rajendranagar, Hyderabad, India

K. Kumar
ICAR-Directorates of Groundnut Research, Junagadh, Gujarat, India



S. R. Mohanty, B. Kollah (eds.), *Greenhouse Gas Regulating Microorganisms in Soil Ecosystems*, https://doi.org/10.1007/978-3-031-70569-4_14

$CO_2$ (50–200 years) (120 years), controlling emission of $CH_4$ seems likely to be favorable in terms of short-term control of global warming (Dedysh et al. 2007).

Next to fertilizer application, microbial activities play a major role in the determination of atmospheric concentrations of greenhouse gases such as methane ($CH_4$), carbon dioxide ($CO_2$), and nitrous oxide ($N_2O$) in the course of earth's history (Singh et al. 2010). Microbes have been involved in the production and removal of GHGs since their evolution in the ocean more than 3.5 billion years ago and after their entry into land about 2 billion years ago (Zimmer 2010). Methanotrophic bacteria buffer the effect of huge amounts of methane emitted in the environment (Singh et al. 2010). For example, they consume huge amounts of methane emitted not only from marine sediments but also from well blowouts like deep water horizon spills (Zimmerman and Labonte 2015). The marine environment is in fact a formidable source of release of methane to the atmosphere, as there is constant leaking of this greenhouse gas. However, such a methane trap has its own unique community of methane-eating microbes; such microbial populations remove around 75% of the arising methane before the gas enters into the atmosphere. These organisms thus provide an important service in maintaining the climate by reducing the emission of greenhouse gases (Trinastic 2015).

## Sources of Emission of Greenhouse Gases

Carbon dioxide is the primary greenhouse gas emitted by human activities. Carbon dioxide enters into the atmosphere mainly through fossil fuel burning (natural gas, coal, and oil), trees, wood products, solid waste, and also by product of some chemical reactions (e.g., cement manufacturing process). Carbon dioxide is removed and sequestered from the atmosphere by plants in the biological carbon cycle. Methane is majorly emitted from coal production and transport, natural gas, and oil. Methane is also released from livestock and other agricultural practices, land use, and during decay of organic waste in municipal solid waste. Nitrous oxide emissions were estimated to occur during agriculture and industrial activities, as well as during combustion of solid waste, fossil fuels, and waste water treatment.

## Impact of Agriculture on Greenhouse Gases

Over the years, agriculture has been a significant contributor to global emissions of $NH_3$, CO, and NOx. Agriculture contributed towards global warming through anthropogenic greenhouse gas emissions and by the conversion of non-agricultural land such as forests into agricultural land (Sarkodie et al. 2019). In the last 150 years, cumulative emissions of $CO_2$ associated with clearing of land for agriculture have been comparable to those from combustion of fossil fuel. However, the latter is the major source of $CO_2$ at present and is projected to become more dominant in the future. Growing rice in waterlogged conditions, ruminant animals, and biomass burning are major agricultural sources of $CH_4$. Oxidation of $CH_4$ in aerobic soils has been decreased by perturbations to natural N cycles. Agricultural contribution towards $N_2O$ emissions has probably been underestimated because of incomplete analysis of increased nitrogen flows in the environment, especially through $NH_3$ volatilization from animal manures, nitrate leaching, and increased use of biological nitrogen fixation (Fig. 1). The agriculture contribution to radiative forcing of climate is analyzed by collecting data from the Intergovernmental Panel on Climate Change (base case) and cases where the global warming potential of methane and agricultural emissions of $N_2O$ are doubled. With these scenarios, agriculture along with land clearing is estimated to contribute around 28–33% of the radiative forcing created over the next 100 year by 1990 anthropogenic emissions of major greenhouse gases.

Agricultural lands, due to their large area and intensive management practices, have a significant influence on the earth's carbon and nitrogen cycles and agricultural activities, leading to the discharge of greenhouse gases. Worldwide, anthropogenic activities are causing a rise in greenhouse gases, specifically carbon dioxide, methane, and nitrous oxide, originating from the agricultural sector. Although

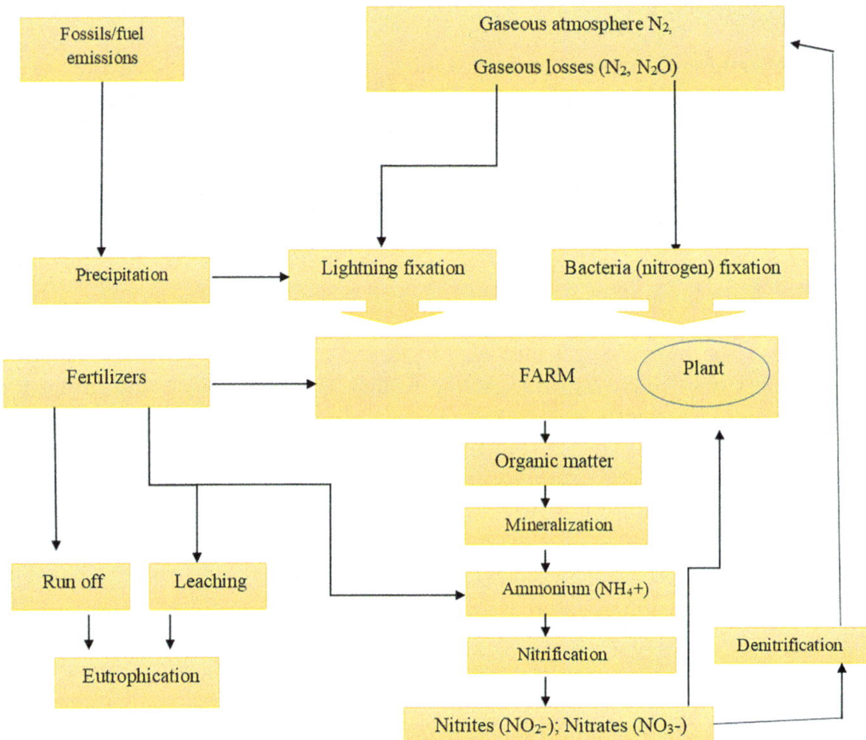


**Fig. 1** Nitrogen cycle in soil-plant-atmosphere continuum (SPAC)

animal manure application to agricultural soils improves soil quality, health, and agricultural production but it has a significant impact on GHGs emissions. But the extent of GHGs release in response to manure application in diverse environmental ecosystems remains uncertain.

## Effect of Greenhouse Gases on Ecosystem

Since the beginning of the industrial revolution, the concentrations of atmospheric greenhouse gases, such as $CO_{2,}$ tropospheric $O_2$, methane ($CH_4$), and nitrous oxides ($N_2O$), have been increasing due to human activities such as deforestation, fossil fuel combustion, and agriculture (Roeckner 1992). Each of these greenhouse gases is involved in trapping heat near the Earth's surface in a process referred to as radiative forcing (Stevenson et al. 1998). Greenhouse gases warm the earth surface and the atmosphere, with significant implications for rainfall, sea level, and the retreat of glaciers. The most abundant greenhouse gases in Earth's atmosphere are: water vapor ($H_2O$), carbon dioxide, methane, nitrous oxide, ozone, and chlorofluorocarbons (CFCs). Among these, carbon dioxide accounts for 76 percent, methane contributes 16 percent, and nitrous oxide contributes around 6 percent to global GHG emissions (EPA, 2017).

**Global Warming** The term "global warming" refers to an aspect of climate change characterized by a rapid increase in the Earth's average surface temperature over the past century, mainly due to the burning of fossil fuels that increase greenhouse gases. Increased GHG concentration in the earth's atmosphere leads to a reduction in outgoing infrared radiation. The earth's climate must be buffered to restore the balance between incoming and outgoing radiation. This climatic change will cause global warming of the Earth's surface and the lower atmosphere. Warming up is the easiest way for the climate to get rid of the extra energy. However, a little increase in temperature will induce so many other changes, such as wind patterns and cloud cover. Some of these changes may lead to increased warming (positive feedback), while others may counteract it (negative feedback). A naturally occurring greenhouse warming is majorly exerted by water vapor, which plays an important role in maintaining the earth's surface temperature at a habitable ~15 °C (59 °F) level for humans and other species. On the contrary, human activities, especially fossil fuel burning, have disrupted the Earth's natural greenhouse. The level of atmospheric carbon dioxide in the atmosphere has increased by more than a third since the industrial revolution. Excessive levels of greenhouse gases increase the greenhouse effect, trapping heat and leading to temperature rise in the Earth's atmosphere. As the temperature increases, more water vapor evaporates from the oceans and other water bodies into the atmosphere, causing a further increase in temperature. From 1905 to 2005, the earth's surface temperature has increased by 0.6–0.9 °C. The Intergovernmental Panel on Climate Change, in their third assessment report, forecast that global mean surface temperature will rise by 1.4 °C to 5.8 °C by the end of

2100. But many uncertainties were associated with this projection, such as emission rates of greenhouse gases, climate feedbacks, and the size of the ocean delay.

**Sea Level Rise** Human activities such as fossil fuel burning have increased emissions of greenhouse gases, leading to warming of the Earth's atmosphere and ocean and raising of sea levels. Some of the GHGs (e.g., methane) have atmospheric lifetimes of decades or less, whereas others (e.g., carbon dioxide) exist for centuries to millennia. Short-lived greenhouse gases contribute to sea-level rise by thermal expansion over much longer periods of time than their atmospheric lifetimes. Sea level rise by global warming occurs due to two different processes. Firstly, warmer temperatures cause sea level to rise through seawater thermal expansion. Secondly, water from melting ice sheets and glaciers also adds water to the sea. It is predicted that the Earth's average sea level will rise by 0.09 to 0.88 meters between 1990 and 2100.

## Water Stress and Water Insecurity

The challenging problem of insufficient water access is particularly pronounced in developing nations. Presently, a staggering 1.1 billion individuals worldwide do not have access to water, and 2.6 billion lack sanitation facilities. The impacts of climate change are anticipated to further exert pressure on water resources. The expanding scarcity of water is amplifying food insecurity and exacerbating the competitions for water, both domestically and internationally. As the global population rises and water consumption increases, the issues related to water are expected to become more pronounced.

**Economic Impact** The environment is largely polluted by increasing urbanization, industrialization, and greenhouse gases emitted from residential areas. This causes the atmosphere to become increasingly polluted. These anomalies threaten life in water, land, and air by destroying nature and changing the climate. Over half of the human population lives within 100 kilometers of the sea. Most of this population lives in urban places that serve as seaports. A measurable increase in sea level will have a significant economic impact on low-lying coastal areas and islands, for instance, increasing erosion rates in beaches along the coastlines and rising sea level displacing fresh groundwater for a substantial distance inland. Changes in temperature and rainfall, as well as the increase in $CO_2$ levels leading to climate change, have substantial impacts on global agriculture. The decline in crop yields and the change in land structure cause people not to receive recompense for their labor and lose their hopes in farming. This leads people to different goals and occupations. Climate change has a wide range of negative impacts, including on health, social mobility, agriculture, the economy, industry, and tourism.

**Agricultural Impact** Agriculture is the backbone of the Indian economy and provides food and livelihood security to the majority of the Indian population. The impact of climate change as anticipated in present times has enormous potential to adversely influence agriculture in a variety of ways. A major part of the arable land in India is rain-fed; the agriculture productivity depends on the quantity of rainfall and distribution pattern. Consequently, alterations in overall precipitation levels or shifts in rainfall patterns not only pose a serious threat to agriculture but also jeopardize the broader aspects of the economy and food security. Beyond a certain range of temperatures, warming enhances the speed of plant growth and development, ultimately resulting in yield reduction. Rising temperatures would increase fertilizer requirements to achieve the same production targets, causing higher GHG emissions, ammonia volatilization, and crop production costs. Warming of more than 3 °C is expected to have an adverse effect on production in all the regions (IPCC Fourth Assessment Report, 2007). Higher temperatures are also associated with the ability of plants to get and utilize the moisture. Evaporation from the soil enhances when temperatures rise and plants increase their transpiration (Cline 2008). Carbon emissions can also help agriculture by enhancing photosynthesis in many important $C_3$ crops such as wheat, rice, and soybeans. Experiments have shown that with higher concentrations of $CO_2$, plants can grow bigger and faster. However, the effects of global warming influence the atmospheric general circulation, changing the global precipitation pattern as well as changing the soil moisture levels across the globe. The Third Assessment Report of the IPCC, 2001, concluded that climate change would hit the poorest countries adversely in terms of shrinking the agricultural production. The report claimed that crop yields would be decreased in most of the tropical and sub-tropical regions due to reduced water availability and a new or changed insect/pest scenario.

**Effects on Aquatic Systems** Aquatic ecosystems are essential components of the global environment. Human well-being and health are sustained by the critical ecosystem services provided by biodiversity, which have been negatively affected by global warming in recent decades (Millennium Ecosystem Assessment, 2005). Apart from the challenges posed by land-use change, water diversion, and pollution, aquatic systems are anticipated to face the added stress of global climate change in the near future. The rise in water temperatures resulting from climate change is poised to modify essential ecological processes and the geographic distribution of aquatic species. Shifts in the seasonal patterns of rainfall and runoff will also transform the hydrological characteristics of aquatic systems, impacting species composition and ecosystem productivity. Increased salinity in estuaries could reduce the abundance of freshwater species and increase the presence of marine species. Aquatic ecosystems have a limited ability to adapt to climate change. Mitigating the effects of climate change on these systems will rely significantly on human activities that reduce other sources of ecosystem stress and increase adaptive capacity (Poff et al. 2002).

**Effects on Hydrological Cycle** Increased temperature near the earth's surface changes surface moisture content and increases the rate of evaporation. This phenomenon increases the moisture content within the atmosphere, leading to intense precipitation events (rainfall or snowfall), increasing the potential for disastrous floods, e.g., urban floods and flash floods, etc. Projected changes in future climate, i.e., a warming climate, impact the spatial and temporal distribution of precipitation events, glaciers, sea level, etc. Melting glaciers along with the augmented precipitation events results in increased runoff, which has the potential to severely impact the water balance across India (Bhunya et al. 2020). In addition, changing climate has the potential to impact groundwater resources due to projected changes in precipitation magnitude and intensity, rate of infiltration, and evapotranspiration.

**Effects on Ecosystems and Biodiversity** Global warming has severe consequences on the habitats of numerous plants and animals, posing a significant threat to the survival of many animal and plant species. While the Earth has witnessed mass extinctions in the past due to natural factors, the anticipated future extinctions of flora and fauna are predominantly influenced by human activities. The rising human population, coupled with pollution and loss of habitat, has set the stage for mass extinctions and substantial changes in the flora and fauna.

**Effects on Agroecosystem** Agroecosystems are sources and sinks of greenhouse gases such as $CO_2$, $CH_4$, and $N_2O$. The major pathways for GHG emissions are gas exchange between the plant-soil ecosystem and the atmosphere (direct emission) and emissions from tile drainage and surface runoff to water bodies (indirect emission). Therefore, GHG emissions are governed by plant growth and microbial activities and influenced by climatic factors (temperature and precipitation) and soil factors (soil temperature and soil moisture). Agricultural management practices, such as tillage, sowing, fertilization, irrigation and drainage, harvesting, and organic matter management, such as manure application and residue burning and incorporation of residues into soil, greatly influence GHG emissions.

## Fertilizers and GHGs

Inorganic fertilizer production, delivery, and use in agriculture fields all contribute to greenhouse gas emissions, particularly $CO_2$ and $N_2O$. Fertilizer manufacturing industries consume about 1.2% of the global energy and release 1.2% of greenhouse gas emissions (Kongshaug 1998). Nitrogen fertilizers are the major catalyst for the crop production of about 50 percent of the food consumed by humans (Smil 2000; Erisman et al. 2008). About 20% of global greenhouse gas (GHG) flux is accounted for by land use change. Fertilizers boost agricultural productivity, encourage crop $CO_2$ uptake, and decrease the need to cultivate new land (deforestation), resulting in fewer GHG emissions as a result of land use change. The $NO_2$ and $CO_2$ emitted during the production of ammonia and nitric acid are the two most important GHG emissions associated with the manufacture of nitrogen fertilizers.

The carbon footprint (CF) of fertilizers varies between the countries due to the difference in the production technology and raw materials used. In China, the average CF of N-based fertilizers (urea, compound fertilizers, and ammonium bicarbonate) is 8.1 kg $CO_2$-eq kg N − 1, 7.4 kg $CO_2$-eq kg $N^{-1}$, and 7.2 kg $CO_2$-eq kg $N^{-1}$, respectively (Zhang et al. 2013). China's average manufacturing GHG emission factor for DAP is 10.3 kg $CO_2$-eq kg DAP-$N^{-1}$ (Chen et al. 2015). In Europe, the average CF from using ammonium nitrate (AN) fertilizer is 5.6 kg $CO_2$-eqv per kg applied N. In India, the estimated average GHG emission during the production and transport of 1 kg of N (DAP), N (urea), $K_2O$ (MOP), and $P_2O_5$ (SSP) is 1.84, 2.02, 0.25, and 0.06 kg $CO_2$ equivalent, respectively (Kool et al. 2012). The region-wise carbon footprint of major N fertilizers is presented in Table 1. In developed countries, like India, urea is the most common source of nitrogen for crop production. Urea offers array advantages in terms of distribution, storage, and handling costs due to its high N concentration (46%) and solid physical form. However, ammonium nitrate-based fertilizers like ammonium nitrate and CAN have proven more effective and environmentally friendly than urea under European climate and soil conditions.

During the manufacturing stages of the fertilizer's life cycle, urea has a lower carbon footprint than ammonium nitrate. This is mainly because urea captures a portion of the $CO_2$ produced during ammonia synthesis. However, as soon as the urea is spread on the field, the captured $CO_2$ is released back into the atmosphere. 2O emissions are a major contributor to ozone depletion due to their long atmospheric lifespan (about 116 years), high global warming potential (310 times that of $CO_2$), and high global climate change potential (290 on a 100-year scale) (Fagodiya et al. 2020). Urea also emits more ammonia into the atmosphere during farming than ammonium nitrate, which increases the risk of not meeting the national limits on emissions of acidifying chemical substances set by the Gothenburg Protocol. The Indian government has introduced a bill to produce 100% of urea with neem oil coating. The chemical compound present in neem oil acts as a nitrification inhibitor (NI) and also acts as a physical barrier, slowing down the speed of urea solubility (Reddy and Prasad 1975). *Nitrosomonas* sp. and *Nitrobacter* sp. are two bacteria

**Table 1** Carbon footprint of N fertilizers (Hoxha and Christensen 2019)

| | Fertilizer | CF (kg $CO_2$ eq/kg N)[a] | | | | CF (kg $CO_2$ eq/kg product)[a] | | | |
|---|---|---|---|---|---|---|---|---|---|
| S. no | Region | Europe | NA | Africa | SEA | Europe | NA | Africa | SEA |
| 1 | AN (35% N) | 3.319 | 6.712 | 6.153 | 6.997 | 1.112 | 2.249 | 2.061 | 2.344 |
| 2 | CAN (27% N) | 3.523 | 6.926 | 6.384 | 7.252 | 0.951 | 1.870 | 1.724 | 1.958 |
| 3 | Urea (46% N) | 3.502 | 3.781 | 3.610 | 3.623 | 1.611 | 1.739 | 1.661 | 1.667 |
| 4 | UAN (30% N) | 3.402 | 5.238 | 4.887 | 5.327 | 1.021 | 1.571 | 1.466 | 1.598 |

Note: *AN* Ammonium nitrate, *CAN* Calcium ammonium nitrate, *UAN* (liquid fertilizer), *CF* Carbon footprint, *NA* North America, *SEA* South East Asia

[a]Incl. $CO_2$ captured in product

that are majorly responsible for nitrification process. Neem-coated urea retards the growth and multiplication of nitrifying bacteria.

Optimizing N-use efficiency is crucial to sustain productivity and profitability to meet the demand. Depending on the edaphic and climatic conditions, improved N management can dramatically cut GHGs emissions. Producers must ensure that the kind, rate, and time of N application do not result in substantial losses owing to volatilization, leaching, or denitrification. The good soil structure improves N-use efficiency and reduces $N_2O$ losses. Adaption of best nitrogen management practices like urea deep placement, use of nitrification inhibitors, urease inhibitors, and slow-release nitrogen fertilizers will reduce the N loss and increase N-use efficiency.

## Greenhouse Gas Emission form Phosphate Fertilizers

For phosphate fertilizers, emissions estimates are dominated by $CO_2$, and most of GHG emissions are due to the use of fossil fuels as an energy source for the various phosphate fertilizer production processes. The net GHG emission from phosphate fertilizer production is primarily determined by the method used in the production of sulfuric acid. Raw material transport is a foremost source of GHG emissions from phosphate fertilizers, accounting for about one-third of total emissions. The carbon footprint of major phosphate fertilizers is presented in Table 2.

### *Greenhouse Gas-Mitigating Microbes*

Microbial processes play a crucial role in global fluxes of biogenic greenhouse gases (methane, nitrous oxide, and carbon dioxide) and are responding rapidly to climate change. The effect of microbial processes on greenhouse gas emissions is unclear, but understanding the mechanisms of microorganisms based on terrestrial greenhouse gas flux is essential for predicting climate change. This involves understanding the complex interactions between microbes and various biotic and abiotic

**Table 2** Europe carbon footprint of phosphate fertilizers (Davis and Haglund 1999)

| Fertilizer | CF (g $CO_2$ eq/kg product)[a] |
|---|---|
| SSP | 291 |
| TSP | 690.6 |
| MAP | 941 |
| DAP | 1077 |

Note: *SSP* Single superphosphate, *DAP* Diammonium phosphate, *TSP* Triple superphosphate, *MAP* Monoammonium phosphate, *CF* Carbon footprint

[a]Incl. $CO_2$ emission during transport

factors. Managing terrestrial microbial processes to reduce greenhouse gas emissions is an exciting prospect for mitigating climate change.

## Methane-Mitigating Microbes

Methanotrophs: Methanotrophs, methane-oxidizing bacteria, are a major group of bacteria primarily responsible for the natural global methane cycle. Methanotrophs play a vital role in lowering methane levels in the atmosphere under certain environmental conditions and factors (Hanson and Hanson 1996). Lactic acid bacteria: Lactic acid bacteria (LAB) are facultative anaerobic, gram-negative, and acid-tolerant bacteria that can reduce methane emissions from livestock farming and are essential for ensuring sustainable food production in the future. One possible approach for lactic acid production is to use LAB to produce lactic acid as an important end product of carbohydrate fermentation (Stilez and Holzapfel 1997). Biochemically, LAB can be divided into homofermenters, which produce lactic acid, and heterofermenters, which produce a variety of fermentation end products, such as ethanol, acetic acid, and $CO_2$. LAB have been used as starter cultures for a wide range of dairy, meat, and plant-based fermentations. Their long history of use in human and animal foods has resulted in most LAB having Qualified Presumption of Safety (QPS) status in the European Union or Generally Recognized as Safe (GRAS) status in the United States. The main LAB genera used as starter cultures are Lactobacillus, Lactococcus, Leuconostoc, and Pediococcus (Bintsis 2018), along with some species of Enterococcus and Streptococcus.

**Sulfate-Reducing Microorganisms (SRM)** Sulfate-reducing microorganisms (SRMs) are robustly favored in competition for substrates with other microbes involved in methanogenic degradation pathways, resulting in the diversion of carbon flow from methane to carbon dioxide (Gauci et al. 2004). This mitigating effect of sulfate reduction on methane flux from wetlands is expected to become even more pronounced in the near future (Gauci et al. 2004). Molecular-based approaches have revealed that members of recognized SRM taxa are found in freshwater wetlands. The existence of SRMs that grow well in low-sulfate environments has been demonstrated by enrichments from lake sediments (Bak and Pfennig 1991; Ramamoorthy et al. 2006), rice paddy fields (Dalsgaard and Bak 1994; Wind and Conrad 1995; Wind et al. 1999), permafrost soil (Vatsurina et al. 2008), and constructed wetlands (Lee et al. 2009). These include species of the genera Desulfovibrio, Desulfobulbus, Desulfotomaculum, and Desulfosporosinus. The latter genes encode the alpha- and beta-subunits of dissimilatory (bi)sulfite reductase, which is used by SRMs for energy conservation.

SRMs have been detected by molecular analyses such as quantitative PCR assays with taxa-specific 16S rRNA gene-based primers (Scheid and Stubner 2001; Stubner 2002, 2004; Stahl et al. 2007) and with the DSR1F-DSR4R primer variants targeting the functional marker genes dsrAB (Wagner et al. 2005).

## Effect of Fertilizer Application on GHG-Mitigating Microbes

Many soil organisms rely on organic matter, such as crop residues and animal manure, while inorganic fertilizers can have a negative impact on soil biology and productivity. Changes in soil nutrient availability can alter the structure and/or activity of microbial communities, and consequently nutrient cycling processes, all of which affect greenhouse gas (GHG) emissions (Blagodatskiy et al. 2008; Blagodatsky and Richter 1998). Generally, nutrients in soluble form are used by both plants and microorganisms. Water-soluble fertilizers can effectively supply nutrients to microbes when present in the optimal amount. Conversely, higher doses of fertilizers can temporarily suppress microbial growth, but this reduction is minor given that a teaspoon of soil can contain anywhere from 100 million to 1 billion individual bacteria. Long-term use of mineral fertilizers has been reported to reduce soil organic carbon content, microbial population, and microbial biomass carbon, despite increasing NPK levels. Inorganic fertilizers can also elevate the possibility of high salinity by increasing soil pH, which modifies the microbial community structure and activity.

Long-term application of organic manure combined with inorganic fertilizer can effectively increase soil nutrient content, improve soil physicochemical properties, increase crop productivity, and thereby affect microorganisms (Oehl et al. 2004; Shi et al. 2008; Feitosa de Souza et al. 2016; Lin et al. 2018). In contrast, long-term use of synthetic nitrogenous fertilizer alone reduces soil pH, microbial biomass, and damages soil microbial structure (Sun et al. 2017; Huang et al. 2018; Li et al. 2019). The combined use of organic manure and mineral fertilizer reduces soil acidification and increases soil organic carbon (SOC), total nitrogen, nitrate nitrogen, total, and available phosphorus levels. The changing physicochemical properties of soil, depending on nutrient management practices, influence microbial growth, activity, and diversity, which in turn affects GHG emissions.

### *Nitrogen Fertilizers Effect on GHG Microbes*

Long-term fertilization altered the physicochemical properties of soil, affecting soil fertility and the abundance and community structure of ammonia-oxidizing archaea (AOA) and ammonia-oxidizing bacteria (AOB). Soil organic carbon (SOC), total nitrogen, nitrate nitrogen, total and available phosphorus, and available potassium all influenced AOA and AOB (Cai et al. 2020). Specific and rare AOA and AOB communities played a more dominant role in nitrification than the whole AOA or AOB community (Cai et al. 2020). Nitrification, the conversion of ammonia to nitrite and eventually to nitrate, is an important mechanism in the agricultural nitrogen cycle. AOA may be able to adapt to oxygenated or hypoxic water conditions (Liu et al. 2015) and may therefore be dominant over AOB in nitrogen cycling under extreme environmental conditions (Erguder et al. 2009).

In diverse environmental conditions, differing $NH_4^+$-N concentrations and levels of total carbon and total nitrogen in soils altered the composition and abundance of AOA and AOB (Huang et al. 2013; Li and Gu 2013; Wang et al. 2013). The application of ammonium-containing N fertilizers often increases CH4 emissions by stimulating plant growth and methanogenic bacterial activity and inhibiting methane-oxidizing bacteria activity (Lindau 1994; Kruger and Frenzel 2003). In contrast, nitrate and sulfate have been reported to inhibit microbial methanogenesis in anoxic soil (Bollag and Czlonkowski 1973; Cappenberg 1975; Westermann and Ahring 1987; Kitada et al. 1993; Achtnich et al. 1995). Rhizobium sp. is an important greenhouse gas-mitigating microbe in legume crops. It reduces the need for externally applied N fertilizers and thereby reduces GHG emissions. When legume crops are grown in soil with suboptimal N levels, externally applied N fertilizers act as a starter dose and stimulate the growth of Rhizobium sp., enhancing nodulation by improving the initial growth of the plants.

## *Phosphate Fertilizers Effect on GHG Microbes*

Methane ($CH_4$) production in flooded rice soils depends on the source of applied phosphorus (P). Applying P through Mussorie rock phosphate or single superphosphate (SSP) inhibits CH4 emissions in flooded rice fields, while applying P through $KH_2PO_4$ or Jordanian rock phosphate stimulates CH4 production in P-deficient soils. When compared to the control, Mussorie rock phosphate prevented CH4 emissions by 70%, while SSP treatment resulted in a substantially larger reduction (1% of the control). The inhibitory effect was linked to the sulfate content of SSP (Adhya et al. 1997). The amount of sulfate in P sources influences the extent of stimulation or inhibition of CH4 flux. Under anaerobic conditions, sulfate acts as an inhibitor of CH4 production by sulfate-reducing bacteria, which use acetate and $H_2$ as energy sources, which are otherwise used as substrates by methanogens (Cappenberg 1975; Winfrey and Zeikus 1977; Achtnich et al. 1995). Therefore, adding phosphorus to P-deficient soils using SSP could reduce $CH_4$ emissions and sustain higher yields.

Acetate can be secreted from rice roots or generated through the fermentation of other substrates, such as sugars or polysaccharides, derived from root exudates or root degradation. Acetate concentrations in the rhizosphere vary depending on the rice variety and show pronounced spatial and seasonal variations in the micromolar to millimolar range (Rothfuss and Conrad 1992; Sigren et al. 1997). This indicates that acetate is an important regulatory factor for CH4 flux by the rice root microflora under in situ conditions. The addition of phosphate ($\geq$20 mM) to rice has been shown to suppress acetotrophic methanogenesis, thereby reducing CH4 flux (Conrad et al. 2000). However, the addition of only 10 mM phosphate increased CH4 flux by a small amount (Conrad et al. 2000). This is likely due to a lack of phosphate availability to rice plants (Lu et al. 1999). Increased phosphate concentrations tend to inhibit methanogens, particularly acetotrophic ones, but the mechanism by which they do so is uncertain. Both phosphorus and potassium fertilizers promote the growth of Rhizobium sp., regardless of the rate and timing of application. Rhizobium sp. are nitrogen-fixing bacteria found in the root nodules of legumes such as alfalfa and soybeans. Phosphorus fertilizer application encourages root growth and nitrogen fixation in legumes. Phosphate fertilizer source, rate of application, and bacterial interactions all affect the dry matter and structure of soil microbial communities, with a higher proportion of P2O5 resulting in increased effects (Gumiere et al. 2019). Thus, phosphorus fertilization increases the growth and activity of greenhouse gas-mitigating microbes in agriculture.

Phosphorus fertilizer application has beneficial effects on the growth and reproduction of ammonia-oxidizing bacteria (AOB) by increasing soil organic matter (Cai et al. 2020). AOB are more sensitive to phosphorus application than ammonia-oxidizing archaea (AOA). In urea-treated soil, the addition of phosphorus fertilizer reduced the nitrification process, with the N:P ratio appearing to be a critical factor restricting nitrification (Ning et al. 2021). Soil phosphorus has a significant impact on nitrous oxide ($N_2O$) fluxes, with higher emissions at lower phosphorus concentrations, indicating that phosphorus is a key driver in the nitrogen cycle (O'Neill et al. 2021). Nitrogen undergoes numerous transformations as it moves through the nitrogen cycle. The primary pathways are nitrification and denitrification, in which nitrogen molecules ($NH_4^+$, organic N) are oxidized to $NO_3$, which can then be reduced to gaseous nitrogen compounds such as $N_2O$ or $N_2$ by denitrification (Signor and Cerri 2013). Therefore, the synergistic effects of nitrogen and phosphorus fertilizer application should be further explored to reduce nitrogen losses from agriculture. Mineralization converts organic forms of nitrogen to inorganic forms of nitrogen, which are available to plants and microorganisms (Signor and Cerri 2013). Microbes require inorganic nitrogen to produce alkaline phosphatase enzymes, which release phosphorus.

## *Potassium Fertilizers Effect on GHG Microbes*

Potassium fertilizer application significantly reduced methane (CH4) emissions from flooded rice fields by inhibiting methanogenic bacteria and stimulating methanotrophic bacterial populations. A study by Babu et al. (2006) found that applying 30 kg K/ha of muriate of potash (MOP) reduced CH4 emissions by 49% compared to no K application. Potassium fertilizer addition also prevented the decline in redox potential of flooded soil and reduced the concentrations of active reducing substances and $Fe_2^+$ in the rhizosphere, thereby decreasing reduction reactions (Chen et al. 1997).

## Strategies to Mitigate Greenhouse Gases

Reducing greenhouse gas (GHG) emissions is a top priority for all nations, as the problem is global and no single country or group of countries can solve it alone. This is why worldwide and regional cooperation is so important and well-supported by the global community. In response to this challenge, the United Nations Framework Convention on Climate Change (UNFCCC) recently entered into force to address the global climate problem. This international agreement, which involves many countries from different regions, aims to reduce harmful levels of anthropogenic GHGs in the atmosphere.

### *Clean Development Mechanism*

The promotion of sustainable development is facilitated by the Clean Development Mechanism (CDM), which encourages the adoption of renewable energy technologies in power generation and $CO_2$ sequestration. Apart from their GHG mitigation potential, renewable energy resources can also help to ensure energy security. The CDM is a key mechanism of the Kyoto Protocol, an international agreement under the United Nations Framework Convention on Climate Change (UNFCCC).

Developing countries have been actively involved in renewable power generation, hosting 53% of global RE power generation in 2009. The CDM was initially designed to strategically reduce emissions from energy generation and consumption to a sustainable level. It was envisaged that developed countries would fund emission reduction mechanisms and provide developing countries with sponsorship for renewable energy programs. However, after more than a decade, the implementation of the CDM has been disappointing, and the global pace of renewable power deployment is not in line with expectations.

### *Financing Low-Carbon Energy*

The combustion of petroleum products releases large quantities of greenhouse gases, including carbon dioxide ($CO_2$). Access to modern energy is essential for development, but it is also important to reduce $CO_2$ emissions from energy production. One way to do this is to finance low-carbon energy projects. This is especially important in countries where oil is the main source of income and energy production. A low-carbon economy can be achieved by harnessing the potential of renewable energy resources. Utilizing renewable energy sources can enhance the energy infrastructure in rural areas, contributing to the alleviation of energy poverty. Decentralized renewable energy systems can provide a strategic solution to energy poverty in developing regions. In some countries, emissions trading schemes (ETS) are used to control and monitor greenhouse gas emissions. This is done by allowing companies to trade emissions allowances. According to the DOC Office of Sustainable Energy and Environmental Programs (2012), reducing greenhouse gas emissions, such as CO2 and CH4, is not the only way to mitigate the greenhouse effect.

## International Organizations Also Recommend the Following Actions

Switch to renewable energy. Use public transportation and other low-emission modes of transportation, such as electric vehicles and bicycles. Educate citizens and governments about the importance of environmental protection. Adopt recycling and circular economy practices. Eat less meat and reduce food waste.

Choose organic foods when possible. Climate action is any policy, program, or measure that works to reduce greenhouse gas emissions, build resilience to climate change, and finance those goals.

The Paris Agreement (2015) was the first and foremost international agreement toward these ends. At COP21, 174 countries and the European Union agreed to work towards the goal of keeping global warming below 2 °C.

## Conclusion

Since the industrial revolution, the concentrations of atmospheric greenhouse gases such as $CO_2$, tropospheric $O_2$, $CH_4$, and $N_2O$ have been increasing. This is due to human activities such as deforestation, fossil fuel combustion, and agriculture. Inorganic fertilizer production, delivery, and use in agriculture also contribute to greenhouse gas emissions, particularly $CO_2$ and $N_2O$. In addition, terrestrial microbial activities also contribute to the atmospheric concentrations of greenhouse gases. Recent research on methanogenesis, fertilizer contributions to GHG emissions, and the role of soils in the $CH_4$ cycle indicates that further investigations are required to acquire more knowledge on global $CH_4$ balance. However, further investigations are required to acquire a more comprehensive understanding of this complex issue. We need to strategically tackle GHG reduction to prevent the degradation of the global environment and protect biodiversity. The potential to mitigate climate change by reducing greenhouse gas emissions through the management of terrestrial microbial processes is a promising vision for the future. This could be achieved by enhancing the population of methanotrophs. Methanotrophs play crucial roles in the global carbon cycle by removing methane from the atmosphere. Further research is needed to strengthen areas like long-term field studies on management strategies to reduce GHG emissions from rice cultivation.

## References

Achtnich C, Bak F, Conrad R (1995) Competition for electron donors among nitrate reducers, ferric iron reducers, sulfate reducers, and methanogens in anoxic paddy soil. Biol Fert Soils 19:65–72

Adhya TK, Pattanaik P, Sathpathy SN, Kumaraswamy S, Sethunathan N (1997) Influence of phosphorus application on methane emission and production in flooded paddy soils. Soil Biol Biochem 30:177–181

Babu YJ, Nayak DR, Adhya TK (2006) Potassium application reduces methane emission from a flooded field planted to rice. Biol Fertil Soils 42(6):532–541

Bak F, Pfennig N (1991) Sulfate-reducing bacteria in littoral sediment of Lake Constance. FEMS Microbiol Ecol 85:43–52

Bhunya PK, Kumar S, Gurrapu S, Bhuyan MK (2020) A review of case studies on climate change impact on hydrologic cycle: an Indian perspective. Int J Sci Res Sci Eng Technol 7(5):249–266

Bintsis T (2018) Lactic acid bacteria as starter cultures: an update in their metabolism and genetics. AIMS Microbiol 4:665–684

Blagodatskiy SA, Avksent'Ev AA, Davydova MA, Blagodatskaya E, Kurakov AV (2008) Nitrous oxide production in soils and the ratio of the fungal to bacterial biomass. Eurasian Soil Sci 41(13):1448–1455

Blagodatsky SA, Richter O (1998) Microbial growth in soil and nitrogen turnover: a theoretical model considering the activity state of microorganisms. Soil Biol Biochem 30(13):1743–1755

Bollag JM, Czlonkowski ST (1973) Inhibition of methane formation in soil by various nitrogen containing compounds. Soil Biol Biochem 5:673–678

Bouwman AF (1996) Direct emission of nitrous oxide from agricultural soils. Nutr Cycl Agroecosyste 46:53–70

Cai F, Luo P, Yang J, Irfan M, Zhang S, An N, Da J, Han X (2020) Effect of long-term fertilization on ammonia-oxidizing microorganisms and nitrification in brown soil of Northeast China. Front Microbiol 11:3579

Cappenberg TE (1975) Relationship between sulphate reducing and methane producing bacteria. Plant Soil 43:125–139

Chen J, Xuan J, Du C, Xie J (1997) Effect of potassium nutrition of rice on rhizosphere redox status. Plant Soil 188:131–137

Chen BM, Wang ZH, Li SX, Wang GX, Song HX, Wang XN (2004) Effects of nitrate supply on plant growth, nitrate accumulation, metabolic nitrate concentration and nitrate reductase activity in three leafy vegetables. Plant Sci 167:635–643

Chen S, Lu F, Wang XK (2015) Estimation of greenhouse gases emission factors for China's nitrogen, phosphate, and potash fertilizers. Acta Ecol Sin 35:6371–6383 (In Chinese)

Cline RW (2008) Global warming and agriculture. Finan Dev:23–27

Conrad R, Klose M, Claus P (2000) Phosphate inhibits acetotrophicmethanogenesis on rice roots. Appl Environ Microbiol 66:828–833

Dalsgaard T, Bak F (1994) Nitrate reduction in a sulfate-reducing bacterium, *Desulfovibriodesulfuricans*, isolated from rice paddy soil: sulfide inhibition, kinetics, and regulation. Appl Environ Microbiol 60:291–297

Davis J, Haglund C (1999) Life Cycle Inventory (LCI) of fertiliser production. Fertiliser products used in Sweden and Western Europe. SIK-Report No. 654. Masters Thesis, Chalmers University of Technology

Dedysh SN, Belova SE, Bodelier PLE, Smirnova KV, Khmelenina VN, Chidthaisong A, Trotsenko YA, Liesack W, Dunfield PF (2007) *Methylocystisheyerispnov.,* a novel type II methanotrophic bacterium possessing 'signature' fatty acids of type I methanotrophs. Int J Syst Evol Microbiol 57:472–479

Erguder TH, Boon N, Wittebolle L, Marzorati M, Verstraete W (2009) Environmental factors shaping the ecological niches of ammonia-oxidizing archaea. FEMS Microbiol Rev 33:855–869

Erisman JW, Sutton MA, Galloway J, Klimont Z, Winiwarter W (2008) How a century of ammonia synthesis changed the world. Nat Geosci 1:636–639
Fagodiya RK, Pathak H, Kumar A, Bhatia A, Jain N, Malyan SK (2020) Global warming impacts of nitrogen use in agriculture: an assessment for India since 1960. Carbon Manag 11(3):291–301
Feitosa de Souza TA, Rodrigues AF, Marques LF (2016) The trend of soil chemical properties, and rapeseed productivity under different long-term fertilizations and stubble management in a Ferralsols of Northeastern Brazil. Org Agric 7:353–363
Gauci V, Matthews E, Dise N, Walter B, Koch D, Granberg G, Vile M (2004) Sulfur pollution suppression of the wetland methane source in the 20th and 21st centuries. Proc Natl Acad Sci 101:12583–12587
Gebert J, Grongroft A, Schloter M, Gattinger A (2004) Community structure in a methanotroph by phospholipid fatty acid biofilter as revealed analysis. FEMS Microbiol Lett 240:61–68
Gumiere T, Rousseau AN, da Costa DP, Cassetari A, Cotta SR, Andreote FD, Gumiere SJ, Pavinato PS (2019) Phosphorus source driving the soil microbial interactions and improving sugarcane development. Sci Rep 9(1):1–9
Hanson RS, Hanson TE (1996) Methanotrophic bacteria. Microbiol Rev 60:439–471
Hoxha A, Christensen B (2019) The carbon footprint of fertiliser production: regional reference values. Proc-Int Fertil Soc 805:1–20
Huang L, Dong H, Wang S, Huang Q, Jiang H (2013) Diversity and abundance of ammonia-oxidizing archaea and bacteria in diverse chinese paddy soils. Geomicrobiol J 31:12–22
Huang X, Liu Y, Li Y, Guo P, Fang X, Yi Z (2018) Foliage application of nitrogen has less influence on soil microbial biomass and community composition than soil application of nitrogen. J Soils Sediments 19:221–231
Kitada K, Ozaki Y, Akiyama Y, Yagi K (1993) Effect of high contents of $NO_3$-N in irrigation water and rice straw application on $CH_4$ emission from paddy fields. J Soil Sci Plant Nutr 64:49–54
Kongshaug G (1998) Energy consumption and greenhouse gas emissions in fertilizer production. IFA technical conference, Marrakech, Morocco, 28 September-1 October, p 18
Kool A, Marinussen M, Blonk H (2012) LCI data for the calculation tool feed print for greenhouse gas emissions of feed production and utilization. In: GHG emissions of N, P and K fertilizer production. Editor B Consultants
Kruger M, Frenzel P (2003) Effects of N-fertilisation on $CH_4$ oxidation and production, and consequences for $CH_4$ emissions from microcosms and rice fields. Glob Chang Biol 9:773–784
Lee YJ, Romanek CS, Wiegel J (2009) *Desulfosporosinusyoungiae* sp. *nov*., a spore-forming, sulfate-reducing bacterium isolated from a constructed wetland treating acid mine drainage. Int J Syst Evol Microbiol 59:2743–2746
Li M, Gu JD (2013) Community structure and transcript responses of anammox bacteria, AOA, and AOB in mangrove sediment microcosms amended with ammonium and nitrite. Appl Microbiol Biotechnol 97:9859–9874
Li N, Kumar P, Lai L, Abagandura GO, Kumar S, Nleya T (2019) Response of soil greenhouse gas fluxes and soil properties to nitrogen fertilizer rates under camelina and carinatanonfood oilseed crops. Bioenergy Res 12:524–535
Lin Y, Slessarev EW, Yehl ST, D'Antonio CM, King JY (2018) Long-term nutrient fertilization increased soil carbon storage in California grasslands. Ecosystems 22:754–766
Lindau CW (1994) Methane emission from Louisiana rice fields amended with nitrogen fertilizers. Soil Biol Biochem 26:353–359
Liu XJ, Zhang Y, Han WX, Tang AH, Shen JL, Cui ZL, Vitousek P, Erisman JW, Goulding K, Christie P, Fangmeier A, Zhang FS (2013) Enhanced nitrogen deposition over China. Nature 494:459–462
Liu S, Hu B, He Z, Zhang B, Tian G, Zheng P (2015) Ammonia-oxidizing archaea have better adaptability in oxygenated/hypoxic alternant conditions compared to ammonia-oxidizing bacteria. Appl Microbiol Biotechnol 99:8587–8596
Lu Y, Wassmann R, Neue HU, Huang C (1999) Impact of phosphorus supply on root exudation, arenchyma formation and methane emission of rice plants. Biogeochemistry 47:203–218

Ning J, Arai Y, Shen J, Wang R, Ai S (2021) Effects of phosphorus on nitrification process in a fertile soil amended with urea. Agriculture 11(6):523

O'Neill RM, Krol DJ, Wall D, Lanigan GJ, Renou-Wilson F, Richards KG, Jansen-Willems AB, Müller C (2021) Assessing the impact of long-term soil phosphorus on N-transformation pathways using 15N tracing. Soil Biol Biochem 152:108066

Oehl F, Sieverding E, Mader P, Dubois D, Ineichen K, Boller T (2004) Impact of long-term conventional and organic farming on the diversity of arbuscularmycorrhizal fungi. Oecologia 138:574–583

Poff NL, Brinson MM, Day JW (2002) Aquatic ecosystems and Global climate change. Millennium Ecosystem Assessment, 2005. Ecosystems and human wellbeing: synthesis. Island Press, Washington, DC

Ramamoorthy S, Sass H, Langner H, Schumann P, Kroppenstedt RM, Spring S, Overmann J, Rosenzweig RF (2006) Desulfosporosinus lacus spnov., a sulfate-reducing bacterium isolated from pristine freshwater lake sediments. Int J Syst Evol Microbiol 56:2729–2736

Reddy RNS, Prasad R (1975) Studies on the mineralization of urea, coated urea, and nitrification inhibitor treated urea in soil. Eur J Soil Sci 26(3):304–312

Roeckner E (1992) Past, present and future levels of greenhouse gases in the atmosphere and model projections of related climatic changes. J Exp Bot 43(8):1097–1109

Rothfuss F, Conrad R (1992) Vertical profiles of $CH_4$ concentrations, dissolved substrates and processes involved in $CH_4$ production in a flooded Italian rice field. Biogeochemistry 18(3):137–152

Sarkodie SA, Ntiamoah EB, Dongmei L (2019) Panel heterogeneous distribution analysis of trade and modernized agriculture on $CO_2$ emissions: the role of renewable and fossil fuel energy consumption. Nat Res Forum 43:135–153

Scheid D, Stubner S (2001) Structure and diversity of gram-negative sulfate-reducing bacteria on rice roots. FEMS Microbiol Ecol 36:175–183

Shi ZH, Chen LD, Cai CF, Li ZX, Liu GH (2008) Effects of long-term fertilization and mulch on soil fertility in contour hedgerow systems: a case study on steeplands from the three gorges area, China. Nutr Cycl Agroecosyst 84:39–48

Signor D, Cerri CEP (2013) Nitrous oxide emissions in agricultural soils: a review. Pesquisa Agropecuária Trop 43:322–338

Sigren LK, Byrd GT, Fisher FM, Sass RL (1997) Comparison of soil acetate concentrations and methane producton, transport, and emission in two rice cultivars. Glob Biogeochem Cycles 11(1):1–14

Singh BK, Bardgett RD, Smith P, Reay DS (2010) Microorganisms and climate change: terrestrial feedbacks and mitigation options. Nat Rev Microbiol 8:779–790

Smil V (2000) Feeding the world: a challenge for the twenty-first century. MIT Press, Cambridge, MA

Stahl DA, Loy A, Wagner M (2007) Molecular strategies for studies of natural populations of sulphate-reducing microorganisms. In: Barton LL, Hamilton WA (eds) Sulphate-reducing bacteria. Cambridge University Press, Cambridge, pp 39–115

Stevenson DS, Johnson CE, Collins WJ, Derwent RG, Shine KP, Edwards JM (1998) Evolution of tropospheric ozone radiative forcing. Geophys Res Lett 25:3819–3822

Stilez ME, Holzapfel W (1997) Lactic acid bacteria of foods and their current taxonomy. Int J Food Microbiol 36(1):1–29

Stubner S (2002) Enumeration of 16S rDNA of *Desulfotomaculum* lineage 1 in rice field soil by real-time PCR with SybrGreenTM detection. J Microbiol Methods 50:155–164

Stubner S (2004) Quantification of gram-negative sulphate-reducing bacteria in rice field soil by 16S rRNA gene-targeted real-time PCR. J Microbiol Methods 57:219–230

Sun YF, Shen JP, Zhang CJ, Zhang LM, Bai WM, Fang Y (2017) Responses of soil microbial community to nitrogen fertilizer and precipitation regimes in a semi-arid steppe. J Soils Sediments 18:762–774

Trinastic J (2015) Methane munching microbes limit global warming. http://www.nature.com/scitable/blog/eyesonenvironment/methanemunchingmicrobes limitglobalwarming. Accessed 15 Sept 2021

Vatsurina A, Badrutdinova D, Schumann P, Spring S, Vainshtein M (2008) Desulfosporosinushippei sp. nov., a mesophilic sulfate-reducing bacterium isolated from permafrost. Int J Syst Evol Microbiol 58:1228–1232

Wagner M, Loy A, Klein M, Lee N, Ramsing NB, Stahl DA, Friedrich MW (2005) Functional marker genes for identification of sulphate-reducing prokaryotes. Methods Enzymol 397:469–489

Wang C, Zhu G, Wang W, Yin C (2013) Preliminary study on the distribution of ammonia oxidizers and their contribution to potential ammonia oxidation in the plant-bed/ditch system of a constructed wetland. J Soils Sediments 13:1626–1635

Westermann P, Ahring BK (1987) Dynamics of methane production, sulfate reduction and denitrification in a permanently waterlogged alder swamp. Appl Environ Microbiol 53:2554–2559

Wind T, Conrad R (1995) Sulfur compounds, potential turnover of sulfate and thiosulfate, and numbers of sulfate-reducing bacteria in planted and unplanted paddy soil. FEMS Microbiol Ecol 18:257–266

Wind T, Stubner S, Conrad R (1999) Sulfate-reducing bacteria in rice field soil and on rice roots. Syst Appl Microbiol 22:269–279

Winfrey MR, Zeikus JG (1977) Effect of sulphate on carbon and electron flow during microbial methanogenesis in fresh water sediments. Appl Environ Microbiol 33:275–281

Zhang WF, Dou ZX, He P, Ju XT, Powlson D, Chadwick D, Norse D, Lu YL, Zhang Y, Wu L, Chen XP, Cassman KG, Zhang FS (2013) New technologies reduce greenhouse gas emissions from nitrogenous fertilizer in China. Proc Natl Acad Sci 110:8375–8380

Zimmer C (2010) The microbe factor and its role in our climate future. http://e360.yale.edu/feature/themicrobefactoranditsroleinourclimatefuture/2279/. Accessed 22 Sept 2021

Zimmerman L, Labonte B (2015) Climate change and the microbial methane banquet. Climate Alert 27(1). http://www.climate.org/publications/Climate%20Alerts/2015summer/climatechangemicrobialmethanebanquet.Html. Accessed 10 Sept 2021

# Microbial Diversity Under Organic Fertilizer Management Systems and Identification of Best Fertilizer Practice

**Himanshu Mahawar, Jairam Choudhary, and Ekta Narwal**

## Introduction

The soil ecosystem consists of a variety of micro- and macrofauna that play an important role in breaking down complex substances from plant and animal decay into simpler forms for use by living or growing plants. Soil organisms act as catalysts for many biogeochemical cycles, which in turn influence soil biotic and abiotic properties. Among them, soil microorganisms are a particularly important group that is essential for soil structure and health. The spatial and temporal variations in the soil microbial diversity govern the sustainability and vital functions of the soil ecosystem. Soil fertility maintenance, carbon storage, and regulation by natural biogeochemical cycles or organic matter degradation are all subject to quantitative and qualitative alterations in soil microbial diversity. Symbiotic nitrogen-fixing microbes, mycorrhizal fungi, phosphate solubilizers, cellulose decomposers, and other microbes have been shown to play a significant role in plant nutrition by refining the mineralization and stabilization of minerals. Organic farming systems are low-input agroecosystem farms that rely on the natural availability of plant nutrients, organic/green manure, microbial biofertilizers, and biological pest control to boost plant productivity and ecosystem functionality (Van Bueren et al. 2002). In contrast, conventional farming systems rely on the intensive application of agrochemicals, such as synthetic fertilizers, fungicides, and pesticides, to increase crop productivity and protect plants from pathogens and insects. Plant roots regulate soil

H. Mahawar (✉)
ICAR-Directorate of Weed Research, Jabalpur, India

J. Choudhary
ICAR-Indian Institute of Farming Systems Research, Modipuram, India

E. Narwal
Shobhit Institute of Engineering and Technology, Modipuram, India



S. R. Mohanty, B. Kollah (eds.), *Greenhouse Gas Regulating Microorganisms in Soil Ecosystems*, https://doi.org/10.1007/978-3-031-70569-4_15

pH, increase soil nutrient levels, and secrete specific compounds that create a rhizosphere bacterial community structure enriched with beneficial microbes (Peiffer et al. 2013). Rhizosphere microorganisms provide valuable benefits to host plants, such as nutrient availability, guard against soil-borne pathogens, and stress resistance (Mendes et al. 2011, 2014).

Farming systems significantly influence the microbial communities, leading to loss of diversity, poor nutrient recycling, crop susceptibility to diseases, insect emergence, and alters soil productivity (Jonason et al. 2011). The rhizosphere is a dynamic interaction between plant roots and microorganisms that is influenced by plant species, soil minerals, and fertilizer application, reflecting crop yield and soil bioactivity in terrestrial agroecosystems (Edmeades 2003; Aira et al. 2010; Laurent et al. 2013; Bai et al. 2015).

## Organic Farming and Soil Microbial Species Richness

Organic farming enhances soil microbiome diversity and abundance, promoting a more diverse ecosystem compared to orthodox farming (Lupatini et al. 2017). The agronomic practices for management in organic farms alter the microbial composition, promoting habitat diversification and increased diversity. Despite challenging measurements, organic systems consistently exhibit higher diversity levels than conventional systems (Hartmann et al. 2015; Gonthier et al. 2014). Stable ecosystem functionality is the outcome of microbial variations, which foster redundant, stable communities with healthy interactions across trophic levels (Wagg et al. 2014). Lupatini et al. (2017) discovered that organic systems promote ecosystem stability by increasing microbial diversity, with predominant bacterial communities including Proteobacteria, Bacteroidetes, Acidobacteria, Actinobacteria, Firmicutes, Verrucomicrobia, Planctomycetes, Gemmatimonadetes, and Armatimonadetes. However, manure addition may encourage their growth, suggesting their function in organic carbon turnover in soil. The higher values of the Shannon index, which examines the distribution and abundance of species, indicate greater diversity in communities. Studies indicate that the diversity of microorganisms may be impacted by the use of organic fertilizers. For example, the linear discriminant analysis indicated that the accumulation of mushroom residue led to a surge in the bacterial diversity in the soil. Organic fertilizers, such as mushroom residue and vermicompost, have been shown to significantly increase the richness of Proteobacteria and Actinobacteria. In comparison to other amendments, the incorporation of vermicompost resulted in an increased quantity of species and a higher Shannon diversity index in community analysis (Shang et al. 2020). DGGE fingerprinting and PLFA analysis have demonstrated that organic fertilization significantly alters bacterial and fungal communities' structure and composition. The greatest diversity index, indicating a rise in microbial biomass and culturable population, was found in soils amended with compost manure and biofertilizers (Zhen et al. 2014). Organic fertilizers, which are high in essential nutrients and organic matter, increase soil vitality and reduce degradation by providing energy for microbial functions, thus increasing soil fertility.

## Factors Regulating Species Richness Under Organic Farming

Microbial species richness, an indicator of soil vigor, is influenced by environmental and anthropogenic factors. These include soil moisture, inorganic and carbonaceous chemicals, SOM, cropping patterns and practices, temperature, rainfall, and humidity. Cumulative research has demonstrated that shifts in microbial abundance, richness, and diversity reflect their spatial dissemination and arrangement in response to environmental factors (Zeglin 2015). In conventional agriculture, tillage practices directly influence the soil structure, plant growth, and crop yield. However, to maintain soil vitality and preserve the physico-chemical qualities essential for crop growth, conservation agriculture widely employs no-tillage techniques. Like any other factor, soil disturbance by tillage significantly alters its structural and biochemical characteristics. Contrariwise, under no-tillage or minimal tillage, significantly more diverse microbial communities have been observed. These findings corroborate the stimulation of redox and hydrolytic enzyme activities, which further up regulate soil nutrient status and crop vigor (Kraut-Cohen et al. 2020). Therefore, tillage practice highly influences the shape and the microbial community's structure, as well as associated ecological, environmental, and agricultural soil parameters. Crop sequencing has an analogous impact on soil microorganisms' structural heterogeneity. According to meta-analysis research, crop diversity rotations are directly proportional to microbial variety and richness (Figuerola et al. 2015; Venter et al. 2016).

Conventional farming techniques used for the management of insects, weeds, and diseases involve the rigorous application of agrochemicals, creating abiotic stress for microbial diversity. The overall variety of microorganisms decreases as a result. Furthermore, monoculture crops and heavy pesticide usage can result in biotic homogenization, which diminishes biodiversity and ecological niches and impairs the resilience and functionality of ecosystems. There are enduring repercussions from these impacts (Stagnari et al. 2014; Figuerola et al. 2015).

## Nutrient Management Systems and Their Effect on Soil Microbial Assemblage

Nutrient management is critical for maintaining the quality and vitality of the soil, and that subsequently affects crop productivity. Nutrients are essential for below-ground organisms' survival and functioning, and their enrichment alters the soil's chemical and physical properties, affecting microbial communities. Nitrogen, regardless of its form or soil edaphic factors, is reported to lower soil microbial gas exchange (Ramirez et al. 2010). Nitrogen has a detrimental effect on microbial biomass, whereas the addition of phosphorus is known to stimulate its growth (Liu et al. 2013). Changes in the composition of bacterial and fungal communities are directly linked to the incorporation of phosphorus. When phosphorus was administered into the soil, bacterial interactions were enhanced more than fungal partnerships, resulting in a greater accumulation of dry matter (Gumiere et al. 2019). Another study has shown that the quantity and composition of bacterial assemblages, particularly Proteobacteria and Actinobacteria, are affected by nitrogen fertilization over phosphorus fertilization (Wang et al. 2018). Fertilizers can modify soil properties and change carbon input from plant waste, impacting bacterial communities (Li et al. 2012; Wang et al. 2016; Kumar et al. 2017). The specific effects on these bacterial communities can vary depending on the type and amount of fertilizer used.

## *Inorganic Fertilizers*

Chemical or synthetic fertilizers like urea, DAP, and muriate of potash are widely used inorganic fertilizers to provide essential nutrients to plants when applied to soil. In modern agriculture, most of the nitrogen, phosphorus, potassium, and other major nutrients are applied to fields through these mineral fertilizers. However, inequitable fertilization practices have drastically altered natural nutrient cycles, leading to soil acidification, degradation, erosion, groundwater contamination, GHG emissions, and other problems. Additionally, inorganic fertilizers can directly deplete available reserves, such as rock phosphate, and indirectly deplete natural resources, such as through the chemical production of nitrogen. Though inorganic fertilizers can enhance the soil's nitrogen, potassium, and phosphorus levels as they can be administered in large quantities, it is important to note that using them in astronomical volumes can have adverse effects. For instance, introducing inorganic nitrogen fertilizers to the soil can result in a reduction in microbial biomass carbon, which is a crucial measure of soil quality. Moreover, the use of inorganic fertilizers can also decrease the population of beneficial microbes such as bacteria and fungi (Nakhro and Dkhar 2010).

## *Organic Fertilizers*

These organic or carbon-based fertilizers are byproducts or end products of natural processes involving animal or plant materials. They contain a significant amount of essential nutrients required for plant growth and can be broadly categorized into two types: (1) natural organic fertilizers, such as cow dung or manure, and (2) processed organic fertilizers, such as humic acid and compost. Organically produced fertilizers are of biotic origin and have several advantages over inorganic fertilizers, including reduced environmental concerns such as groundwater contamination, eutrophication, and soil erosion. Furthermore, in soil, the organically bound nutrients are more stable than inorganic nutrients, making them less susceptible to leaching or volatilization (Jacoby et al. 2017). These characteristics make organic fertilizers more critical for the well-being of the soil, as they act as slow-release fertilizers that maintain the nutrient balance of soils, a crucial prerequisite for healthy plant growth. Microbial activity is the primary indicator of the benefits of organic fertilization. Organic manure application increases soil OC reserves and improves soil physical attributes like cation exchange capacity, which buffers soil pH (Ojeniyi et al. 2007). Considering all these aspects and the significance of ecologically sustainable soil health, this discussion focuses on the abundance, structure, and functioning of microorganisms in soil as influenced by organic fertilizers. These fertilizers aid native microbial communities by regulating their community and structure. However, more research will be necessary to fully comprehend the effects of organic fertilizers on crop growth, production, and soil microbial diversity.

## *Natural Organic Fertilizers*

Farmyard manure (FYM) is derived from the decomposition of cattle shed wastes, including dung, urine, fodder, and roughage. When incorporated into the soil, properly decomposed FYM improves water-holding capacity and preserves soil structure. The nutrient-rich environment created by FYM stimulates the native microbial population, increasing soil microbe biomass and activity. In a study comparing FYM and inorganic fertilizer, FYM was shown to enhance the functioning of the soil's enzyme glucosidase, which is important for C-cycling. Conversely, inorganic fertilizer reduced the action of the soil enzyme arylsulphatase, which is important for sulfur cycling and an important indicator of the fertility of the soil. These findings demonstrate that FYM not simply restores but also enhances the soil's biological properties (Holík et al. 2019). Although FYM is cheap and easy to produce, it must be applied repeatedly to obtain optimal benefits, given its relatively low nutrient content.

Animal manure is an excellent source of readily accessible nutrients. Cow dung has been utilized as an organic amendment in Indian farming systems for centuries. Goat manure is a good source of essential nutrients with a relatively high organic content. In a study, goat manure enhanced the nutrient availability of N, P, K, Mg, and Ca and increased soil pH at an application rate of 10.0 t ha-1. This resulted in increased leaf nutrient content and vigor in pepper plants (Awodun et al. 2007). Another study investigated the effect of goat manure and sugarcane straw on soil nutrient transformation and soil microbial population. Goat manure enhanced the nutrient availability of C, N, and P at both 50 and 100 days after application, as estimated by the activities of major soil enzymes such as urease, β-glucosidase, and cellulases. Plant growth-promoting and lignocellulose-degrading rhizobacteria (PGPR) bacteria increased quantitatively, which was consistent with this. High throughput coupled with distance-based redundancy analysis and screening-based analysis of soil metagenome revealed that soils amended with goat manure alone and together with sugarcane straw had a greater number of Proteobacteria and Bacteroids (Tayyab et al. 2018). Proteobacteria are metabolically called copiotrophs since they can flourish in nutrient-rich conditions. The incorporation of swine manure also resulted in an upsurge in their relative abundance (Xun et al. 2016). With their comparatively high content of organic matter, fertilization with these manures can be a good source of organic amendment, especially for acidic soils and soils low in native organic content. This highlights their importance in soil remediation.

The agro-industrial sector generates waste, including crop residues and animal manure, including poultry manure. Poultry litter is an effective organic fertilizer used in organic farming to enhance microorganism diversity (Faissal et al. 2017). Shannon index diversity was greater in poultry manure-amended soils compared to control soils, with a substantial rise in soil enzyme activities. In comparison to other manures, poultry manure contains a relatively higher nitrogen content and, accordingly, has been discovered to stimulate the population of nitrifiers and increase microbial biomass (Mierzwa-Hersztek et al. 2018). This could prove beneficial for reducing ammonia emissions while boosting the amount of plant-available nitrogen. Hoover et al. (2019) reported that the problem of nitrate leaching was reduced and lowest in poultry manure-amended soils, with a simultaneous rise in soil organic accumulation. However, because of its higher C:N ratio, poultry manure mineralizes rapidly, resulting in heat generation. Consequently, it's not suggested for use during the summer season (Shaji et al. 2021).

Peanut hulls have traditionally been used by farmers to enhance soil quality, particularly in paddy fields affected by salinity. A team of researchers conducted experiments to evaluate the effectiveness of peanut hulls in mitigating saline conditions. They discovered that applying peanut hulls at a rate of 4 t $ha^{-1}$ significantly increased mycorrhizal root colonization. However, excess doses in saline conditions had negative effects. Therefore, peanut hulls can be suggested as a very good organic amendment option for saline soils (Fall et al. 2018). Green manuring and residue from crop application have long been used to enhance soil qualities and crop productivity. Leguminous crops, including *Cajanuscajan*, *Glysine max*, *Vignaradiata*, and *Cyamopsistetragonoloba*, as well as non-leguminous crops such as *Crotalaria juncea* and *Sesbania* species, are incorporated as cover crops to optimize soil quality and fertility. Studies have demonstrated that green manuring can increase microbial biomass nitrogen (MBN) and carbon (MBC) by 2.30–145.07% and 172%, respectively, and accelerate soil microbial enzyme activity. However, not much is known regarding how green manuring regulates the soil microbial assemblage (Ye et al. 2014).

## *Processed Organic Fertilizers*

Compost, vermicompost, seaweed extracts, cottonseed meal, humic acid, etc., are examples of processed organic fertilizers. They are mostly derived from plant litter and crop residues. Similarly, animal-based processed organic fertilizers include bone meal, fish meal, manure compost, blood meal, and wood ash. Natural organic fertilizers involve the direct use of plant or animal materials for fertilization, while processed organic fertilizers undergo a decomposition process, either induced or natural. Both types of fertilizers have their advantages and disadvantages. However, they have both been shown to strongly influence the soil microbial community. For example, kitchen waste compost stimulates the actinomycetes population in the soil, followed by bacterial growth, due to its neutral to alkaline pH (Arslan et al. 2008). It also shifts the pH from neutral to alkaline. However, high microbial respiration initially leads to the accumulation of anaerobic microbial populations, which can generate a foul odor.

FAME analysis revealed substantial changes in the soil microbial community composition under two water conditions: paddy and pepper (Lee et al. 2019). Irrespective of water saturation conditions, the total PLFA correlated positively with the fungi-to-bacteria ratio and compost application rates. The fungal population is essential, as it is the main driver of recalcitrant carbon cycling. Compost is derived by decomposing animal manure or plant waste. Hence, it's better than naturally applied compost because it checks the proliferation of pathogenic microbes and the germination of weed seeds. However, compost tends to have a slightly lower nutrient content than manure. (Shaji et al. 2021).

Bone meal, a phosphorus-rich ground mixture from slaughterhouse waste, is utilized as an organic fertilizer to enhance soil nutrient status and crop growth, boosting microbial biomass and enzymatic activities, thus increasing nutrient availability. The use of bone meal as fertilizer in farming has generated debate. This is a result of its association with the dissemination of long-term conditions such as spongiform encephalopathies. Concerns regarding the development of chronic problems and the development of microbiota that cause zoonotic diseases are also raised by using organic fertilizers obtained from animal sources (Goss et al. 2013; Mondini et al. 2008). Furthermore, the same manure composts into an entirely distinct product. To boost soil nutritional value and subsequent crop productivity, the adoption of bioinoculants—also referred to as effective microorganisms or agriculturally important microorganisms, or AIMs—has gained traction along with the application of natural or processed organic fertilizers.

## Effect of Bioinoculants on Microbial Diversity and Functions in Soil

Biofertilizers are preparations containing live or latent agriculturally beneficial microorganisms that, when added to soil, can improve or enhance soil health and plant vigor. Nisha et al. (2007) discovered that the autotrophic nature of three native cyanobacterial isolates in the biofertilizer used for organically deficient semi-arid soil resulted in a considerable increase in the total organic carbon content. Wang et al. (2021) studied the consequences of biological fertilizers on the structural and functional rhizospheric microbial community of Zea mays. They discovered that *Bacillus amyloliquefaciens* and *B. licheniformis* increased total N, P, K, and SOM, indicating they stimulated the plant to secrete root exudates, thereby increasing the accessibility of soil nutrients.

The four most common bacterial phyla found are Proteobacteria, Acidobacteria, Actinobacteria, and Gemmatimonadetes. The most prevalent bacterial species are *Aciditerrimonasferrireducens*, *Gemmatimonasaurantiaca*, and *Pyrinomonasmethylaliphatogenes*. Significant improvements have been detected in the metabolic processes of bacteria regarding energy, sugars, and amino acids. In soil, microorganisms primarily use organic matter as substrates and sources of energy. According to Kordelas et al. (2014), the microbial community's composition and activities are dictated by the chemical makeup of carbon in the soil. As was previously said, applying biofertilizer alters the physicochemical characteristics of the soil; these modifications frequently have a direct connection to the makeup of the soil microbial population (Zhou et al. 2015; Sun et al. 2016; Neufeld et al. 2017).

## Effect of Soil Microbial Community on Soil Organic Carbon

Long-term agronomic practices like chemical fertilizers, pesticides, heavy machinery, monoculture, and reduced organic amendments can negatively impact soil properties, leading to soil organic carbon loss and reducing agricultural production sustainability (Lal et al. 2011). SOM is central to all microbial processes, such as nutrient cycling, decomposition, and mineralization. Therefore, alternative management practices such as the use of biofertilizers and organic manures can help to alleviate the negative impacts of conventional chemical farming and maintain or build up SOC to ensure sustainable production over time. The amount, formation, and consumption of dissolved SOM are closely related to soil microbial activity. Bioavailability and association with different dissolved substances are modified by microbial actions, which also aid in the dissolution of organic substances and modify their chemical structure. The chemical structure of DOM determines both the extent of microbial decomposition and its susceptibility to it. Dissolved organic carbon compounds can be both substrates and products in the microbiological transformation process of organic matter, leading to the generation of complex compounds (Kalbitz et al. 2003; Marschner and Kalbitz 2003; Kiikkilä et al. 2014). The most microbiologically stable organic matter reservoir in soil, known as humus, is composed of fulvic acid, humic acids, and humin. The most inert component of humified materials is called humin. Using the carbon ratio of humic acid to fulvic acid (CHA/CFA) as a turnover indicator, the degree of the humification of organic matter in soil has been determined (Yang et al. 2004). The loss of SOM due to limited biomass addition and $CO_2$, $CH_4$, and soil-dissolved carbon emissions has led to an increased focus on soil carbon fixation, or retention (Smith 2004). Plants fix carbon, which decomposes and humifies the soil. Effective microorganism application can boost soil health. Although biofertilizers have long been recognized, additional study is still required to fully understand how they affect the biological characteristics, physicochemical attributes of soil, and crop productivity (Dinesh et al. 2010; Mayer et al. 2010; Piotrowska et al. 2012). According to a study by Debska et al. (2016), after three years of biofertilizer treatment, TOC increased by 6.3%, whereas dissolved organic carbon decreased by 0.19%. This suggests biofertilizer application improves the proportion of stable organic matter in the soil, a crucial parameter for boosting soil carbon capture and reducing decomposition. Microbial biomass carbon (MBC) is a crucial factor indicating the distinctness among organic and conventional regions, and a rise in MBC could indicate changes in organic matter's nutrient-supplying capacity compared to total organic matter. Singh et al. (2015) found that organic substances significantly enhanced β-glucosidase activity, resulting in improved soil C availability and a higher microbial population.

## Efficient and Effective Organic Fertilizer Management System(s)

Organic fertilizers offer additional organic carbon as an energy source, along with essential micronutrients. This could help with sustaining microbial diversity through the species-energy hypothesis, which predicts that more species can be sustained in an ecosystem with greater energy availability. Organic amendments stimulate microbial biomass more than inorganically applied NPK fertilizers. Given the highly varied agricultural and environmental conditions, there's a definite need for an integrated fertilization management approach, or "organic fertilizer cocktail," to nourish the soil. A blend of organic amendments like compost, animal or plant-based manure, and bioinoculants could serve this purpose to improve soil health metrics more efficiently. Similarly, the substitution of crop cover with additional organic inputs like manure might be of vital importance (Tully and McAskill 2020).

The introduction of nutrient-rich organic amendments rebuilds nutrient-depleted soils and provides organic carbon, the backbone of organic supplements, as an energy source for microorganisms to carry out their job more efficiently and effectively. Additionally, the simultaneous application of bioinoculants of agriculturally beneficial microbes can further accelerate microbial processes and enhance nutrient availability for plants. Bioinoculants and organic manures are suitable alternatives for improving soil biological diversity and productivity. Several investigations have demonstrated that bio-organic fertilizers significantly alter the microbial communities in the soil and boost crop yields (Ding et al. 2013; Liu et al. 2015). Exudates from host plants' roots regulate the organization of microbial communities as a result of plant growth-promoting microbes in biofertilizers or manures. These exudates include amino acids, sugars, organic acids, vitamins, and polymers. In an experiment by Valarini et al. (2003), only soil subjected to a combination of crop residues, animal manure, and efficient microbes (30 L $ha^{-1}$) showed a substantial rise in the amount of SOC after three months. It was discovered that the application of efficient microorganisms in the soil significantly enhanced the percentage of organic carbon because of the rapid breakdown of manure and crop waste. Wu et al.

(2005) discovered a 75% boost in soil carbon that was enriched with N-fixers, K and P solubilizers, AM fungi species, and organic fertilizer in comparison to the control. Singh et al. (2015) found that combining vermicompost, crop waste, and biofertilizer effectively increases MBC and SOC. Soil bacteria rely on β-glucosidase hydrolysis for energy, which is essential for the carbon cycle and organic material breakdown.

Besides using a combined organic fertilization approach, the fate of organic modifications in agriculture is influenced by several other factors. The nutrient content of manures varies based on their origin. For example, swine manure has higher proportions of overall SOC and SOM, besides the highest C:N ratio, while cattle manures are enriched with more organic phosphorus, and manure from poultry has a high nitrogen content. Some manures are suitable for acidic soils and others for saline soils. Therefore, the effectiveness of manure depends on the native nutrient makeup of the soil, its type, and the crop towards which the organic amendment is intended. Both manure and compost can significantly alter microbial diversity, but they affect the soil differently due to their inherent properties. Whether or not manure should be composted is another key concern for farmers and agricultural researchers. Composting modifies the physicochemical structure of manure, which could have a more beneficial or detrimental effect than manure alone. Properly composted manure has reduced water content, overall volume, and bulk density, and it has also increased dry matter content. These are advantages from a handling and ease-of-application standpoint. However, the major properties that make compost superior to manure are the elimination of pathogenic microbes, weed seeds, and parasites. Additionally, compost and manure have been shown to have a similar influence on soil nutrient status. Composting, however, can result in higher nutrient losses. Several processes, like ammonia volatilization, denitrification, and leaching, can promote nitrogen depletion from manure. When manure is properly composted, the nitrification process increases the volume of plant-available forms of nitrogen, including nitrates immobilized as organic fertilizers, in the compost compared to fresh manure. Compost releases organic nutrients instead of nitrogen, reducing the problem of nitrogen leaching or surface runoff, unlike fresh manure. This is advantageous to the environment because it reduces the problem of nitrogen leaching or surface runoff. To fully capitalize on the advantages of organic modifications and promote ecosystem services, it's crucial to develop a holistic strategy that can accurately define the utility of all such amendments.

## Conclusion

Organic fertilizers in agrarian systems not only conserve finite nutrient resources but also mitigate soil changes caused by crop production practices. Naturally available assets, largely neglected or not properly managed, provide an ample source of micro- and macronutrients. Of these, carbon composition is key to maintaining soil's physicochemical and biological integrity. The soil microflora composition, a heterogeneous entity, serves as the fundamental unit for maintaining the soil's structure and carrying out essential functions in a conserved and sustainable ecosystem. The range of organic modifications in the soil, as demonstrated by activated microbial populations and enzymatic activities, has an important effect on its richness and abundance. Thus, soil biological activity and health are intrinsically shaped by the community buildup of microorganisms in the soil. Despite suggestions that organic amendments influence copiotrophic metabolism by providing plants with a readily available pool of nutrients, the efficacy of a particular organic amendment is subject to several abiotic and biotic factors. Therefore, in light of the variability in soil disturbances, a combination of organic fertilizers may be the most effective course of action. The findings highlight that one or another form of organic fertilizer group (manures, compost, crop residue, etc.) can be deployed to manage depleted or degraded soil. Their integration with bioinoculants will generate an additional synergy essential for conserving soil health while preserving environmental dignity. The enduring multiplicity resulting from several soil diversity meta-analyses suggests that further in-depth research is necessary to properly comprehend the implications of organic fertilization on microbial assemblages and subsequent ecosystem functions. The organically operated system offers great nuance; hence, further studies ought to concentrate on augmenting these practices to improve soil health. Lastly, the dearth of adequate information to generate standard risk assessment procedures for individual organic amendment performance warrants the establishment of standard operating procedures (SOPs) for their proper evaluation.

## References

Aira M, Gomez-Brandon M, Lazcano C, Baath E, Dominguez J (2010) Plant genotype strongly modifies the structure and growth of maize rhizosphere microbial communities. Soil Biol Biochem 42:2276–2281

Arslan E, Obek E, Kirbag S, Ipek U, Topal M (2008) Determination of the effect of compost on soil microorganisms. Int J Sci Technol 3:151–159

Awodun MA, Omonijo LI, Ojeniyi SO (2007) Effect of goat dung and NPK fertilizer on soil and leaf nutrient content, growth and yield of pepper. Int J Soil Sci 2:142–147

Bai L, Cui J, Jie W, Cai B (2015) Analysis of the community compositions of rhizosphere fungi in soybeans continuous cropping fields. Microbiol Res 180:49–56

Dębska B, Jacek D, Anna P, Magdalena B (2016) The impact of a bio-fertilizer on the soil organic matter status and carbon sequestration-results from a field-scale study. J Soil Sediment 10:2335–2343

Dinesh R, Srinivasan V, Hamza S, Manjusha A (2010) Short-term incorporation of organic manures and biofertilizers influences biochemical and microbial characteristics of soils under an annual crop [Turmeric (*Curcuma longa* L.)]. Bioresour Technol 101:4697–4702

Ding C, Shen Q, Zhang R, Chen W (2013) Evaluation of rhizosphere bacteria and derived bio-organic fertilizers as potential biocontrol agents against bacterial wilt (*Ralstoniasolanacearum*) of potato. Plant and Soil 366:453–466

Edmeades DC (2003) The long-term effects of manures and fertilizers on soil productivity and quality: a review. Nutr Cycl Agroecosyst 66:165–180

Faissal A, Ouazzani N, Parrado JR, Dary M, Manyani H, Morgado BR, Barragán MD, Mandi L (2017) Impact of fertilization by natural manure on the microbial quality of soil: molecular approach. Saudi J Biol Sci 24:1437–1443

Fall D, Bakhoum N, Fall F, Diouf F, Ndiaye C, Faye MN, Hocher V, Diouf D (2018) Effect of peanut shells amendment on soil properties and growth of seedlings of *Senegaliasenegal* (L.) Britton, *Vachelliaseyal* (Delile) P. Hurter, and *Prosopisjuliflora* (Swartz) DC in salt-affected soils. Ann For Sci 75:1–11

Figuerola ELM, Guerrero LD, Turkowsky D, Wall LG, Erijman L (2015) Crop monoculture rather than agriculture reduces the spatial turnover of soil bacterial communities at a regional scale. Environ Microbiol 17:678–688

Gonthier DJ, Ennis KK, Farinas S, Hsieh HY, Iverson AL, Batary P, Rudolphi J, Tscharntke T, Cardinale BJ, Perfecto I (2014) Biodiversity conservation in agriculture requires a multi-scale approach. Proc R Soc B Biol Sci 281:20141358

Goss MJ, Tubeileh A, Goorahoo D (2013) A review of the use of organic amendments and the risk to human health. Adv Agron 12:275–379

Gumiere T, Rousseau AN, da Costa DP, Cassetari A, Cotta SR, Andreote FD, Gumiere SJ, Pavinato PS (2019) Phosphorus source driving the soil microbial interactions and improving sugarcane development. Sci Rep 9:1–9

Hartmann M, Frey B, Mayer J, Mader P, Widmer F (2015) Distinct soil microbial diversity under long-term organic and conventional farming. Multidiscip J Microb Ecol 9:1177–1194

Holík L, Hlisnikovský L, Honzík R, Trögl J, Burdová H, Popelka J (2019) Soil microbial communities and enzyme activities after long-term application of inorganic and organic fertilizers at different depths of the soil profile. Sustain For 11:3251

Hoover NL, Law JY, Long LAM, Kanwar RS, Soupir ML (2019) Long-term impact of poultry manure on crop yield, soil and water quality, and crop revenue. J Environ Manage 252:109582

Jacoby R, Peukert M, Succurro A, Koprivova A, Kopriva S (2017) The role of soil microorganisms in plant mineral nutrition-current knowledge and future directions. Frontiers. Plant Sci 8

Jonason D, Andersson GKS, Ockinger E, Rundlof M, Smith HG, Bengtsson J (2011) Assessing the effect of the time since transition to organic farming on plants and butterflies. J Appl Ecol 48:543–550

Kalbitz K, Schwesig D, Schmerwitz J, Kaiser K, Haumaier L, Glaser B, Ellerbrock R, Leinweber P (2003) Changes in properties of soilderived dissolved organic matter induced by biodegradation. Soil Biol Biochem 35:1129–1142

Kiikkilä O, Kanerva S, Kitunen V, Smolander A (2014) Soil microbial activity in relation to dissolved organic matter properties under different tree species. Plant and Soil 377:169–177

Kordelas L, Rebmann V, Ludwig AK, Radtke S, Ruesing J, Doeppner TR, Epple M, Horn PA, Beelen DW, Giebel B (2014) MSC-derived exosomes: a novel tool to treat therapy-refractory graft-versus-host disease. Leukemia 28:970–973

Kraut-Cohen J, Zolti A, Shaltiel-Harpaz L, Argaman E, Rabinovich R, Green SJ, Minz D (2020) Effects of tillage practices on soil microbiome and agricultural parameters. Sci Total Environ 705:135791

Kumar U, Shahid M, Tripathi R, Mohanty S, Kumar A, Bhattacharyya P, Lal B, Gautam P, Raja R, Panda BB, Jambhulkar NN (2017) Variation of functional diversity of soil microbial community in sub-humid tropical rice-rice cropping system under long-term organic and inorganic fertilization. Ecol Indic 73:536–543

Lal R (2011) Sequestering carbon in soils of agro-ecosystems. Food Policy 36:533–539

Laurent P, Raaijmakers JM, Philippe L, Putten WH, Der V (2013) Going back to the roots: the microbial ecology of the rhizosphere. Nat Rev Microbiol 11:789–799

Lee CH, Park SJ, Hwang HY, Kim MS, Il Jung H, Luyima D, Hong SY, Oh TK, Kim SH (2019) Effects of food waste compost on the shift of microbial community in water saturated and unsaturated soil condition. Appl Biol Chem 62:1–7

Li R, Khafipour E, Krause DO, Entz MH, de Kievit TR, Fernando WD (2012) Pyrosequencing reveals the influence of organic and conventional farming systems on bacterial communities. PloS One 7:e51897

Liu L, Sun C, Liu S, Chai R, Huang W, Liu X, Tang C, Zhang Y (2015) Bioorganic fertilizer enhances soil suppressive capacity against bacterial wilt of tomato. PloS One 10:e0121304
Liu L, Zhang T, Gilliam FS, Gundersen P, Zhang W, Chen H, Mo J (2013) Interactive effects of nitrogen and phosphorus on soil microbial communities in a tropical forest. PloS One 8
Lupatini M, Korthals GW, de Hollander M, Janssens TK, Kuramae EE (2017) Soil microbiome is more heterogeneous in organic than in conventional farming system. Front Microbiol 7:2064
Marschner B, Kalbitz K (2003) Controls of bioavailability and biodegradability of dissolved organic matter in soils. Geoderma 113:211–235
Mayer J, Scheid S, Widmer F, Fließbach A, Oberholzer HR (2010) How effective are 'Effective microorganisms® (EM)'? Results from a field study in temperate climate. Appl Soil Ecol 46:230–239
Mendes LW, Kuramae EE, Navarrete AA (2014) Taxonomical and functional microbial community selection in soybean rhizosphere. Multidiscip J Microb Ecol 8:1577
Mendes R, Kruijt M, De Bruijn I, Dekkers E, van der Voort M, Schneider JH, Piceno YM, DeSantis TZ, Andersen GL, Bakker PA, Raaijmakers JM (2011) Deciphering the rhizosphere microbiome for disease-suppressive bacteria. Science 332:1097–1100
Mierzwa-Hersztek M, Klimkowicz-Pawlas A, Gondek K (2018) Influence of poultry litter and poultry litter biochar on soil microbial respiration and nitrifying bacteria activity. Waste Biomass Valori 9:379–389
Mondini C, Cayuela ML, Sinicco T, Sánchez-Monedero MA, Bertolone E, Bardi L (2008) Soil application of meat and bone meal. Short-term effects on mineralization dynamics and soil biochemical and microbiological properties. Soil Biol Biochem 40:462–474
Nakhro N, Dkhar MS (2010) Populations and biomass carbon in paddy field soil. Agron J 9:102–110
Neufeld KR, Grayston SJ, Bittman S, Krzic M, Hunt DE, Smukler SM (2017) Long-term alternative dairy manure management approaches enhance microbial biomass and activity in perennial forage grass. Biol Fertil Soils 53:613–626
Nisha R, Kaushik A, Kaushik CP (2007) Effect of indigenous cyanobacterial application on structural stability and productivity of an organically poor semi-arid soil. Geoderma 138:49–56
Ojeniyi SO, Akanni DI, Awodun MA (2007) Effect of goat manure on some soil properties and growth yield and nutrient status of tomato. J Agric Sci 15:396–405
Peiffer JA, Spor A, Koren O, Jin Z, Tringe SG, Dangl JL, Buckler ES, Ley RE (2013) Diversity and heritability of the maize rhizosphere microbiome under field conditions. Proc Natl Acad Sci U S A 110:6548–6553
Piotrowska A, Długosz J, Zamorski R, Bogdanowicz P (2012) Changes in some biological and chemical properties of an arable soil treated with the microbial biofertilizer UGmax. Pol J Environ Stud 21:455–463
Ramirez KS, Craine JM, Fierer N (2010) Nitrogen fertilization inhibits soil microbial respiration regardless of the form of nitrogen applied. Soil Biol Biochem 42:2336–2338
Shaji H, Chandran V, Mathew L (2021) Organic fertilizers as a route to controlled release of nutrients. In: Controlled release fertilizers for sustainable agriculture. Academic Press, pp 231–245
Shang L, Wan L, Zhou X, Li S, Li X (2020) Effects of organic fertilizer on soil nutrient status, enzyme activity, and bacterial community diversity in Leymuschinensis steppe in Inner Mongolia, China. PloS One 15:1–18
Singh G, Kumar D, Sharma P (2015) Effect of organics, biofertilizers and crop residue application on soil microbial activity in rice–wheat and rice-wheat mungbean cropping systems in the Indo-Gangetic plains. Cogent Geosci 1(1):1085296
Smith P (2004) Carbon sequestration in croplands: the potential in Europe and the global context. Eur J Agron 20:29–236
Stagnari F, Perpetuini G, Tofalo R, Campanelli G, Leteo F, Della Vella U, Schirone M, Suzzi G, Pisante M (2014) Long-term impact of farm management and crops on soil microorganisms assessed by combined DGGE and PLFA analyses. Front Microbiol 5:644

Sun R, Dsouza M, Gilbert JA, Guo X, Wang D, Guo Z, Ni Y, Chu H (2016) Fungal community composition in soils subjected to long-term chemical fertilization is most influenced by the type of organic matter. Environ Microbiol 18:5137–5150

Tayyab M, Islam W, Arafat Y, Pang Z, Zhang C, Lin Y, Waqas M, Lin S, Lin W, Zhang H (2018) Effect of sugarcane straw and goat manure on soil nutrient transformation and bacterial communities. Sustainability (Switzerland) 10:1–21

Tully KL, McAskill C (2020) Promoting soil health in organically managed systems: a review. Org Agric 10:339–358

Valarini PJ, Díaz MC, Gascó JM, Guerrero F, Tokeshi H (2003) Assessment of soil properties by organic matter and EM Microorganisms incorporation. Rev Bras Ciênc Solo 27:519–525

Van Bueren EL, Struik PC, Jacobsen E (2002) Ecological concepts in organic farming and their consequences for an organic crop ideotype. NJAS: Wageningen Journal of. Life Sci 50:1–26

Venter ZS, Jacobs K, Hawkins HJ (2016) The impact of crop rotation on soil microbial diversity: A meta-analysis. Pedobiologia 59:215–223

Wagg C, Bender SF, Widmer F, Van Der Heijden MGA (2014) Soil biodiversity and soil community composition determine ecosystem multifunctionality. Proc Nat Acad Sci USA 111:5266–5270

Wang L, Yang F, Yuan J, Raza W, Huang Q, Shen Q (2016) Long-term application of bioorganic fertilizers improved soil biochemical properties and microbial communities of an apple orchard soil. Front Microbiol 7:1893

Wang Q, Wang C, Yu W, Turak A, Chen D, Huang Y, Ao J, Jiang Y, Huang Z (2018) Effects of nitrogen and phosphorus inputs on soil bacterial abundance, diversity, and community composition in chinese fir plantations. Front Microbiol 9:1–10

Wang J, Liu L, Gao X, Hao J, Wang M (2021) Elucidating the effect of biofertilizers on bacterial diversity in maize rhizosphere soil. PloS One 16:e0249834

Wu SC, Cao ZH, Li ZG, Cheung KC, Wong MH (2005) Effects of biofertilizer containing N-fixer, P and K solublizers and AM fungi on maize growth: A greenhouse trial. Geoderma 125:155–166

Xun W, Xiong W, Huang T, Ran W, Li D, Shen Q, Li Q, Zhang R (2016) Swine manure and quicklime have different impacts on chemical properties and composition of bacterial communities of an acidic soil. Appl Soil Ecol 100:38–44

Yang ZH, Singh BR, Sitaula BK (2004) Soil organic carbon fractions under different land uses in Mardi Watershed of Nepal. Commun Soil Sci Plant Anal 35:615–629

Ye X, Liu H, Li Z, Wang Y, Wang Y, Wang H, Liu G (2014) Effects of green manure continuous application on soil microbial biomass and enzyme activity. J Plant Nutr 37:498–508

Zeglin LH (2015) Stream microbial diversity in response to environmental changes: Review and synthesis of existing research. Front Microbiol 6:1–15

Zhen Z, Liu H, Wang N, Guo L, Meng J, Ding N, Wu G, Jiang G (2014) Effects of manure compost application on soil microbial community diversity and soil microenvironments in a temperate cropland in China. PloS One 9:e108555

Zhou J, Guan D, Zhou B, Zhao B, Ma M, Qin J, Jiang X, Chen S, Cao F, Shen D, Li J (2015) Influence of 34-years of fertilization on bacterial communities in an intensively cultivated black soil in northeast China, vol 90. Soil Biology, p 42

# Climate Resilient Agriculture: The Perspective of Conservation Agricultural Practice

**Yanglem Sofia Devi and M. Premi Devi**

## Introduction

The report by the United Nations in 2022 highlighted the distressing global hunger crisis affecting nearly 828 million individuals in 2021. This represents about 10% of the world's population (UN Report 2022). Besides, the crisis is compounded by projections indicating the necessity of a 70% increase in food production by 2050 to sustain an estimated 9 billion people (FAO 2009). The current agricultural landscape faces challenges arising from climate change, including abiotic and biotic stresses that hinder crop cultivation. These challenges are fueled by agriculture's susceptibility to climate variations, resulting in adverse impacts such as rising temperatures, erratic weather patterns, and the explosion of resilient pests and diseases (UN Report 2022).

Addressing these pressing issues necessitates a shift towards sustainable agricultural practices. CA, characterized by principles such as minimal soil disturbance, permanent soil cover, and crop diversification, offers a holistic approach to farming that mitigates climate risks (FAO 2017). The transition from conventional agriculture is driven by various factors, including the detrimental effects of intensive tillage, inadequate residue management, and natural resource depletion (Sangar and Abrol 2005; Jat et al. 2020b). CA not only safeguards soil health and biodiversity but also enhances soil quality and carbon sequestration while optimizing crop yields (FAO 2017; Murungu 2012; Lal and Stewart 2010; Brouder and Gomez-Mcpherson 2004). Furthermore, CA's integration with precise water and nutrient management enhances resilience to climate variability while reducing production costs and labor

Y. S. Devi
Krishi Vigyan Kendra (CAU-Imphal), East Garo Hills, Meghalaya, India

M. P. Devi (✉)
College of Agriculture (CAU-Imphal), Kyrdemkulai, Meghalaya, India



S. R. Mohanty, B. Kollah (eds.), *Greenhouse Gas Regulating Microorganisms in Soil Ecosystems*, https://doi.org/10.1007/978-3-031-70569-4_16

requirements (Sidhu et al. 2019). Its versatility across various crops and compatibility with integrated pest management make it an attractive option for farmers seeking sustainable solutions (Brown et al. 2017). Hence, CA is a promising solution for climate-resilient agriculture in the face of climate change.

## Present Scenario of Agriculture with Climate Change

Climate influences global agricultural production, with escalating concerns over its impacts on livelihoods. Climate change threatens food security, characterized by the dual challenges of escalating food prices and diminishing food production (Lobell et al. 2011; Seo and Mendelsohn 2008; Cline 2007).

The consequences of climate change are already manifesting globally, with events such as heat waves and catastrophic floods wreaking havoc on agricultural systems and worsening food insecurity. For instance, the summer of 2010 witnessed heat waves in key agricultural regions like Russia, Ukraine, and Kazakhstan, leading to yield losses and a hike in food prices. Similarly, devastating floods overcame several European countries in July 2021, contributing to widespread damage to agricultural infrastructure. Moreover, record-high temperatures in 2022, particularly in France and the UK, coupled with wildfires and drought, have severely impacted crop cultivation, particularly in leading corn-producing nations (World Economic Forum 2022). Regions like West Africa have been identified as climate change hotspots, where adverse climatic conditions continue to hinder crop yields and production contributing to food insecurity (Sultan and Gaetani 2016; Palazzo et al. 2017). Countries such as China and Pakistan have also dealt with severe climate-related catastrophes in 2022, further challenging food production and livelihoods (Mallapaty 2022).

Agricultural landscapes reliant on rainfed systems face heightened vulnerability, despite being responsible for generating a significant portion of staple foods. India, for instance, with over 60% of its agricultural area under rainfed cultivation, faces acute vulnerability to climate change impacts (Asha latha et al. 2012: Zaman et al. 2017).

## Impact of Climate Change on Agriculture

The effects of climate change are altering ecosystems, human well-being, and agricultural systems. While elevated levels of carbon dioxide may offer some benefits to certain crops, the concurrent rise in temperature and variability in rainfall patterns are anticipated to profoundly impact food production (Thomas and Rosegrant 2015).

Warmer and drier summers are likely to result in increased fallowing of land in rainfed areas traditionally cultivated annually. Furthermore, climate change has the potential to disrupt food supply chains and alter trade patterns, contributing to heightened food prices and market instability (FAO 2015). The socioeconomic repercussions of climate change on agriculture are substantial. The projection suggest that it could push anywhere from 3 million to 16 million people into extreme poverty, primarily due to its adverse effects on agricultural productivity and resultant food prices (Hallegatte et al. 2016; Hallegatte and Rozenberg 2017).

## Impact on Crop and Crop Yield

While most regions are expected to suffer negative impacts, areas experiencing increased precipitation and situated at higher altitudes may witness positive effects, as warmer temperatures render them more conducive to agriculture (Thomas and Rosegrant 2015). However, projections for South Asian countries indicate a substantial decline of approximately 30% in crop yields by the mid-twenty-first century. Studies have forecasted significant farm revenue losses ranging from 9% to 25%. This is due to diminished yields in dryland wheat, rice, maize, and soybeans in response to temperature increases of 2–3.5 °C. The increase in temperature results in a global decline of 17% in yields by 2050 compared to a scenario with a stable climate (Asha latha et al. 2012; Zhao et al. 2017).

Major agricultural producers in temperate zones, such as the European Union and the United States, particularly vulnerable to adverse climate impacts during critical stages like flowering, have exhibited some resilience under unfavorable conditions (Muller and Elliott 2015). However, the effects of climate change on other crops, including roots, tubers, pulses, vegetables, fruits, and horticultural products, which are essential for nutrition and livelihoods, have received less attention (HLPE 2012).

Climate change has induced earlier flowering and maturity in perennial horticultural crops like grapes and apples globally (Porter et al. 2014). While crops like cassava may benefit from increased temperatures and $CO_2$ levels due to their high optimum temperature for growth and photosynthesis, temperate fruits like apples, cherries, and nuts, with specific chilling temperature requirements, are expected to suffer. Likewise, most wine-producing regions are projected to experience reduced suitability for grapevines. In regions like Brazil, sugarcane and coffee cultivation zones are anticipated to shift to more favorable areas, while coffee-growing nations such as Costa Rica, Nicaragua, and El Salvador might face a 40% decrease in suitability (Porter et al. 2014).

Research conducted at the Indian Agricultural Research Institute indicates greater susceptibility of rabi crops compared to kharif crops to climate change impacts. In India, a 1.5 °C temperature increase and a 2 mm reduction in precipitation could lead to rice yield reductions ranging from 3% to 15% (Ahluwalia and Malhotra 2006). Consequently, reports suggest a potential 20% income reduction among dryland farmers in Andhra Pradesh, a 25–30% decrease in sugarcane yields in Maharashtra, and a 12% decline in rice yields in Odisha due to flooding events.

## Impact on Livestock Production

The livestock industry contributes to 75% of historical deforestation in the Brazilian Amazon rainforest. It also plays a significant role in nearly a third of biodiversity loss, as well as air and water pollution (Steinfeld et al. 2006). The UN Food and Agriculture Organization (FAO 2013) suggests that the livestock sector produces14.5% of global greenhouse gas emissions, encompassing $CO_2$, $N_2O$, $CH_4$, and $NH_3$ (Leytem et al. 2011; Russell 2014), thus significantly exacerbating global warming.

Moreover, climate change poses substantial threats to livestock production, both directly and indirectly (FAO 2015). The consequences include heightened competition for land, water, and food resources, worsening food insecurity precisely when it is most critical (Thornton and Herrero 2010). Potential impacts encompass alterations in the production and quality of feed crops and forage, water availability, animal growth, milk production, disease prevalence, reproductive rates, and biodiversity (Chapman et al. 2012; Polley et al. 2013; Henry et al. 2012; Nardone et al. 2010).

## Impact on Pest and Diseases

Climate change elevates the risk posed by pests and diseases to crops, potentially affecting even regions previously deemed immune. The population densities of pathogens and insect pests are linked to temperature and humidity, implying that an uptick in these parameters will likely alter their prevalence. The adverse impacts of pests and diseases can manifest both biologically and institutionally.

The surge in pests and diseases may offset the direct benefits of climate change, such as extended growing seasons in high-latitude areas. In these regions, the climatic conditions will not only favor crop growth but also facilitate the proliferation of weeds and pests. Consequently, there may be a heightened reliance on pesticides, which, in turn, could entail significant risks to human health and the environment. Additionally, there is the concern of pesticide residues contaminating food supplies, potentially reaching unsafe levels.

## Coverage of Conservation Agriculture in the World

Conservation agriculture is a strategy to mitigate erosion by minimizing tillage and maintaining soil coverage with crop residues. No-till farming, pioneered in the USA during the 1960s (Table 1), marked a significant advancement in this approach (Despsch 2004). Over the past decade, the global area under no-till cultivation has surged by 93%, reaching 507.6 million acres, with an annual adoption growth rate of approximately 25 million acres (Paukner 2022). The primary focus of most

**Table 1** Area coverage under conservation agriculture in the world

| Sl. No. | Country | % of global area |
|---|---|---|
| 1 | USA | 23 |
| 2 | Brazil | 20 |
| 3 | Argentina | 19 |
| 4 | Canada | 12 |
| 5 | Australia | 11 |
| 6 | China | 4 |
| 7 | Russia | 3 |
| 8 | Paraguay | 2 |
| 9 | Kazakhstan | 1 |
| 10 | India | 1 |
| 11 | Uruguay | 1 |

Source: Brown et al. (2017)

resource conservation techniques revolves around enhancing the efficiency of pivotal natural resources, namely water and soil. However, it's worth noting that some of these methods can also impact the efficiency of other production resources and inputs such as labor, farm power, and fertilizer.

## Perspectives of Minimum Soil Disturbance

Conservation agriculture (CA) minimizes mechanical soil disturbance, which involves directly planting seeds into undisturbed soil to prevent or mitigate soil compaction. This practice can be categorized into two main approaches: no-till, where there is complete absence of tillage, and reduced or minimum tillage, which entails an intermediate level of soil disturbance. Traditional tillage methods, such as intensive plowing and disking, have been found to negatively impact both crop productivity and soil properties globally.

Research by Rusinamhodzi et al. (2011) revealed that in regions with an annual rainfall of 600 mm, no-tillage practices resulted in higher yields compared to conventional practices in areas receiving 1000 mm of rainfall. Changes in land use, particularly the conversion of natural vegetation to agricultural land, have depleted global soil carbon (C) stocks and contributed to increased $CO_2$ emissions (Lal 2011).

No-till farming reduces greenhouse gas (GHG) emissions from farming operations by over 75% and has proven to be 10–20% more profitable compared to conventional burning and tilling methods (CGIAR 2022).

However, Powlson et al. (2014) argue that while no-till farming is beneficial for soil quality and adaptation to climate change, its role in mitigating climate change may be overstated. Reduced tillage practices can also contribute to GHG emission reductions by decreasing fossil fuel use in land preparation and enhancing carbon sequestration (Mangalassery et al. 2014).

Reducing tillage methods may increase emissions of nitrous oxide ($N_2O$), which has a higher warming potential than $CO_2$ and methane ($CH_4$) (Hemle et al. 2008; Six et al. 2004). Conservation agriculture practices (no-till farming) without straw mulching and crop rotation can lead to reduced crop yields compared to when implemented in combination (Pittelkow et al. 2015b). Despite these complexities, reduced tillage and no-till practices can enhance climate resilience by improving soil health, increasing water holding capacity, and reducing soil erosion rates compared to conventional tillage methods (Gathala et al. 2020; Seitz et al. 2019; Murungu 2012; Bertrand et al. 2022).

## Perspectives of Permanent Soil Cover

Conservation agriculture (CA) aims to mitigate climate change while enhancing agricultural efficiency. CA preserves soil through permanent cover, which serves to combat soil erosion and runoff loss. Soil erosion depletes fertile topsoil and also results in the loss of crucial organic matter and nutrients (Zheng 2005). By implementing permanent soil cover using crops, cover crops, or mulch, CA helps maintain soil moisture, reduce evaporation losses, prevent erosion, and alleviate compaction (Lalljee 2013). Moreover, it fosters soil health by bolstering microbial populations, minimizing nutrient leaching, managing weeds, and ultimately boosting yields (Kaye and Quemada 2017; Mhlanga et al. 2015).

CA facilitates carbon sequestration and also mitigates greenhouse gas emissions from soil (Archarya et al. 2018). Under CA, it's crucial to maintain at least 30% surface cover with crop residues, cover crops, or biomass sourced through agroforestry practices, thereby reducing reliance on burning crop residues, which is known to produce significant greenhouse gas emissions (Lohan et al. 2018). Additionally, retaining crop residues serves as a renewable energy source, enriches soil with organic matter, and creates a conducive microclimate for agricultural ecosystems (Jat et al. 2020a).

Strategies such as straw mulching have shown promising results in maximizing "crop yield per drop and bag" (Gan et al. 2013; Li et al. 2013). However, it's essential to address potential drawbacks, such as increased methane emissions, when applying rice straw to paddy fields. Strategies to mitigate these emissions involve promoting the oxidative decomposition of straw both in and out of the field, alongside enhancing carbon stock in the soil through microbial interventions (Kumar et al. 2019).

Cover crops represent another valuable aspect of CA, contributing to soil health improvement and nitrogen fixation, thereby reducing nitrogen leaching and synthetic fertilizer input (Cline 2007; Kaye and Quemada 2017). They serve as a natural and cost-effective climate solution, capturing atmospheric carbon dioxide and enhancing soil health and crop resilience to climate change (Greene 2022). Meta-analyses have shown that cover crops can increase carbon sequestration and mitigate warming, making them valuable tools for climate change adaptation (McDaniel et al. 2014). Additionally, cover crops play a crucial role in weed management by creating a physical barrier that impedes weed seedling emergence, thus reducing weed population and diversification (Mhlanga et al. 2015).

## Perspectives of Species Diversification

Crop diversification holds significant promise for both climate change adaptation and mitigation efforts. By cultivating a variety of crops through rotations or intercropping, farmers can tap into diverse ecological benefits while enhancing resilience to climate variability.

Diversifying cropping patterns helps farmers mitigate the risks associated with climate change, such as crop failure due to extreme weather events or pest and disease outbreaks (Joshi 2005). Different crops have varying water and nutrient requirements, and rotating or intercropping them allows for more efficient use of resources and reduces soil degradation over time (Singh 2001; Singh et al. 2004). Furthermore, crop diversification promotes ecological balance, enhances agricultural diversity, improves crop quantity and quality, and mitigates yield losses caused by pests, diseases, and weeds (Mousavi and Eskandari 2011).

Integrating intercropping into CA systems significantly increases soil organic carbon (SOC) levels over time, although the extent of this improvement depends on various factors, including agroecological conditions and biomass production (Rusinamhodzi et al. 2012). Additionally, crop rotations within CA help break pest and disease cycles, enhance soil quality through biological nitrogen fixation, and bolster organic matter content (Jordan and Hutcheon 2003; Baudron et al. 2011). Such practices lead to improved productivity, profitability, and resilience to climate stressors (Gangwar and Ram 2005; Rusinamhodzi et al. 2012; Smith and Read 2008).

Specific crop diversification strategies tailored to different agricultural contexts offer further opportunities for climate-resilient agriculture. For instance, replacing rice-wheat systems with maize-wheat or rice-maize systems can enhance adaptation to water stress and profitability in water-stressed areas (Akanda 2011; Islam et al. 2019). Introducing millets, known for their resilience to heat and drought, in drought-prone regions can improve crop selection under moisture stress conditions (Satapathy et al. 2011). Additionally, incorporating legumes into cropping systems not only enhances nitrogen fixation but also contributes to greenhouse gas mitigation efforts (Lemke et al. 2007).

## Combination of the Principles

Conservation agriculture (CA) presents enhanced agricultural productivity while promoting sustainability. While the average yield of CA systems may not significantly differ from conventional agriculture, the integration of all three CA principles—no-tillage, residue retention, and crop diversification—can lead to notable yield increases, particularly under conditions of limited precipitation and without herbicide application (Corbeels et al. 2020). For instance, combining no-tillage and residue retention has been shown to boost crop yields by 6.0% (Xiao et al. 2021), while the adoption of no-tillage alongside straw mulching and crop rotation resulted in a 7.3% yield increase (Pittelkow et al. 2015b). Furthermore, residue retention plays a crucial role in reducing sediment loss by 55% compared to 26.0% under no-tillage alone (Xiao et al. 2021). These practices contribute to improved soil health, water infiltration, and organic carbon buildup, enhancing overall agricultural sustainability (Stanojevic 2021).

However, the widespread adoption of CA faces various barriers that hinder its implementation. These barriers include initial yield decreases, challenges in delivering relevant information, competition for crop residues by livestock, high initial investment costs, and increased labor demands (Pittelkow et al. 2015a; Brown et al. 2017; Murungu 2012; Michler et al. 2019). Overcoming these barriers requires concerted efforts from various stakeholders. Government support, policymaker engagement, adequate funding for transitional needs, and farmer and public awareness-raising initiatives are essential for successful CA implementation on a large scale (Lee and Gambiza 2022).

Moreover, leveraging collective knowledge from diverse scientific communities worldwide is crucial for evaluating the practicality and efficacy of CA principles in different contexts. By integrating insights from various disciplines, stakeholders can identify innovative approaches to enhance food productivity while safeguarding natural resources for long-term economic, human, and ecological sustainability.

## Conclusion

Rainfed farming, covering 80% of the world's cropland and contributing over 60% of global food production, is highly susceptible to climate conditions. With the expanding global population, there is a growing strain on diminishing resources, compounded by unpredictable climate patterns, posing significant challenges to agricultural systems. Conservation agriculture (CA) emerges as a climate-smart agricultural strategy aimed at bolstering resilience to climate change while enhancing the capacity of cropping systems for mitigation. By efficiently utilizing resources, CA conserves them for future generations, offering a sustainable alternative to conventional agriculture. The adoption of CA practices, such as no-till agriculture, crop residue retention, and crop rotation, leads to a substantial reduction in fossil fuel consumption, minimizing greenhouse gas emissions. Furthermore, decreased agrochemical usage in CA contributes to lower pollution levels in the air, soil, and water, promoting environmental sustainability. To maximize outcomes and yields, it's imperative to implement all three principles of CA synergistically. CA presents a holistic farming approach with crops that require fewer resources or inputs, making it particularly beneficial for small and marginal farmers grappling with the impacts of climate change. By embracing CA, farmers can adapt to changing climate conditions while ensuring sustainable agricultural practices for future generations.

## References

Ahluwalia VK, Malhotra S (2006) Environmental science. Anne Books India, New Delhi

Akanda AI (2011) Rethinking crop diversification under changing climate, hydrology and food habit in Bangladesh. J Agric Environ Int Dev (JAEID) 104:3–23

Archarya CL, Bandyopadhyay KK, Hati KM (2018) Mulches: Role in climate resilient agriculture. In: Hillel D, Rosenzweig C, Powlson DS, Scow KM, Singer MJ, Sparks DL, Hatfeild J (eds) Encyclopedia of soils in the environment, Earth systems and environmental sciences. Elsevier, Amsterdam, pp 521–532

Asha latha KV, Gopinath M, Bhat ARS (2012) Impact of climate change on rainfed agriculture in India: A case study of Dharwad. Int J Environ Sci Dev 3(4):368–371

Baudron F, Tittonell P, Corbeels M et al (2011) Comparative performance of conservation agriculture and current smallholder farming practices in semiarid Zimbabwe. Field Crop Res 132:117–128

Bertrand S, Roberts AS, Walker E (2022) No-till farming improves soil health & mitigates climate change, Agriculture and climate series, Washington, DC. https://www.eesi.org/articles/view/no-till-farming-improves-soil-health-and-mitigates-climate-change/

Brouder SM, Gomez-Mcpherson H (2004) The impact of conservation agriculture on smallholder agriculral yields: ascoping review of the evidence. Agric Ecosyst Environ 187:11–32

Brown B, Llewellyn R, Nuberg I (2017) Global learnings to inform the local adaptation of conservation agriculture in Eastern and Southern Africa. Glob Food Sec 17:213–220

CGIAR (2022) Zero tillage to reduce air pollution in India. https://www.cgiar.org/innovations/zero-tillage-to-reduce-air-pollution-in-india/. Assessed 1 Sept 2022

Chapman SC, Chakraborty S, Dreccer MF, Howden SM (2012) Plant adaptation to climate change: opportunities and priorities in breeding. Crop Pasture Sci 63:251–268

Cline WR (2007) Global warming and agriculture: Impact estimates by country. Peterson Institute of International Economics, NW, Washington, DC

Corbeels M, Naudin K, Whitbread AM et al (2020) Limits of conservation agriculture to overcome low crop yields in sub-Saharan Africa. Nat Food 1(7):447–454

Despsch R (2004) History of crop production, with and without tillage. Lead Edge 3(1):150–154

FAO (2009) How to feed the world in 2050. FAO CA website. http://www.fao.org/wsfs/forum2050

FAO (2013) Climate-smart agriculture: sourcebook. FAO, Rome. http://www.fao.org/3/a-i3325e/

FAO (2015) Climate change and food security: risks and responses. https://www.fao.org/3/i5188e/I5188E

FAO (2017) Conservation agriculture – revised edition online FAO. http://www.fao.org/3/a--i7480e.pdf/

Gan Y, Siddique KHM, Turner NC et al (2013) Ridge-furrow mulching systems- an innovation techniques for boosting crop productivity in semiarid rainfed environments. Adv Agron 118:429–476. https://doi.org/10.1016/B978-0-12-405942-9.00007-4

Gangwar B, Ram B (2005) Effect of crop diversification on productivity and profitability of rice (*Oryzasativa*)–wheat (*Triticumaestivum*) system. Indian J Agric Sci 75:435–438

Gathala MK, Laing AM, Tiwari TP et al (2020) Energy-efficient, sustainable crop production practices benefit smallholder farmers and the environment across three countries in the eastern Gangetic Plains, South Asia. J Clean Prod 246:118982

Greene GM (2022) Cover crops play a starring role in climate change mitigation. US Department of Agriculture (USDA). www.farmers.gov/blog/cover-crops

Hallegatte S, Rozenberg J (2017) Climate change through a poverty lens. Nat Clim Chang 7(4):250–256
Hallegatte S, Bangalore M, Bonzanigo L (2016) Shock waves: managing the impacts of climate change on poverty. The World Bank, Washington, DC
Hemle S, Anken T, Leifeld J et al (2008) The effect of the tillage system on soil organic carbon content under moist, cold-temperate conditions. Soil Tillage Res 98:94–105
Henry B, Charmley E, Eckard R et al (2012) Livestock production in a changing climate: adaptation and mitigation research in Australia. Crop Pasture Sci 63:191–202
HLPE (2012) Food security and climate change. A report by the High Level Panel of Experts on Food Security and Nutrition of the Committee on World Food Security, Rome
Islam S, Gathala MK, Tiwari TP et al (2019) Conservation agriculture based sustainable intensification: increasing yields and water productivity for smallholders of the Eastern gangetic Plains. Field Crop Res 238:1–7
Jat HS, Choudhary KM, Nandal DP et al (2020a) Conservation agriculture based sustainable intensification of cereal systems leads to energy conservation, higher productivity and farm profitability. Environ Manag 65:774–786
Jat HS, Datta A, Choudhary M et al (2020b) Conservation agriculture: factors and drivers of adoption and acalable innovative practices in Indo-Gangetic plains of India – a review. Int J Agric Sustain 18(1):1–16
Jordan VWL, Hutcheon JA (2003) Influence of cultivation practices on arable crop diseases. In: Titi AE (ed) Soil Tillage in Agroecosystems. CRC Press, Boca Raton
Joshi P (2005) Crop diversification in India: nature, pattern, and drivers. New Delhi India: Asian Development Bank
Kaye JP, Quemada M (2017) Using cover crops to mitigate and adapt to climate change. A review. Agron Sustain Dev 37(4):1–17
Kumar U, Panneerselvan P, Priya H (2019) Microbial technologies for land rejuvenation and climate resilient crop productivity. A manual on climate resilient agricultural technologies for future. Published by ICAR-NRRI, Cuttack
Lal R (2011) In: Hillel D, Rosenzweig C (eds) Handbook of climate change and agroecosystem: impacts, adaptation and mitigation. Imperial College Press, pp 287–305
Lal R, Stewart B (2010) Food security and soil quality. CRC Press, Boca Raton
Lalljee B (2013) Mulching as a mitigation agricultural technology against land degradation in the wake of climate change. Int Soil Water Conserv Res 1(3):68–74
Lee M, Gambiza J (2022) The adoption of conservation agriculture by smallholder farmers in southern Africa: a scoping review of barriers and enables. J Rural Stud 92:214–225
Lemke RL, Zhong Z, Campbell CA, Zentner R (2007) Can pulse crops play a role in mitigating greenhouse gases from North American Agriculture? Agron J 99(6):1719–1725
Leytem AB, Dungan RS, Bjorneberg DL, Koehn AC (2011) Emissions of ammonia, methane, carbon dioxide, and nitrous oxide from dairy cattle housing and manure management systems. J Environ Qual 40:1383–1394
Li SX, Wang ZH, Li SQ et al (2013) Effect of plastic sheet mulch, wheat straw mulch and maize growth on water loss by evaporation in dryland areas of China. Agric Water Manag 116:39–49
Lobell DB, Schlenker W, Costa-Roberts J (2011) Climate trends and global crop production since 1980. Science 333:616–620
Lohan SK, Jat HS, Yadav AK (2018) Burning issues of paddy residue management in north-west states of India. Renew Sust Energ Rev 81:693–706
Mallapaty S (2022) Why Pakistan's floods so extreme this year? Nature. https://www.nature.com/articles/d41586-022-02813-6. Assessed 2 Sept 2022
Mangalassery S, Sjogersten S, Spakes DL (2014) To what extent can zero tillage lead to a reduction in greenhouse gas emissions from temperate soils? Sci Rep 4(4586):1–8
McDaniel M, Tiemann L, Grandy AS (2014) Does agricultural crop diversity enhance soil microbial biomass and organic matter dynamics? A meta-analysis. Ecol Appl 24:560–570

Mhlanga B, Cheesman S, Maasdorp B et al (2015) Weed community responses to rotations with cover crops in maize-based conservation agriculture systems of Zimbabwe. Crop Prot 69:1–8
Michler JD, Baylis K, Arends-Kuenning M, Mazvimavi K (2019) Conservation agriculture and climate resilience. J Environ Econ Manag 93:148–169
Mousavi RS, Eskandari H (2011) A general overview on intercropping and its advantages in sustainable agriculture. J Appl Environ Biol Sci 1:482–486
Muller C, Elliott J (2015) The Global Gridded Crop Model intercomparison: approaches, insights and caveats for modelling climate change impacts on agriculture at the global scale. In: Elbehri A (ed) Climate change and food systems: global assessments and implications for food security and trade. FAO, Rome
Murungu FS (2012) Conservation agriculture for smallholder farmers in the Eastern Cape Province of South Africa: recent developments and future prospects. Afr J Agric Res 7(38):5278–5284
Nardone A, Ronchi B, Lacetera N et al (2010) Effects of climate change on animal production and sustainability of livestock systems. Livest Sci 130:57–69
Palazzo A, Vervoort J, Mason-D'Croz D et al (2017) Linking regional stakeholder scenarios and shared socioeconomic pathways: quantified West African food and climate futures in a global context. Glob Environ Chang 45:227–242
Paukner M (2022) No-till farming 101, soil health. https://www.no-tillfarmer.com/topics/88-no-till-farming-101. Assessed 5 Sept 2022
Pittelkow CM, Linguist BA, Lundy ME et al (2015a) When does no-till yield more? A global meta-analysis. Field Crop Res 183:156–168
Pittelkow CM, Liang X, Linguist BA et al (2015b) Productivity limits and potentials of the principles of conservation agriculture. Nature 517:365–368
Polley HW, Briske DD, Morgan JA (2013) Climate change and North American rangelands: trends, projections, and implications. Rangel Ecol Manag 66:493–511
Porter JR, Xie L, Challinor AJ et al (2014) Food security and food production systems. In: Climate change 2014: impacts, adaptation, and vulnerability. Part A: global and sectoral aspects. Contribution of working group II to the fifth assessment report of the intergovernmental panel on climate change. Cambridge University Press, Cambridge/New York, pp 485–533
Powlson DS, Stirling CM, Jat ML et al (2014) Limited potential of no-till agriculture for climate change mitigation. Nat Clim Chang 4(8):678–683
Rusinamhodzi L, Corbeels M, Wijk MV et al (2011) A meta-analysis of long-term effects of conservation agriculture on maize grain yield under rain-fed conditions. Agron Sustain Dev 31:657–673
Rusinamhodzi L, Corbeels M, Nyamangara J et al (2012) Maize-grain legume intercropping is an attractive option for ecological intensification that reduces climatic risk for smallholder farmers in central Mozambique. Field Crop Res 136:12–22
Russell S (2014) Everything you need to know about agricultural emissions. http://www.wri.org/blog/2014/05/everything-you-need-know-about-agricultural-emissions/ Accessed 24 Nov 2018
Sangar S, Abrol IP (2005) Conservation agriculture for transition to sustainable agriculture. Curr Sci 88:686–687
Satapathy S, Porsche I, Kunkel N et al (2011) Adaptation to climate change with a focus on rural areas and India. Deutsche Gesellschaft fur Internationale Zusammenarbeit (GIZ) GmbH, India Project on Climate Change Adaptation in Rural Areas of India. GIZ and GoI
Seitz S, Goebes P, Puerta VL (2019) Conservation tillage and organic farming reduce soil erosion. Agron Sustain Dev 39(4):1–10
Seo N, Mendelsohn R (2008) A Ricardian analysis of the impact of climate change on South American farms. Chil J Agric Res 68(1):69–79
Sidhu HS, Jat ML, Singh T et al (2019) Sub-surface drip fertigation with conservation agriculture in a rice-wheat system: a breakthrough for addressing water and nitrogen use efficiency. Agric Water Manag 216:273–283
Singh V (2001) Production sustainability and economic viability of different cropping systems under various levels of irrigation. Indian J Agric Sci 71:576–580

Singh RD, Khan A, Sikka A (2004) Sustainable agricultural development in Bihar. Agric Situat India 22:799–805

Six J, Ogle SM, Jay Breidt J (2004) The potential to mitigate global warming with no-tillage management is only realized when practised in the long term. Glob Chang Biol 10:155–160

Smith SE, Read DJ (2008) Mycorrizal symbiosis, 3rd edn. Academic/Elsevier, Amsterdam

Stanojevic AB (2021) Conservation agriculture and its principles. Ann Environ Sci Toxicol 5(1):018–022. https://doi.org/10.17352/aest.000031

Steinfeld H, Gerber P, Wassenaar T et al (2006) Livestock's long shadow. FAO, Rome. Accessed 20 Nov 2018

Sultan B, Gaetani M (2016) Agriculture in West Africa in the twenty-first century: climate change and impacts scenarios and potential for adaptation. Front Plant Sci 7:1262

Thomas TS, Rosegrant MW (2015) Chapter 5: Climate change impact on key crops in Africa: Using crop models and general equilibrium models to bound the predictions. In: Elbehri A (ed) Climate change and food systems: global assessments and implications for food security and trade. Food and Agricultural Organization of the United Nations (FAO), Rome, pp 146–175

Thornton PK, Herrero M (2010) The Inter-linkages between rapid growth in livestock production, climate change, and the impacts on water resources, land use, and deforestation. World Bank Policy Research Working Paper, WPS 5178. World Bank, Washington, DC

UN Report (2022) Global hunger numbers rose to as many as 828 million in 2021. In: Media release 6 July 2022, Rome/New York https://www.who.int/news/item/06-07-2022-un-report%2D%2Dglobal-hunger-numbers-rose-to-as-many-as-828-million-in-2021/

World Economic Forum (2022) 5 things to know about Europe's scorching heatwave. https://www.weforum.org/agenda/2022/07/heatwaves-europe-climate-change. Accessed 21 July 2022

Xiao L, Kuhn NJ, Zhao R, Cao L (2021) Net effects of conservation agriculture principles on sustainable land use: a synthesis. Glob Chang Biol 27(24):6321–6330. https://doi.org/10.1111/gcb.15906

Zaman AM, Molla MK, Pervin IA et al (2017) Impacts on river systems under 2°C warming: Bangladesh Case Study. Clim Serv 7:96–114

Zhao C, Liu B, Piao S (2017) Temperature increase reduces global yields of major crops in four independent estimates. Proc Natl Acad Sci 114(35):9326–9331

Zheng FL (2005) Effects of accelerated soil erosion on soil nutrient loss after deforestation on the Loess Plateau. Pedosphere 15(6):707–715

# Carbon Sequestration Strategies in Response to Fertilizer Application

**Ryusuke Hatano, Ikabongo Mukumbuta, and Mariko Shimizu**

## Introduction

Carbon (C) sequestration is the process of removing C from the atmosphere and capturing it in plants and soils. Soil C sequestrations include increasing the soil C content, increasing its's distribution within the depth of soil profiles, and ensuring that it is protected from microbial decomposition (Lal 2004). In agroecosystems, the C budget or C cycle is used to estimate C flows (inputs and outputs) and C transformations from one form to another. Carbon capture from the atmosphere is primarily through photosynthesis by plants (Janzen 2004; Trumbore 2006). Approximately half of this C from photosynthesis is released back into the atmosphere through plant respiration, and the other half is locked in the plant tissues. The C in plants is then added to soils as plant residue, litter, or as roots. In agricultural soils, C is also added through the addition of organic fertilizers such as animal manure. The C output (loss) from soils is largely through autotrophic and heterotrophic respiration. In addition to soil respiration, soil C is lost through leaching, erosion, and removal through harvested biomass. Autotrophic root respiration is part of the normal plant growth and maintenance of roots and does not result in depletion of soil C (Cahoon et al. 2016; Kuzyakov and Larionova 2005) because C lost through root respiration is always smaller than C assimilated into the plant tissues (Ferréa et al. 2012). Heterotrophic respiration, on the other hand, is the microbial decomposition of pre-formed C contained in soil organic matter (SOM) (Balogh

R. Hatano (✉)
Research Faculty of Agriculture, Hokkaido University, Sapporo, Japan
e-mail: hatano@chem.agr.hokudai.ac.jp

I. Mukumbuta
Faculty of Veterinary Medicine, Hokkaido University, Sapporo, Japan

M. Shimizu
Civil Engineering Research Institute for Cold Region, Sapporo, Japan



S. R. Mohanty, B. Kollah (eds.), *Greenhouse Gas Regulating Microorganisms in Soil Ecosystems*, https://doi.org/10.1007/978-3-031-70569-4_17

et al. 2016) and results in the loss of soil C. In soils, C is largely stored as SOM. Soil organic matter is the cornerstone of soil function and the most important measure of soil health, including the enhancement of water retention, permeability, buffer power, C sequestration from the atmosphere, and storage and supply of plant nutrition (Lal 2016). Soil organic matter is a product of the net C budget, i.e., the difference between C input (photosynthesis) and C output (respiration by plants and microbes). Therefore, maintenance of SOM is strongly dependent on the ecosystem C and nutrient management practices.

Soil organic matter is produced as a result of net ecosystem production (NEP) due to the difference between net primary production in aboveground and belowground by plants (ANPP and BNPP) and heterotrophic respiration by soil microbes (Rh), that is:

$$\text{NEP} = \text{ANPP} + \text{BNPP} - \text{Rh} \tag{1}$$

In a natural ecosystem, when plant litter is added to the soil, it is shredded and tilled by soil animals and decomposed into organic matter that is easily decomposed by microbes, and organic matter that is difficult to decompose, such as lignin, remains and polymerizes to form humus. Humus has acidic functional groups such as carboxyl group and phenol hydroxyl group, forms organo-mineral compound with clay mineral, and stabilizes soil structure by bridging silt and sand. That is, SOM is produced through soil genesis.

In an agroecosystem, harvest is always conducted, and manure application is also usually conducted. Therefore, net ecosystem C balance (NECB) is obtained as net biome production (Schulze et al. 2000). The NECB of agricultural land is obtained by adding C input (C input) through organic fertilizer application and C output (C output) by harvesting and erosion. However, C loss through erosion is usually not included in the NECB calculation because it is usually comparatively very small (Olson 2010).

Therefore, NECB is calculated as:

$$\text{NECB} = \text{ANPP} + \text{BNPP} - \text{Rh} + \text{C input} - \text{C output} \tag{2}$$

Plant residue is expressed as:

$$\text{Residue} = \text{NPP} - \text{Harvest} \tag{3}$$

So, if erosion is negligible, NECB can also be expressed as:

$$\text{NECB} = \text{Residue} - \text{Rh} + \text{C input} \tag{4}$$

Based on the above relational expressions, the components of NECB in agricultural ecosystems have been measured using various methods and techniques. The eddy covariance method can measure NEP directly. The chamber method is commonly used to measure soil respiration, with the Rh measured by the chamber method in the area where root respiration is removed through root exclusion plots, which is also called the trenching method. Soil produces other greenhouse gases (GHGs) such as methane ($CH_4$) and nitrous oxide ($N_2O$) together with carbon dioxide ($CO_2$), and these gases are also measured by chamber methods.

Net GHG balance (NGB) is estimated by using global warming potential (GWP), which is the total $CO_2$ equivalents of $CO_2$, $CH_4$, and $N_2O$ emissions and is calculated using the $CO_2$ conversion coefficients, 1 for $CO_2$, 28 for $CH_4$, and 265 for $N_2O$ (IPCC 2014). $CO_2$ emission is given as a negative value of NECB:

$$GWP = GWP\,CO_2 + GWPCH_4 + GWPN_2O \quad (5)$$

Net GHG balance is obtained as negative of GWP:

$$NGB = -GWP \quad (6)$$

$CH_4$ is largely produced under anaerobic conditions as the final step in anaerobic decomposition of SOM, and hence $CH_4$ emission is dominant in oxygen-limited ecosystems such as wetlands. In upland soils, $CH_4$ is usually taken up and consumed through oxidation. On the other hand, $N_2O$ emission is strongly influenced by the application of nitrogen (N) fertilizers, either in chemical/inorganic or organic forms. $N_2O$ is produced both under aerobic conditions through nitrification and under anaerobic conditions through denitrification. Figure 1 shows the schematic illustration of NECB and NGB in relation to emission and uptake of GHG gases, harvest, and organic fertilizer application. How to manage N fertilization and C input is key to achieving adequate soil C sequestration and ensuring sustainable crop production.

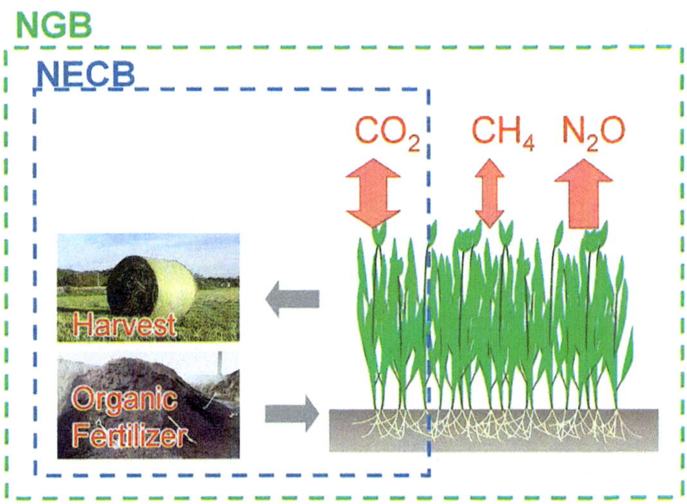


**Fig. 1** Schematic illustration of net ecosystem carbon balance (NECB) and net greenhouse gas balance (NGB) in agroecosystem

## Technologies of Soil C Sequestration and Its Impact on Ecosystems

In order to recover SOM degradation and increase SOM content, it is necessary to prevent excessive decomposition of SOM and to increase the input of organic matter to the soil. Excessive decomposition of SOM can be prevented by adopting conservation farming methods such as no-tillage, reduced-tillage, and mulch. On the other hand, in order to increase the input of organic matter to the soil, it is necessary to increase the application of organic fertilizers and the retention of crop residues. Recently, biochar addition has shown high potential to increase soil C. Integrated farming practices such as integrating crops and livestock, agroforestry, improvement of pastures, and conservation of urban ecosystems can increase soil C sequestration. Technologies such as organic farming, agroforestry, ecosystem restoration, and infrastructure development have been developed for these practices (Lal 2016), and these are further important to maintain regional circular economies under the COVID-19 pandemic (Lal et al. 2020). Figure 2 shows that these technologies improve soil C and contribute to reducing global warming potential and environmental burdens while enhancing plant production. Increasing the global soil C content in top 30 cm soil by 4 per mille (0.4%) can absorb all of the 4 Gt C remaining in the atmosphere annually. If the C content of agricultural soils can be successfully increased by 4 per mille in top 1 m soil, SOC sequestration reaches 2–3 Gt C $y^{-1}$ (Minasny et al. 2017).

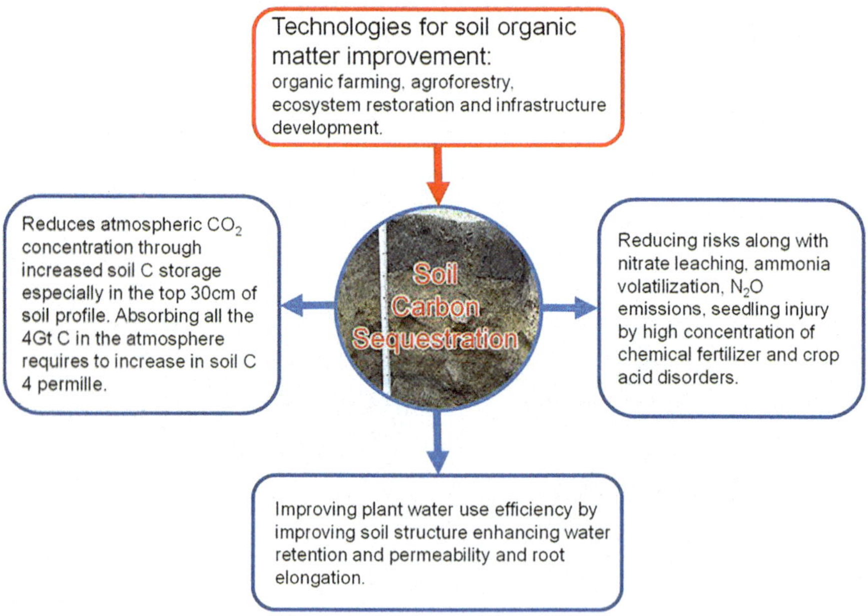


**Fig. 2** Technologies of soil carbon sequestration and its impact on soil ecosystems

The use of organic fertilizers increases the content of SOM, improves soil structure, enhances soil water retention and permeability, increases root elongation, improves the utilization efficiency of rainfall, and therefore improves plant growth (Wang et al. 2016). Therefore, it is effective for C sequestration of agricultural land soil.

Since the nutrient supply from organic fertilizers occurs during mineralization, the nutrient release is likely to be synchronized with plant nutrient requirements compared to inorganic fertilizers and therefore suppresses $NO_3$-N leaching and $N_2O$ emissions (Nagatake et al. 2018). This becomes more important during drought (Ullaha et al. 2020). Since N transformation strongly depends on soil moisture content. Nitrogen mineralization and nitrification increase with increasing water-filled pore space (WFPS) up to 60%, and beyond that, they decrease, and when WFPS is more than 80%, denitrification becomes predominant (Linn and Doran 1984).

Plant growth also depends on soil moisture. Rooting is maximized when the soil matrix potential is −10 kPa (Iijima and Kato 2007). When the soil is too moist, there is a shortage of air, and when the soil is too dry, the hardness of the soil increases in addition to the lack of water resulting in suppressed root elongation. Organic matter addition helps to maintain soil moisture. In addition, the application of organic fertilizer increases cation exchange capacity (CEC) of soil (Iwasaki et al. 2017), and because of increased humus, buffers against soil acidification, thereby preventing the decrease of soil pH. Through the above mechanisms, the application of organic fertilizer greatly contributes to increasing C input and reducing C output, in addition to directly adding C contained in the organic fertilizers.

## Effect of Organic Fertilizer on NECB and NGB

Crop growth is limited by the least supplied/available plant nutrient. Organic fertilizers do not always supply elements in an ideal balance for crop growth. In general, the content and mineralization rate of potassium (K) from livestock excreta organic fertilizer is higher than that of phosphorus (P) and N. Therefore, if the organic fertilizer is the only source of N and P, the K supply will usually be excessive. Therefore, fertilizer application designs based on organic fertilizer should ensure that the mineralized amount of each individual element (i.e., N, P, and K) does not exceed the elemental requirements of plants for any of these elements. Therefore, it is usually necessary to supplement the missing elements with chemical fertilizers (Asaka 2021).

The effects of this supplemental fertilizer application on crop growth and greenhouse gas emissions of three organic fertilizers (manure, slurry, and digestive liquid) were assessed for three years in managed grassland in an Andosol in southern Hokkaido (Kitamura et al. 2021). The amount of N from chemical fertilizer in organic fertilizer treatments decreased by 10% in manure, 19.7% in slurry, and 29.7% in digestive liquid compared with chemical fertilizer alone treatment, but the grass biomass yield was not significantly different among fertilizer treatments. The three-year NECB was significantly lower in organic fertilizer treatments than in chemical fertilizer-only treatment due to the addition of organic C from organic fertilizer. The three-year NGB in the organic fertilizer treatments reduced by 16.5% in slurry, 27.0% in digestive liquid, and 36.2% in manure compared with the chemical fertilizer-only treatment, and NECB accounted for more than 90% of the NGB. $N_2O$ emissions accounted for less than 8%, and the order was slurry > digestive liquid > manure.

An investigation of the appropriate manure application rate to compensate for negative NGB was conducted in four managed grasslands sites with different climates from Hokkaido to Kyushu in Japan by measuring NECB, $N_2O$, and $CH_4$ emissions for three years. Annual NECB and NGB were found to have the following significant positive correlation with the manure C application rate (C manure) (Hirata et al. 2013):

$$\text{NECB} = 1.0 \times \text{C} - \text{manure} - 2.4,\ r^2 = 0.88,\ P < 0.01$$

$$\text{NGB} = 1.2 \times \text{C} - \text{manure} - 4.0,\ r^2 = 0.87,\ P < 0.01$$

From the above regression equations, it is possible to estimate the manure C application rate required to make NECB and NGB positive, as shown in Table 1 and Fig. 3. The y-intercept of each regression equation is an approximation of NECB and NGB when no manure was applied to the field, which was −2.4 t $CO_2$–C eq $ha^{-1}$ $y^{-1}$ for NECB and − 4.0 t $CO_2$–C eq $ha^{-1}$ $y^{-1}$ for NGB. The measured value of NECB in the chemical fertilizer plot ranged from −0.08 to −3.46 t $CO_2$–C eq $ha^{-1}$ $y^{-1}$, while the NGB ranged from −0.27 to −4.10 t $CO_2$–C eq $ha^{-1}$ $y^{-1}$. The y-intercepts obtained from the regression equations were within these ranges. On the other hand, the x-intercept of each regression equation is an approximation of the amount of manure C required to reduce the NECB and NGB to zero, and this was 2.5 and 3.5 t C $ha^{-1}$ $y^{-1}$, respectively. Assuming that the C content per fresh weight of manure is 10%, it is necessary to add 35 t fresh $ha^{-1}$ $y^{-1}$ of manure in order to eliminate GHGs.

The difference between the 2.5 t C $ha^{-1}$ $y^{-1}$ required for NECB to reduce to zero and the 3.5 t C $ha^{-1}$ $y^{-1}$ required for NGB to reduce zero was solely caused by $N_2O$ emission since the soil was a $CH_4$ sink (Shimizu et al. 2013). Therefore, the difference of 1 t C $ha^{-1}$ $y^{-1}$ caused by $N_2O$ emissions must be supplemented by soil C

**Table 1** Effect of manure application on net ecosystem carbon balance (NECB) and net greenhouse gas balance (NGB)

| Manure | NECB | NGB |
|---|---|---|
| t C $ha^{-1}$ $y^{-1}$ | | t $CO_2$–C eq $ha^{-1}$ $y^{-1}$ |
| 0 | −2.4 | −4.0 |
| 2.5 | 0 | −1.2 |
| 3.5 | 1.0 | 0 |

Hirata et al. (2013)

The values were estimated from the regression equations obtained from the three years study in four grasslands with chemical fertilizer and manure treatments in Japan

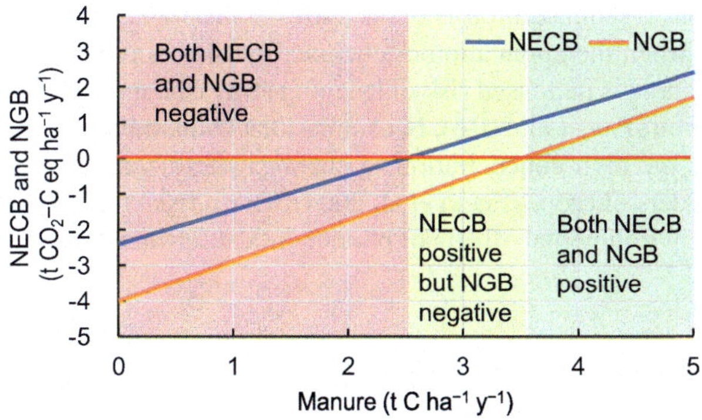


**Fig. 3** Relationship between manure application rate and net ecosystem carbon balance (NECB) and net greenhouse gas balance (NGB). It is necessary to achieve soil C sequestration that makes not only NECB but also NGB positive

sequestration. The C content of the four grasslands at a depth of 30 cm ranged from 85 to 170 t C $ha^{-1}$ $y^{-1}$. This implies that applying compost with zero greenhouse gas emissions required an increase in soil C of 0.6% to 1.2% (that is 6 to 12 per mille). This means that $N_2O$ emissions must be compensated by increasing soil C sequestration, and therefore, fertilization management practices that suppress $N_2O$ emissions are important.

## Effect of Inorganic Fertilizer on NECB and NGB

While the role of organic fertilizers in increasing soil C sequestration is well documented, the role of inorganic fertilizers is not. Inorganic fertilizers do not directly increase soil C; however, they are necessary in many parts of the world, especially in infertile soils, to increase productivity and biomass production (Follett 2001; Lal 2004). Increased biomass production in turn can lead to increased amounts of residue left in the field and therefore increase NECB. This is, however, dependent on residue retention being practiced. Cenini et al. (2015) reported increased soil C in fertilized compared to unfertilized soils 19 years after the set-up of experiments. Other studies have equally reported the importance of N fertilization. Follett (2001) estimated that N application rates of 67–134 kg N $ha^{-1}$ $y^{-1}$ resulted in 140–182 kg $ha^{-1}$ $y^{-1}$ SOC when compared with treatments without N fertilization. The addition of other nutrients such P, K, and trace elements is equally important in improving the general fertility of soil and can potentially help to increase the soil's ability to sequester C (Follett 2001). However, as indicated above, N fertilization generally leads to increased $N_2O$ emissions and could consequently reverse any gains made in SOC sequestration by increasing the net GHG emissions and GWP. Additionally, the process of inorganic fertilizer production releases significant amounts of $CO_2$ into the atmosphere and therefore indirectly negates the benefits of increased N fertilization on C sequestration (Follett 2001; Schlesinger and Amundson 2018). Additionally, when the application of chemical fertilizers is not synchronized with plant growth, there is increased risk of leaching (Hayashi and Hatano 1999), ammonia volatilization (Pan et al. 2016), $N_2O$ emissions (Mukumubuta et al. 2017), seedling injury by the high concentration of chemical fertilizers (Stevens et al. 2011), and acid disorders of crops (Fueki et al. 2004). These risks further add to the potential negative environmental effects of poor chemical fertilizer use.

## Factors Controlling $N_2O$ Emissions

Soil $N_2O$ is produced as a by-product by the nitrification process and as an intermediate product by the denitrification process. There is a significant relationship between $N_2O$ emission and N application rate and N mineralization or their total (Mu et al. 2009). Nitrification requires aerobic conditions and NH4-N, while denitrification requires anaerobic conditions and $NO_3$-N and organic C (Davidson 1992; Bremner 1997). Nitrification and denitrification also produce NO as a by-product and intermediate product. Since nitrification and denitrification are affected by soil moisture (Linn and Doran 1984), NO and $N_2O$ production are also affected by soil moisture (Bouwman 1998). NO increases as WFPS increases to 60% and then drops to almost zero at 70% of WFPS. On the other hand, $N_2O$ production starts at 30% of WFPS, peaks at 60–70% of WFPS, and decreases when WFPS exceeds 80%. In addition, N2 is produced when WFPS exceeds 70% and becomes dominant when WFPS exceeds 80%. These gas productions are strongly associated with the N conversion process in soil in relation to WFPS, and the N conversion process can be estimated from the $N_2O$-N/NO-N production ratio. NO production predominates in the nitrification process in dry soil condition, and the $N_2O$-N/NO-N ratio is 1 or less. On the other hand, $N_2O$ production predominates in the denitrification process in wet soil conditions, with an $N_2O$-N/NO-N ratio greater than 100 (Lipschultz et al. 1981). Therefore, reduction of $N_2O$ emission from denitrification is an important strategy, especially that management of N fertilizers, organic matter, and water is thought to be very important for denitrification.

In the onion fields in Hokkaido in Northern Japan, the occurrence of precipitation at the time of harvest led to large amounts of $N_2O$ emissions, accounting for about 70% of the annual $N_2O$ emissions (Kusa et al. 2002). This is largely due to the high amount of fresh residue, leaves, and roots retained in the field at the time of harvest in this area. At this time, the $N_2O$-N/NO-N ratio exceeded 100, indicating that denitrification was predominant. This indicates that the supply of residual organic matter at times of high rainfall, combined with high/excess $NO_3$-N, stimulated denitrification. In other words, it is important to reduce the amount of fertilizer applied so that there is little or no $NO_3$-N remaining in the soil at the time of harvest. If too much fertilizer is applied, regardless whether it is in inorganic or organic form, it will leave a large amount of $NO_3$-N and significantly increase $N_2O$ emissions. A comparison of chemical fertilizer treatment and cow dung manure treatment in a managed grassland showed increased $N_2O$ emissions in the manure treatment (Shimizu et al. 2010). However, both the treatments of chemical fertilizer and manure showed a similar relationship between $N_2O$ flux and soil moisture and soil temperature. The peaks of $N_2O$ flux were observed at 70–80% of WFPS and at more than 18 °C soil temperature. In addition, the $N_2O$-N/NO-N ratio was high when the $N_2O$ flux was high in both chemical fertilizers and manure treatments, indicating that denitrification was the main process of $N_2O$ production. Furthermore, the surplus amount of inorganic N, which is calculated as (chemical N fertilizer application + soil N mineralization - plant N uptake), showed a significant correlation with $N_2O$ emission. From these results, it is clear that the surplus/excess amount of inorganic N is a significant cause of soil $N_2O$ emission.

Manure releases N as it decomposes the decomposition rate is relatively low for cow dung manure, at an annual average rate of less than 20% of the total N content of the manure. However, cow dung manure has a high concentration of K, which has a higher release rate. Manure is applied as much as the amount of NPK mineralized from manure does not exceed the plant requirement, and therefore, the shortfall is usually supplemented with chemical fertilizer. However, since organic fertilizers continue to release nutrients in subsequent years (second year onwards), continuous application of manure can reduce the amount of chemical fertilizer supplements. Fig. 4 shows that with the reduction of chemical N fertilizer supplements in the manure plot, $N_2O$ emissions in the manure plot became lower than that in chemical fertilizer plots (Shimizu et al. 2013). In addition, the manure treatment increased soil pH to above 5, whereas the chemical fertilizer treatment reduced soil pH to below 5 (Jin et al. 2010; Mukumbuta et al. 2017).

Annual $N_2O$ emissions measured in grasslands and cornfields in Hokkaido are significantly negatively correlated with the average soil pH when N is applied to the fields with soil pH less than 6, with or without manure application (Mukumbuta et al. 2017). Incubation experiments with the grassland soil also found no $N_2O$ release when soils were limed to pH above 7 (Mukumbuta et al. 2018). This is because the optimum pH for incomplete denitrifying bacteria that cause $N_2O$ release is lower than the optimum pH of complete denitrifying bacteria, pH 6.5 (Nie et al. 2016). Therefore, maintaining or increasing soil pH to above 6.5 or higher can reduce $N_2O$ emissions. Maintaining soil pH above 6 by applying organic fertilizers, or by liming, is one of the potential strategies for reducing $N_2O$ emissions.

Due to it's bulky nature, it is often difficult to handle and manage organic matter (such as animal manure) application in soils. In Japan, manure pellets are being developed to improve the handling of manure as organic farming becomes more common. However, the pure manure pellets have been reported to increase $N_2O$ emissions due to stimulated denitrification inside the pellets (Hayakawa et al. 2009). However, addition of about 2% mineral N to manure pellets has been shown to significantly reduce $N_2O$ emissions from manure pellets due to inhibition of denitrification by the mineral N (Yamane et al. 2017).

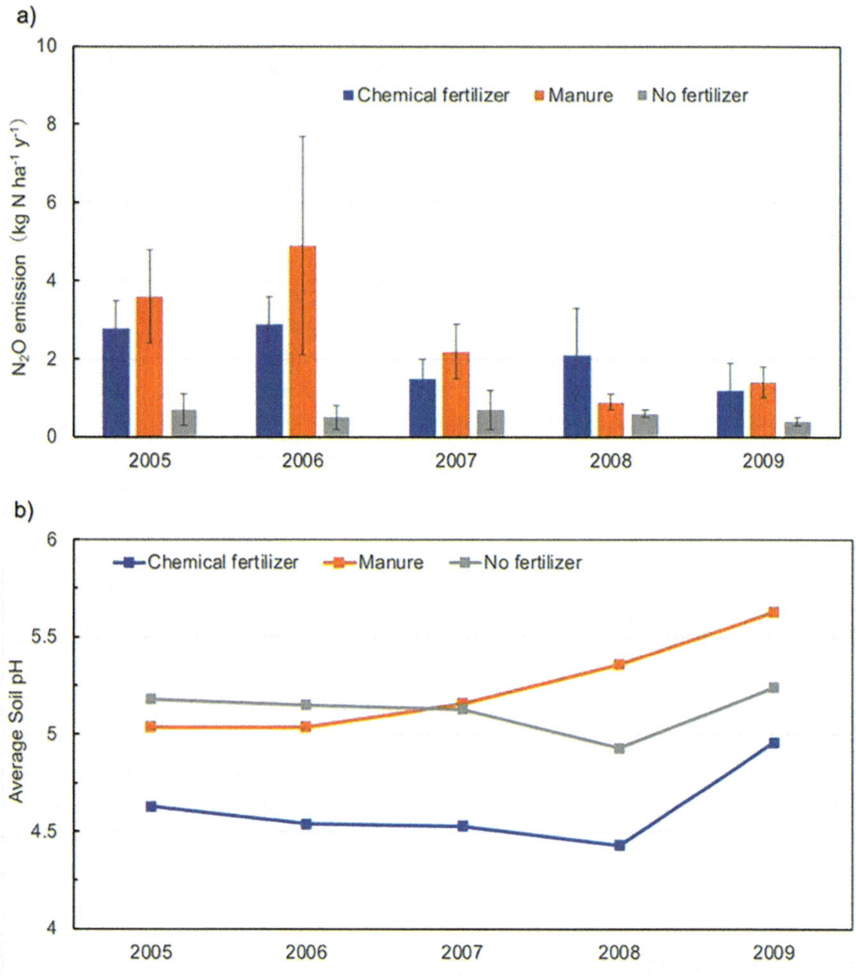


**Fig. 4** Effect of continuous manure application on $N_2O$ emission (**a**), and soil pH (**b**) in managed grassland in South Hokkaido. Chemical fertilizer and manure were applied according to the recommendation from local government (modified from Shimizu et al. 2013 and Mukumbuta et al. 2017). Manure application increased soil pH and hence could reduce $N_2O$ from denitrification

## Conclusion

In crop cultivation, crop residues remain after harvesting, while at the same time SOM is decomposed. If the C output by the SOM decomposition is larger than the C input by the crop residue, soil C is depleted. Therefore, the application of organic fertilizer is indispensable for soil C sequestration. It is also possible to reduce the amount of chemical fertilizer applied and maintain crop yields by applying appropriate organic fertilizers. This is because N is mineralized from organic fertilizer and supplied to crops. However, the increase in soil mineral N pool and the supply of organic matter promote denitrification enhance $N_2O$ emissions. Therefore, it is necessary to achieve C sequestration that offsets both soil C depletion and $N_2O$ emissions with organic fertilizer application. This amount was estimated to be 3.5 t C $ha^{-1}$ $y^{-1}$ as the amount of C input from manure in Japanese managed grassland. This amount corresponds to an increase in soil C of 6–12 per mille within the top 30 cm soil depth. $N_2O$ emission can be greatly reduced by continuous use of organic fertilizer in place of inorganic fertilizer. It is increasingly important to consider soil management with organic fertilizers that reduce chemical fertilizers without reducing yields.

## References

Asaka D (2021) Fertilizer recommendations. In: Hatano R, Shinjo H, Takata Y (eds) The soils of Japan. Springer Nature Singapore Pte Ltd.

Balogh J, Papp M, Pintér K, Fóti S, Posta K, Eugster W, Nagy Z (2016) Autotrophic component of soil respiration is repressed by drought more than the heterotrophic one in dry grasslands. Biogeosciences 13:5171–5182

Bouwman AF (1998) Nitrogen oxides and tropical agriculture. Nature 392:866–867

Bremner JM (1997) Source of nitrous oxide in soils. Nutr Cycl Agroecosys 49:7–16

Cahoon SMP, Sullivan PF, Gamm C, Welker JM, Eissenstat D, Post E (2016) Limited variation in proportional contributions of auto- and heterotrophic soil respiration, despite large differences in vegetation structure and function in the low arctic. Biogeochemistry 127:339–351

Cenini VL, Fornara DA, McMullan G, Ternan N, Lajtha K, Crawley MJ (2015) Chronic nitrogen fertilization and carbon sequestration in grassland soils: evidence of a microbial enzyme link. Biogeochemistry 126:301–313

Davidson EA (1992) Source of nitric oxide and $N_2O$ following wetting of dry soil. Soil Sci Soc Am J 56:95–102

Ferréa C, Zenone T, Comolli R, Seufert G (2012) Estimating heterotrophic and autotrophic soil respiration in a semi-natural forest of Lombardy, Italy. Pedobiologia (Jena) 55:285–294

Follett RF (2001) Soil management concepts and carbon sequestration in cropland soils. Soil Tillage Res 61:77–92

Fueki N, Tani M, Higashida S, Nakatsu S (2004) Effect of soil acidity and nitrification of fertilizer introduced by row application on sugar beet growth in several soil types. Soil Sci Plant Nutr 50:321–329

Hayakawa A, Akiyama H, Sudo S, Yagi K (2009) $N_2O$ and NO emissions from an Andisol field as influenced by pelleted poultry manure. Soil Biol Biochem 41:521–529

Hayashi Y, Hatano R (1999) Annual nitrogen leaching in subsurface-drained water from a clayey aquic soil growing onions in Hokkaido, Japan. Soil Sci Plant Nutr 45:451–459

Hirata R, Miyata A, Mano M, Shimizu M, Arita T, Kouda Y, Matsuura S, Niimi M, Saigusa T, Mori A, Hojito M, Kawamura O, Hatano R (2013) Carbon dioxide exchange at four intensively managed grassland sites across different climate zones of Japan and the influence of manure application on ecosystem carbon and greenhouse gas budgets. Agric For Meteorol 177:57–68

Iijima M, Kato J (2007) Combined soil physical stress of soil drying, anaerobiosis and mechanical impedance to seedling root growth of four crop species. Plant Produc Sci 10:451–459

IPCC (2014) In: Core Writing Team, Pachauri RK, Meyer LA (eds) Climate Change 2014. Synthesis Report. Contribution of Working Groups I, II and III to the Fifth Assessment Report of the Intergovernmental Panel on Climate Change. IPCC, Geneva

Iwasaki S, Endo Y, Hatano R (2017) The effect of organic matter application on carbon sequestration and soil fertility in upland fields of different types of andosols. Soil Sci Plant Nutr 63:200–220

Janzen HH (2004) Carbon cycling in earth systems—a soil science perspective. Agric Ecosys Environ 104:399–417

Jin T, Shimizu M, Marutani S, Desyatkin AR, Iizuka N, Hata H, Hatano R (2010) Effect of chemical fertilizer and manure application on $N_2O$ emission from reed canary grassland in Hokkaido, Japan. Soil Sci Plant Nutr 56:53–65

Kitamura R, Sugiyama C, Yasuda K, Nagatake A, Yuan Y, Du J, Yamaki N, Taira K, Kawai M, Hatano R (2021) Effects of three types of organic fertilizers on greenhouse gas emissions in a grassland on andosol in southern Hokkaido, Japan. Front Sustain Food Syst 5:649613

Kusa K, Sawamoto T, Hatano R (2002) Nitrous oxide emissions for six years from a gray lowland soil cultivated with onions in Hokkaido, Japan. Nutr Cycl Agro Ecosyst 63:239–247

Kuzyakov Y, Larionova AA (2005) Root and rhizomicrobial respiration: a review of approaches to estimate respiration by autotrophic and heterotrophic organisms in soil. J Plant Nutr Soil Sci 168:503–520

Lal R (2004) Soil carbon sequestration to mitigate climate change. Geoderma 123:1–22

Lal R (2016) Soil health and carbon management. Food Energy Secur 5:212–222

Lal R, Brevik EC, Dawson L, Field D, Glaser B, Hartemink AE, Hatano R, Lascelles B, Monger C, Scholten T, Singh BR, Spiegel H, Terribile F, Basile A, Zhang Y, Horn R, Kosaki T, Sánchez LBR (2020) Managing soils for recovering from the COVID-19 pandemic. Soil Systems 4:46

Linn DM, Doran JW (1984) Effect of water filled pore space on carbon dioxide and nitrous oxide production in tilled and non-tilled soils. Soil Sci Soc Am J 48:1267–1272

Lipschultz F, Zafiriou OC, Wofsy SC, McElroy MB, Valois FW, Watson SW (1981) Production of NO and $N_2O$ by soil nitrifying bacteria. Nature 294:641–643

Minasny B, Malone BP, McBratney AB, Angers DA, Arrouays D, Chambers A, Chaplot, Vincent, Chen ZS, Cheng K, Das BS, Field DJ, Gimona A, Hedley CB, Hong SY, Mandal B, Marchant BP, Martin M, McConkey BG, Mulder VL, O'Rourke S, Richer-de-Forges AC, Odeh I, Padarian J, Paustian K, Pan GX, Poggio L, Savin I, Stolbovoy V, Stockmann U, Sulaeman Y, Tsui CC, Vagen TG, van Wesemael B, Winowiecki L (2017) Soil carbon 4 per mille. Geoderma 292:59–86

Mu ZJ, Huang A, Kimura SD, Jin T, Wei S, Hatano R (2009) Linking $N_2O$ emission to soil mineral N as estimated by $CO_2$ emission and soil C/N ratio. Soil Biol Biochem 41:2593–2597

Mukumbuta I, Shimizu M, Jin T, Nagatake A, Hata H, Kondo S, Kawai M, Hatano R (2017) Nitrous and nitric oxide emissions from a cornfield and managed grassland: 11 years of continuous measurement with manure and fertilizer applications, and land-use change. Soil Sci Plant Nutr 63:185–199

Mukumbuta I, Uchida Y, Hatano R (2018) Evaluating the effect of liming on $N_2O$ fluxes from denitrification in an andosol using the acetylene inhibition and N-15 isotope tracer methods. Biol Fertil Soils 54(1):71–81

Nagatake A, Mukumbuta I, Yasuda K, Shimizu M, Kawai M, Hatano R (2018) Temporal dynamics of nitrous oxide emission and nitrate leaching in renovated grassland with repeated application of manure and/or chemical fertilizer. Atmos 9:485

Nie YX, Li L, Isoda R, Wang MC, Hatano R, Hashidoko Y (2016) Physiological and genotypic characteristics of nitrous oxide ($N_2O$)-emitting pseudomonas species isolated from dent corn Andisol farmland in Hokkaido, Japan. Microbes and Environments 31(2):93–103

Olson KR (2010) Impacts of tillage, slope, and erosion on soil organic carbon retention. Soil Sci 175:562–567

Pan B, Lam SK, Mosier A, Luo Y, Chen D (2016) Ammonia volatilization from synthetic fertilizers and its mitigation strategies: a global synthesis. Agric Ecosyst Environ 232:283–289

Schlesinger WH, Amundson R (2018) Managing for soil carbon sequestration: Let's get realistic. Global Chang Biol 25:386–389

Schulze ED, Wirth C, Heimann M (2000) Managing forests after Kyoto. Science 289:2058–2059

Shimizu M, Marutani S, Desyatkin AR, Jin T, Nakano K, Hata H, Hatano R (2010) Nitrous oxide emissions and nitrogen cycling in managed grassland in southern Hokkaido, Japan. Soil Sci Plant Nutr 56:676–688

Shimizu M, Hatano R, Arita T, Kouda Y, Mori A, Matsuura S, Niimi M, Jin T, Desyatkin AR, Kawamura O, Hojito M, Miyata A (2013) The effect of manure application on $CH_4$ and $N_2O$ emission from managed grasslands in Japan. Soil Sci Plant Nutr 59:69–86

Stevens WB, Evans RG, Jabro JD, Iversen WM (2011) Sugarbeet productivity as influenced by fertilizer band depth and nitrogen rate in strip tillage. J Sugar Beet Res 48:137–155

Trumbore S (2006) Carbon respired by terrestrial ecosystems –recent progress and challenges. Global Chang Biol 2:141–153. https://doi.org/10.1111/j.1365-2486.2005.01067.x

Ullah MR, Corneo PE, Dijkstra FA (2020) Inter-seasonal nitrogen loss with drought depends on fertilizer management in a seminatural Australian grassland. Ecosystems 23:1281–1293

Wang X, Jia Z, Liang L, Yang B, Ding R, Nie J, Wang J (2016) Impacts of manure application on soil environment, rainfall use efficiency and crop biomass under dryland farming. Sci Rep 6:20994

Yamane T, Komada M, Koga N, Nishimura S, Kato N (2017) Effects of urea and lime nitrogen in pelleted cattle manure compost on nitrous oxide emissions from soils. Japan J Soil Sci Plant Nutr 88:413–419

# Organic Farming: Approach and Strategies for Sustainable Agriculture

**M. Homeshwari Devi, Santosh R. Mohanty, and Bharati Kollah**

## Introduction

The sector of organic farming is witnessing rapid growth in developing nations. In 2020, Africa saw the most significant increase in organic agricultural land, expanding by 22% compared to 2019. Similarly, Asia experienced a noteworthy rise in its organic land area, with a 12% increase from the previous year. The global demand for organic foods is escalating quickly, with its sales surpassing $106 billion in 2019. The countries with the highest sales of organic foods that year included the United States ($42.0 billion), Germany ($13.0 billion), France ($11.5 billion), China ($10.0 billion), and Italy ($7.1 billion). India has become the leading producer of organic agricultural products, with an organic farming area reaching 7.4 million hectares in the 2022–2023 period, marking a substantial rise from 5.0 million hectares in 2018–2019. The principal organic crops in India include rice, wheat, pulses, cotton, oilseeds, along with fruits and vegetables. The export of organic products from India has seen a rapid increase, with over US$1.1 billion worth of organic goods exported in 2021–2022, which is a 40% increase from the year before. The main markets for Indian organic exports are the United States, the European Union, and Switzerland.

Table 1 details the extent of organic farming, including areas under conversion and wild organic coverage across Indian states for the 2020–2021 period. The combined organic and in-conversion area in India amounted to 2.6 million hectares, with an additional 1.6 million hectares dedicated to wild collection, totaling 4.33 million hectares of organic territory. Madhya Pradesh is at the forefront with the largest organic cultivation area of 1.63 million hectares, including areas of wild organic collection. Rajasthan and Maharashtra follow with 0.48 million and 0.37 million hectares, respectively. Madhya Pradesh not only has the most extensive organic area

M. H. Devi (✉) · S. R. Mohanty · B. Kollah
Indian Institute of Soil Science, Bhopal, Madhya Pradesh, India



S. R. Mohanty, B. Kollah (eds.), *Greenhouse Gas Regulating Microorganisms in Soil Ecosystems*, https://doi.org/10.1007/978-3-031-70569-4_18

**Table 1** State-wise area under organic farming 2020–2021

| Sl No. | States | Organic area (in ha) | Conversion area (in ha) | Organic + Conversion area (in ha) | Organic area under wild (in ha) | Total area (in ha) |
|---|---|---|---|---|---|---|
| 1 | Madhya Pradesh | 540,994 | 479,024 | 1,020,018 | 617,712 | 1,637,730 |
| 2 | Maharastra | 219,659 | 152,063 | 371,723 | 76 | 371,798 |
| 3 | Rajasthan | 177,600 | 121,087 | 298,686 | 183,176 | 481,862 |
| 4 | Gujarat | 72,318 | 75,548 | 147,866 | 0 | 147,866 |
| 5 | Karnataka | 61,116 | 33,934 | 95,050 | 79,373 | 174,424 |
| 6 | Odisha | 78,148 | 14,547 | 92,695 | 3612 | 96,307 |
| 7 | Sikkim | 74,647 | 1082 | 75,730 | 0 | 75,730 |
| 8 | Uttarakhand | 31,557 | 43,270 | 74,826 | 7384 | 82,210 |
| 9 | Uttar Pradesh | 53,195 | 14,248 | 67,443 | 91,865 | 159,308 |
| 10 | Jharkhand | 0 | 53,262 | 53,262 | 28,400 | 81,662 |
| 11 | Kerala | 25,657 | 19,414 | 45,070 | 3294 | 48,364 |
| 12 | Meghalaya | 34,816 | 3560 | 38,376 | 0 | 38,376 |
| 13 | Andhra Pradesh | 22,343 | 14,458 | 36,801 | 0 | 36,801 |
| 14 | Tamil Nadu | 14,086 | 17,543 | 31,629 | 9990 | 41,619 |
| 15 | Jammu & Kashmir | 19,029 | 11,591 | 30,620 | 162,150 | 192,770 |
| 16 | Bihar | 5 | 29,898 | 29,903 | 0 | 29,903 |
| 17 | Chhattisgarh | 14,744 | 8465 | 23,210 | 263,475 | 286,685 |
| 18 | Assam | 6719 | 11,752 | 18,471 | 0 | 18,471 |
| 19 | Nagaland | 7385 | 7405 | 14,790 | 0 | 14,790 |
| 20 | Arunachal Pradesh | 265 | 12,849 | 13,114 | 0 | 13,114 |
| 21 | Mizoram | 40 | 12,998 | 13,039 | 0 | 13,039 |
| 22 | Manipur | 4419 | 8306 | 12,725 | 2000 | 14,725 |
| 23 | Goa | 9244 | 3389 | 12,632 | 5590 | 18,222 |
| 24 | Himachal Pradesh | 9109 | 2745 | 11,854 | 191,882 | 203,736 |
| 25 | Telangana | 4724 | 2142 | 6866 | 0 | 6866 |
| 26 | Tripura | 204 | 6318 | 6521 | 0 | 6521 |
| 27 | West Bengal | 5463 | 840 | 6303 | 14,700 | 21,003 |
| 28 | Haryana | 3346 | 1557 | 4903 | 0 | 4903 |
| 29 | Punjab | 880 | 1142 | 2022 | 16,616 | 18,638 |
| 30 | Lakshadweep | 896 | 0 | 896 | 0 | 896 |
| 31 | LADAKH | 0 | 818 | 818 | 0 | 818 |
| 32 | Pondicherry | 3 | 21 | 24 | 0 | 24 |
| 33 | New Delhi | 1 | 4 | 5 | 0 | 5 |
| | Total | 1,492,611 | 1,165,278 | 2,657,889 | 1,681,296 | 4,339,185 |

Source: APEDA (2021)

but also leads in organic farm production, yielding 1.39 million metric tons (MT) in the 2020–2021 period, with Maharashtra (0.77 million MT) and Karnataka (0.35 million MT) following behind, as reported by APEDA in 2021. Table 1 displays the crop-wise production data under organic farming for the same period.

## Components and Principles of Organic Farming

According to Krishnaprabu (2019), specific components are utilized together because their compatibility enhances a synergistic effect, aiming to reduce dependence on external agricultural inputs. These components include:

- *Seeds:* The use of certified organic or non-genetically modified (GM) seeds is preferred over GM seeds. These seeds may be treated with cow urine, animal dung, biofertilizers, or biofungicides.
- *Biomanures:* These include farmyard manure (FYM), which contains dung, urine, straw, and farm waste, along with concentrated organic manures derived from various sources like non-edible oil cakes, edible oil cakes, fish meal, bones, and manures from poultry, sheep, and goat. Biomanures enrich the soil with source nutrients and organic matter, enhancing soil conservation. Incorporating green manure from plants such as dhaincha, berseem, sunhemp, cowpea, greengram, glyricidia, or sesbania also boosts soil fertility and its physical and chemical characteristics.
- *Compost:* Traditional compost is made through the aerobic decomposition of organic waste, including animal dung, plant debris, crop residues, and fodder residues. Vermicompost is produced by earthworms, and the addition of Trichoderma or Panchagavya can accelerate decomposition and increase the compost's nutritive value.
- *Crop diversification:* Practices like crop rotation, intercropping/mixed cropping, and cover cropping are implemented regularly. Trap crops are also grown along field borders to manage pest and disease incidence.
- *Soil solarization:* Mulching with polyethylene and dried weeds or crop residues reduces water evaporation and improves soil moisture retention.
- *Biofertilizers and microbial inoculants:* These include nitrogen fixers (e.g., Azotobacter, Azolla, Rhizobium, Azospirillum, blue-green algae), phosphate solubilizing bacteria and fungi, and phosphate mobilizers (vesicular arbuscular mycorrhizae), aiding in reducing the need for chemical fertilizers.
- *Microbial pesticides:* Instead of synthetic pesticides, microbial pesticides containing microorganisms like bacteria (e.g., *Bacillus subtilis, B. thuringiensis, Pseudomonas fluorescens*) and fungi (*Trichodermaviride, T. harzianum*) are used.

Chaudhary (2020) and Soumya (2015) further elucidate the principles of organic farming as health, ecology, fairness, and care. The health principle emphasizes the interconnectedness of ecosystem health with that of individuals and communities. The principle of ecology integrates organic agriculture within living ecological systems, focusing on ecological processes and recycling for nourishment and well-being. The fairness principle advocates for equity, respect, justice, and stewardship in all human relations within organic agriculture. Lastly, the principle of care highlights the need to respond adaptively to both internal and external conditions within the dynamic system of organic agriculture.

## Benefits of Organic Farming

The emphasis on food quality and safety is becoming increasingly prominent among consumers worldwide. There's a growing appreciation for organically produced foods, attributed to their health and nutritional advantages. Beyond personal health benefits, organic agriculture contributes to environmental conservation and offers significant socioeconomic benefits at the national level. India, with its rich tradition and immense potential for organic cultivation, was initially slow in embracing organic farming practices due to various challenges. Nevertheless, it has recently seen a surge in organic agriculture, positioning itself as one of the leading organic producers globally (Das et al. 2020). The primary goal of organic farming is to safeguard human health and preserve, or even improve, the quality of our natural resources, ensuring the environment remains viable for future generations and economically sustainable. A key advantage of organic farming is the improvement of soil quality, which involves enhancing its chemical, physical, and biological aspects and the interplay between them (Escobar and Hue 2007). Below are some of the principal advantages associated with organic farming.

### *Nutritional Benefits and Health Safety*

The rising popularity of organically farmed fresh produce has sparked interest among both consumers and producers in the nutritional differences between organic and conventionally farmed foods (Magnusson et al. 2003; Brandt and MØlgaard 2001). Foods grown organically, such as leafy vegetables and root crops, tend to have higher dry matter content than those grown conventionally (AFSSA 2003; Woese et al. 1997; Bourn and Prescott 2002). While organic grains and their derivatives may have lower protein levels compared to their conventional counterparts, the proteins they do contain are of higher quality, boasting superior amino acid profiles. For instance, organic wheat has been found to contain 25% to 30% more lysine than conventional wheat (Woese et al. 1997; Brandt et al. 2000). Additionally, organically farmed fruits and vegetables are often reported to taste better than those from conventional agriculture. This enhanced flavor profile is attributed to their longer growth periods, during which they are not subjected to artificial enhancers, allowing their natural sugars to fully develop and resulting in a more flavorful and nutritious product (Elayaraja and Vijai 2020).

## *Environmental Impact*

Organic agriculture is a practice aimed at sustainability and environmental preservation, avoiding synthetic pesticides to prevent harm to water, soil, and living organisms (Oquist et al. 2007). By limiting chemical inputs, it addresses critical environmental issues like soil erosion and pollution of air and water. This method also safeguards soil health, water quality, air quality, and biodiversity (Elayaraja and Vijai 2020). Some important environmental impacts are listed below:

### Soil Quality Improvements

According to Clark et al. (1998) and Petersen et al. (1999), organic farming improves soil fertility over time. Yet, in the initial transition phase of about 3 years, organic methods might temporarily decrease crop yields due to changes in biological activity and the delayed availability of nitrogen for plants (Petersen et al. 1999). Recent studies have confirmed these findings, with some suggesting that the negative effects on crop production during the transition period may be more severe than previously thought. For instance, research conducted by the Rodale Institute in 2021 revealed that in the early years of transitioning, organic crop yields could be 20–30% lower compared to conventional farming. Nonetheless, organic yields often meet or exceed those of conventional methods after a period of 3–5 years.

### Soil Biological Properties

Increased soil microbial activity and biological processes are one of the most important benefits of organic farming (Gunapala and Scow 1998; Scow et al. 1994; Werner 1997). A comprehensive review of 56 studies revealed that organic farming systems exhibit increases of 32–84% in microbial biomass carbon, nitrogen, total phospholipid fatty acids, and the activities of enzymes such as dehydrogenase, urease, and protease compared to conventional farming (Bruni et al. 2021). Additionally, organic farming has been shown to boost soil biodiversity significantly. In Denmark, shifting to entirely organic farming was found to raise microbial biomass by 77%, springtail population by 37%, and earthworm densities by 154% (Axelsen and Elmholt 1998). Moreover, organic farming can mitigate the prevalence of pests and diseases. For instance, research from California demonstrated that organic fields hosted 20–30% fewer plant parasitic nematodes than their conventional counterparts (Scow et al. 1994).

### Soil Physical Properties

Organic amendments, such as animal and green manures, are known to enhance soil physical characteristics by reducing bulk density, boosting water retention, and facilitating better infiltration rates (Petersen et al. 1999; Werner 1997; Lee et al. 2006; Sadanandan and Hamza 2006). A recent meta-analysis of 134 studies found that organic farming systems had 10–20% higher soil organic matter content, 5–10% higher water holding capacity, and 5–10% higher infiltration rates than conventional systems (Bruni et al. 2021). Organic manures also improve soil structure and aggregation, which leads to increased pore space and aeration. This creates a more favorable environment for plant root growth and development (Werner 1997). Overall, organic manures have a number of benefits for soil physical properties, including improved water holding capacity, infiltration rates, and pore space. These benefits can lead to increased plant growth and resilience to drought.

### Soil Chemical Properties

Numerous studies indicate that organic farming leads to increases in soil organic matter (SOM) content and humic substances. In the initial phase of transitioning from conventional to organic systems, a gradual and consistent augmentation in SOM is typically observed (Clark et al. 1998; Kuo et al. 1997). A recent meta-analysis of 56 studies found that organic farming systems had 32–84% greater SOM content than conventional systems, and this increase was most pronounced in the first 10 years of the transition (Bruni et al. 2021). SOM is a critical component of healthy soil and plays several important roles, including increasing water-holding capacity, improving drainage, reducing erosion, providing nutrients for plants, and sequestering carbon. Humic substances are a complex group of organic compounds formed from the decomposition of organic matter. They are an important component of SOM, and they play a number of important roles in soil health, including: improving soil structure, increasing nutrient availability, chelating heavy metals, and enhancing microbial activity. The enhancement of SOM and humic substances can be attributed to several practices, including the application of organic fertilizers like compost and manure that contribute organic matter to the soil; reduced tillage, which aids in maintaining SOM and humic substances; and crop diversification, which boosts the quantity of organic matter reintroduced into the soil. These improvements in SOM and humic substances within organic farming systems offer various advantages for crop production.

### Promotion of Biodiversity

Organic farms typically have higher levels of biodiversity than conventional farms, due to practices such as crop rotation, cover cropping, and the use of organic manures and pesticides. Crop rotation helps to promote biodiversity by providing a variety of habitats for different plants and animals. It also helps to reduce pests and diseases, which can further improve biodiversity. Cover cropping helps to protect the soil from erosion and improve soil fertility. It also provides food and shelter for beneficial insects and other wildlife. Organic manures and pesticides are less harmful to wildlife than synthetic fertilizers and pesticides. They can also help to improve soil health and promote biodiversity. A recent meta-analysis of 234 studies found that organic farms had 30% more species of plants and animals than conventional farms (Benton et al. 2003). Organic farms also had higher populations of beneficial insects, such as ladybugs and parasitic wasps (Tuck et al. 2014).

### Consume Less Energy

Organic farming typically consumes less energy and generates fewer waste products compared to conventional farming, both per unit of land and per unit of yield (Stolze et al. 2000; Hansen et al. 2001). The lower energy requirement is largely due to the avoidance of synthetic fertilizers and pesticides, which demand considerable energy for production and distribution. A recent meta-analysis of 33 studies found that organic farming systems used 30–40% less energy than conventional systems (Pimentel et al. 2005), attributed to factors such as reduced reliance of synthetic inputs, greater incorporation of renewable energy sources like solar and wind, and more efficient management of water and resources. Moreover, organic farming has been shown to contribute to lower greenhouse gas emissions. In essence, organic farming offers a more energy-efficient and low-emission alternative to conventional farming, underscoring its role in fostering sustainable agriculture practices.

### Climate Change

Organic farming plays a role in combating climate change through the reduction of greenhouse gas emissions and sequestering carbon. One of the main ways that organic farming reduces greenhouse gas emissions is by reducing the use of synthetic fertilizers. Synthetic fertilizers are a major source of nitrous oxide ($N_2O$), a greenhouse gas that is 265–298 times more potent than carbon dioxide ($CO_2$) over a 100-year timescale (United States Environmental Protection Agency 2015). Organic farming also sequesters carbon in the soil. This is because organic farming practices, such as crop rotation and cover cropping, help to build SOM. Soil organic matter is a rich source of carbon, and it helps to keep carbon out of the atmosphere. A recent meta-analysis of 56 studies found that organic farming systems had 32–84% greater soil organic matter content than conventional systems (Bruni et al. 2021). Another study found that organic farming systems emitted 25–30% less greenhouse gases than conventional systems.

### *Socioeconomic Impact*

Organic farming generally requires more labor than conventional farming, which can lead to more job creation in rural areas. For example, a study by Halberg (2008) found that organic farms in the United States employed 25% more workers per acre than conventional farms. Similarly, a study by Bhutani et al. (2018) found that organic farms in India employed 15–20% more workers per acre than conventional farms. Organic farming can also promote women's employment, as many of the tasks involved in organic farming, such as weeding and harvesting, are traditionally done by women. Organic products typically cost more than conventional products, due to factors such as the cost of organic certification, the higher cost of organic inputs, and the lower yields of organic crops. Nevertheless, numerous consumers are prepared to pay higher prices for organic items, motivated by the perceived benefits to health and the environment. According to a study by Winter and Davis (2006), organic products in the United States cost 10–40% more than conventional products. A study by Mukherjee et al. (2018) found that organic products in India cost 20–30% more than conventional products. Despite the higher cost, the demand for organic products is growing rapidly. This is due to increasing awareness of the potential health and environmental benefits of organic farming.

## Organic Agriculture and Sustainable Development

Organic farming offers a more economical and sustainable option compared to conventional farming methods in various aspects.

### *Ecological Sustainability*

Organic farming practices are known to improve ecological sustainability in a number of ways, including: Improved soil health: organic farming practices such as crop rotation, organic manure, and cover crops help to improve soil structure, fertility, and water retention. This reduces soil erosion and compaction and makes the soil more resilient to pests and diseases. Increased biodiversity: organic farming practices such as natural pest control and crop rotation promote biodiversity on farms. This includes beneficial insects, birds, and other wildlife that help to control pests and diseases (Benton et al. 2003). Conservation of plant genetic resources: crop rotation and the cultivation of a diversity of crops help to conserve plant genetic resources. This is important for maintaining food security and adapting to climate change (FAO 2020). Nutrient recycling: organic farming practices such as composting and livestock integration help to recycle nutrients on farms. This reduces the need for external inputs, such as synthetic fertilizers (Reganold and Wachter 2016).

Use of renewable energies: organic farms often integrate renewable energies, such as solar and wind power, into their operations. This reduces their reliance on fossil fuels and greenhouse gas emissions (Eyhorn et al. 2019). Diversified income streams: Organic farms often produce a variety of products, including crops, livestock, and timber. Diversifying assists in minimizing risk and enhancing income stability.

### *Social Sustainability*

Organic agriculture contributes to social well-being in a number of ways, including: Reduced environmental impacts: organic farming practices help to reduce soil erosion, water contamination, and greenhouse gas emissions. They also help to promote biodiversity and conserve plant genetic. Improved food security: organic farming can help to improve food security by increasing farm productivity and reducing vulnerability to weather vagaries (Gomez and Thivant 2015). Increased rural livelihoods: organic farming can help to improve rural livelihoods by providing farmers with access to higher-value markets and increasing their incomes. Promotion of social justice: organic farming can help to promote social justice by respecting indigenous knowledge and practices, promoting gender equality, and empowering rural communities (Pandey and Singh 2012).

## Challenge of Organic Farming

While organic food and farming offer numerous advantages, they still occupy a minor share of the world's total agricultural land. For instance, in India, only about 3.18% of the agricultural land is devoted to organic farming (Willer et al. 2021), highlighting potential gaps in policy direction and farmer engagement with organic practices. Nonetheless, there is a shift towards more sustainable agricultural policies in various countries, with incentives such as conversion subsidies being introduced to motivate farmers towards organic practices (Mzoughi 2011). Transitioning from conventional to organic farming usually spans 3 years (FAO 2015; Swezey et al. 1998; Tranter et al. 2009), a period which can see a notable dip in crop yields (Meena 2010; Pacini et al. 2003), potentially impacting farmer incomes negatively (Lapple and Van Rensburg 2011). To counterbalance these challenges, governmental support in the form of conversion subsidies is crucial (Rezvanfar and Olhan 2011; Soltani et al. 2014). The environmental concerns, subsidies, and market access were key determinants to switch to organic farming. The choice to adopt organic farming was positively affected by farmer knowledge, credit availability, and government assistance (Azam and Shaheen 2019).

## Constraints of Organic Farming

- *Reduced crop yields in the initial years:* According to Savage (2016), farmers adopting conventional organic farming may experience significant yield losses (up to 71%) in the initial years. This is because the soil needs time to adjust to organic farming practices, and organic farmers may not have the same level of experience and expertise as conventional farmers.
- *High cultivation costs:* Organic farming often requires higher inputs than conventional farming, such as labor and organic fertilizers. This can lead to higher cultivation costs, especially in the early years of conversion.
- *Uncertain crop yields:* The time required to achieve crop sustainability under organic farming practices can vary depending on a number of factors, such as soil type, climate, and management practices. This uncertainty can make it difficult for organic farmers to plan and budget for their operations.

In addition to these constraints, Indian organic growers may also face challenges such as:

- *Lack of access to organic markets:* There is a limited market for organic products in India, and organic farmers may have difficulty finding buyers for their products.
- *Lack of government support:* The Indian government does not provide as much support to organic farmers as it does to conventional farmers. Such circumstances may pose challenges for organic farmers striving to compete in the market.

## Opportunities of Organic Farming

**Carbon sequestration** Transitioning from natural ecosystems to farmland significantly reduces the soil's organic carbon content—by as much as 60% in temperate areas and over 75% in tropical agricultural soils (Lal 2010). This conversion highlights the pivotal role that transforming plant-captured carbon into stable soil organic carbon plays in counteracting climate change, as this form of carbon can persist for long periods, from centuries to millennia (Lal et al. 1998). For instance, enhancing the soil carbon content in underproductive agricultural soils by one ton can lead to yield boosts of approximately 20–40 kg per hectare for wheat, 10–20 kg per hectare for corn, and 0.5–1 kg per hectare for cowpeas (Lal 2010), illustrating a tangible connection between carbon sequestration and agricultural productivity.

Enhancing crop production is the foremost objective of agricultural methods, but the focus has recently shifted towards integrating farming approaches that promote carbon storage and ensure the long-term viability of farming ecosystems (Smith et al. 2007). Utilizing organic soil additives like crop residue, FYM, leguminous cover crops, and compost has been recognized for improving both crop yields and the enduring health of soil fertility (Pandey and team 2008), as well as contributing to a greater accumulation of carbon in the soil. Research in Punjab, India, has demonstrated that the use of organic materials can notably lower the emission of methane from paddy fields (Khosa et al. 2010). Consequently, the implementation of recommended management practices (RMPs) can greatly diminish these environmental impacts. The application of these RMPs serves as an economically viable strategy for combating climate change and simultaneously securing food availability in India.

## *Cost Benefits Issue*

Cost-benefit analysis (CBA) serves as a crucial framework for evaluating economic decisions, particularly in weighing the financial implications of various choices. It proves especially significant in gauging the economic viability of organic versus conventional farming practices. Research from Karnataka, India, demonstrated that organic farming yielded higher net profits than conventional methods, attributed largely to premium pricing for organic products, even though the average cultivation cost per acre for paddy was marginally lower (Suresh and Kunnal 2004). Similarly, research conducted by the Central Institute for Cotton Research in Nagpur, India, highlighted that organic farming reduced cultivation costs by approximately 21%, primarily due to the absence of chemical fertilizers and insecticides (IGNOU 2007). Mehmood et al. (2011) assessed the profitability ratios for organic versus inorganic wheat and carrot farming on a per-acre basis, finding ratios of 1:1.08 and 1:1.01 for organic crops, and 1:1.52 and 1:1.44 for inorganic crops, respectively. This indicates that organic farming can match or surpass conventional farming profitability in the short term. Additionally, Ramesh et al. (2010) found that organic farming delivered farmers a net profit margin 22% higher than conventional methods, despite a 9.2% decrease in productivity. This advantage was credited to a combination of reduced production costs by 11.7% and the ability to command a 20–40% higher price for certified organic products. Beyond immediate financial returns, organic farming presents numerous environmental advantages, such as enhanced soil quality, diminished water contamination, and bolstered biodiversity. These ecological benefits may translate into sustained cost reductions and societal savings over the long haul.

## *Employment Opportunities*

Organic farming typically requires more labor than conventional farming, due to the need for manual tasks such as weeding and harvesting. This can provide employment opportunities for rural communities, especially for those with less skilled labor. A study by Pimental et al. (2005) estimated that organic farming requires about 15% more labor than conventional farming. This suggests that organic farming could play a significant role in reducing unemployment in developing countries.

## Perspectives in Future Agriculture

Most Indian farmers are marginal and small, with low incomes (Singh 2000). Government initiatives are essential for the continuation of agriculture as a profession. A study by Azam and Shaheen (2019) found that government involvement and support are essential for promoting organic farming among small farmers, who often lack their own farmland. Furthermore, there is a pressing need for increased training and awareness efforts aimed at newer and less experienced farmers, focusing on the marketing, economic, and social aspects of organic agriculture. In rural areas of India, a significant barrier to farmer education on subjects like negotiation, demand forecasting, and setting prices is illiteracy. Providing education and training to these underrepresented communities could open doors to new possibilities like collective farming, more effective utilization of government schemes, and the exploration of sustainable agricultural practices, including organic farming.

Organic farming is known for being work-intensive, yet it offers significant benefits for rural job creation and the enhancement of environmental health. By incorporating livestock management into organic systems, farmers can generate their own compost and organic inputs. Additionally, the growing demand for organic textiles supports the case for organic cultivation, which has proven advantageous for silk production (Babu and Dandin 2009). Transitioning to organic methods can be a stepwise process, bolstered by a robust research and development framework. Such a transition encourages the adoption of sustainable farming practices well-suited to the Indian agricultural landscape, potentially leading to food self-sufficiency for the country. It is important for policymakers to advocate for organic farming as a pathway to improve living standards, soil health restoration, economic growth, and environmental protection. Offering subsidies to farmers who shift towards organic methods and utilize eco-friendly inputs can facilitate this transition. Furthermore, contract farming presents a viable option by guaranteeing organic farmers a market for their (Gahukar 2007).

Information on organic farming techniques for various crops is insufficient. Advancing research in creating suitable crop varieties or hybrids, enhancing plant nutrition, and developing comprehensive pest management strategies could boost the demand for organic products in both domestic and international markets. During the Bhoodan Movement, the government allocated unproductive lands, which could be repurposed for organic cultivation to expand the food crop sector. Initiatives for seed production could be launched by entities like the National Seed Corporation, agricultural universities, and private firms. Introducing a distinct minimum support price (MSP) for organic goods would inspire farmers to increase organic cultivation. For effective management, especially concerning seed production, it might be necessary to delineate organic farming zones distinctly from conventional ones. Regular updates to Indian organic standards, ensuring they match international benchmarks, are crucial for maintaining the global marketability of Indian organic products. Presently, India exports organic items to 58 countries, with the United States, Europe, and Canada as the primary buyers (APEDA 2021). The Participatory Guarantee System (PGS), a farmer-focused certification approach, has significantly contributed to the growth of organic farming in India, expanding it 25 times over the last 7 years through the collective efforts of farmers, NGOs, government actions, and market dynamics (Kulkarni 2011). Enhancing the PGS could further promote organic farming, making it a more sustainable and economically viable option for the future.

## Government Initiatives

As the demand for organic products raised both locally and globally, it became imperative to verify that products marketed as "organic" met established organic farming standards and that their production processes were independently certified. Initiated by the Indian government's Ministry of Commerce and Industry in 2001, the National Programme for Organic Production (NPOP) marked India's first venture into quality assurance in organic agriculture. The NPOP established a framework for certifying agencies' accreditation and the implementation of certification schemes via its certified bodies, ensuring regular monitoring and effective system operation. In 2004, the integration of NPOP within the scope of the Foreign Trade Development and Regulation (FTDR) Act required that organic products must be certified under the NPOP for exportation (Chaudhary 2020).

- The Paramparagat Krishi Vikas Yojana (PKVY), an initiative by the Indian government, encourages organic farming on a cluster basis, including certification through the Participatory Guarantee System (PGS). This program supports the creation of clusters along with offering training, certification, and marketing assistance. Farmers receive financial aid of Rs. 50,000 per hectare over a period of 3 years, with 62% of this amount (Rs. 31,000) allocated as a subsidy for organic inputs (Elayaraja and Vijai 2020).
- The Rashtriya Krishi Vikas Yojana (RKVY) is a government of India scheme that provides assistance for the promotion of organic farming on various components, subject to the approval of the State Level Sanctioning Committee (Elayaraja and Vijai 2020).
- The One District-One Product (ODOP) program is a flagship initiative of the Government of Uttar Pradesh that aims to promote and develop indigenous and specialized products/crafts of each district in the state. The program aims to generate employment at the district level and improve the earnings of small and marginal farmers. The presence of aggregators is essential to the success of the ODOP program. Aggregators are businesses that collect and consolidate products from multiple suppliers. This allows them to achieve economies of scale,

which can benefit both the suppliers and the consumers. Aggregators can help ODOP farmers in a number of ways, including: providing them with access to a larger market, helping them to get better prices for their products, providing them with training and support on production and marketing techniques, investing in infrastructure and technology to improve the efficiency of the value chain, and by working with aggregators, ODOP farmers can increase their income and improve their livelihoods (Drishti 2020).

- The National Mission on Oilseeds and Oil Palm (NMOOP) offers a subsidy of 50%, capped at Rs. 300 per hectare, to support various components. This financial assistance covers the use of biofertilizers including Rhizobium culture, phosphate solubilizing bacteria (PSB), zinc solubilizing bacteria (ZSB), Azotobacter, mycorrhiza, and vermicompost (Elayaraja and Vijai 2020).
- The Organic Farming Action Programme 2017–2020. The Organic Farming Action Programme 2017–2020 (OFAP) was a 4-year initiative launched by the Indian government to promote and significantly develop organic farming. The OFAP aimed to achieve this through a number of priority measures, including: increasing the area under organic cultivation, improving the productivity and quality of organic produce, strengthening the organic value chain, and creating awareness about organic farming and its benefits. The OFAP was implemented by the Department of Agriculture and Cooperation under the Ministry of Agriculture and Farmers Welfare. The program provided financial assistance to farmers for the conversion to organic farming, as well as for the purchase of organic inputs and certification. The OFAP also supported training and capacity-building programs for farmers and other stakeholders in the organic sector. The OFAP was successful in increasing the area under organic cultivation in India from 5.16 lakh hectares in 2015–2016 to 11.55 lakh hectares in 2019–2020. The OFAP also helped to improve the productivity and quality of organic produce and to strengthen the organic value chain. The program also raised awareness about organic farming and its benefits among consumers (Elayaraja and Vijai 2020).

## Conclusion

Organic farming stands as a pivotal approach to rejuvenating and maintaining the health of the soil, critical for achieving sustainability. Practices including the rotation of crops, the application of green manures, biofertilizers, composting, and the use of biological methods for pest control facilitate the production of nutritious and high-quality food devoid of synthetic fertilizers, pesticides, or hormones. With a rising consumer demand for food that is both healthy and safe, organic agriculture is gaining traction. As the world's leading producer of organic products, India's commitment to organic agriculture could significantly contribute to fostering a future that is nutritionally rich, ecologically sound, and economically vibrant.

## References


AFSSA (2003) Report on Evaluation of the nutritional and sanitary quality of organic foods (Evaluation nutritionnelle et sanitaire des aliments issus de l'agriculturebiologique, in French). AFSSA, p 164. http://www.afssa.fr. Accessed 3 August 2018

Agricultural and Processed Food Products Export Development Authority (APEDA) (2021) Consolidated organic agricultural statistics for the year 2020–21, Statistics for the year 2020–21. Ministry of Commerce and Industry, Government of India. https://apeda.gov.in/apedawebsite/organic/data.htm

Axelsen JAA, Elmholt S (1998) Scenarium om 100% okologiskjordnurgi Denmark, A3.4, Jordbundensbiologi. Rapport for Bichel-Udvalget

Azam MS, Shaheen M (2019) Decisional factors driving farmers to adopt organic farming in India: a cross-sectional study. Int J Soc Econ 46(4):562–580. https://doi.org/10.1108/IJSE-05-2018-0282

Babu CM, Dandin SB (2009) Organic farming for mulberry: an overview. Indian J Seric 48(2):100–110

Benton TG, Vickery AJ, Wilson JD (2003) Farmlanf biodiversity: is habitat heterogeneity the key? Trends Ecol Evol 18(4):182–188

Bhutani S, Kahlon S, Monika (2018) Organic farming in India: an alternative agricultural system. Amity J Agribusiness 3(1):37–49

Bourn D, Prescott J (2002) A comparison of the nutritional value, sensory qualities, and food safety of organically and conventionally produced foods. Crit Rev Food Sci Nutr 42:1–34

Brandt K, MØlgaord JP (2001) Organic agriculture: does it enhance or reduce the nutritional value of plant foods? J Sci Food Agric 81:924–931

Brandt DA, Brand TS, Cruywagen CW (2000) The use of crude protein content to predict concentrations of lysine and methionine in grain harvested from selected cultivars of wheat, barley and triticale grown in Western Cape region of South Africa. S Afr J Anim Sci 30:22–259

Bruni E, Guenet B, Huang Y, Clivot H, Virto I, Farina R et al (2021) Additional carbon inputs to reach a 4 per 1000 objective in Europe: feasibility and projected impacts of climate change based on century simulations of long-term arable experiments. Biogeosciences 18(13):3981–4004. https://doi.org/10.5194/bg-2020-489

Chaudhary D (2020) Organic farming in India: need and future aspects. MuktShabd J 9(5):317–327. IssnNo : 2347-3150

Clark MS, Horwath WR, Shennan C, Scow KM (1998) Changes in soil chemical properties resulting from organic and low-input farming practices. Agron J 90:662–671

Das S, Chatterjee A, Pal TK (2020) Organic farming in India: a vision towards a healthy nation. Food Qual Saf 4:69–76. https://doi.org/10.1093/fqsafe/fyaa018

Drishti IAS (2020) Organic farming in India. Retrieved from https://www.drishtiias.com/daily-updates/daily-news-analysis/organic-farming-in-india

Elayaraja M, Vijai C (2020) Organic farming in India: benefits and challenges. Eur J Mol Clin Med 7(11):3021–3029. ISSN 2515–8260

Escobar MEO, Hue NV (2007) Current developments in organic farming. Recent Res Devel Soil Sci 2:29–62. ISBN: 81-308-0151-5

Eyhorn F, Muller A, Reganold JP et al (2019) Sustainability in global agriculture driven by organic farming. Nat Sustain 2:253–255. https://doi.org/10.1038/s41893-019-0266-6

FAO (2015) Training manual for organic agriculture, technologies and practices for smallholder farmers (TECA). Food and Agriculture Organization of the United Nations, pp 23–25

FAO (2020) The world of organic agriculture-statistics and emerging trends 2020. https://shop.fibl.org/CHen/mwdownloads/download/link/id/1294/?ref=1

Gahukar RT (2007) Contract farming for organic crop production in India. Curr Sci 93(12):1661–1663

Gomez I, Thivant L (2015) Training manual for organic agriculture. Technologies and practices for smallholder farmers (TECA team)- Research and Extension division (DDNR), Food and Agriculture Organization of the United nation (FAO)

Gunapala N, Scow KM (1998) Dynamics of soil microbial biomass and activity in conventional and organic farming systems. Soil Biol Biochem 30:805–816

Halberg N (2008) Energy use and green house gas emission in organic agriculture. In: Proceedings of international conference organic agriculture and climate change. 17–18 April 2008, ENITA of Clermont, France, pp 1–6

Hansen B, Alroe HJ, Kristensen ES (2001) Approaches to assess the environmental impact of organic farming with particular regard to Denmark. Agric Ecosyst Environ 83:11–26

IGNOU (2007) BAPI-003 Economics and marketing of organic produce. http://vedyadhara.ignou.ac.in/wiki/images/b/bb/BAPI-003-01.pdf (Last accessed on September 2011)

Khosa MK, Sidhu BS, Benbi DK (2010) Effect of organic materials ans rice cultivars on methane emission from rice field. J Environ Biol 31:281–285

Krishnaprabu S (2019) Organic farming in India: concept, applications and perspectives. J Phys Conf Ser 1362:012020. https://doi.org/10.1088/1742-6596/1362/1/012020

Kulkarni N (2011) Organic farming picks up in India. Biospectrum 9(5):20–24

Kuo S, Sainju UM, Jellum EJ (1997) Winter cover crop effects on soil organic carbon and carbohydrate in soil. Soil Sci Soc Am J 61:145–152

Lal R (2010) Soil carbon sequestration impacts on global climate change and food security. Science 304:1623–1627

Lal R, Kimble J, Follett R (1998) Land use and soil C pools in terrestrial ecosystems. In: Lal R, Kimble JM, Follett RF, Stewart BA (eds) Management of carbon sequestration in soil. CRC Press, Boca Raton, pp 1–10

Lapple D, Van Rensburg T (2011) Adoption of organic farming: are there differences between early and late adoption? Ecol Econ 70(7):1406–1414

Lee YH, Lee SM, Sung JK, Kim YH, Choi DH, Ryu GH (2006) The effects of applied organic materials for improving soil physico-chemical properties in paddy fields. In: 18th World Congress of Soil Science. Philadelphia, PA, USA

Magnusson MK, Arvola A, Hursti UK, Aberg L, Sjoden PO (2003) Choice of organic foods is related to perceived consequences for human health and to environmentally friendly behaviour. Appetite 40:109–117

Meena BS (2010) Socio-economic study of organic farming in irrigated north western plain zone of Rajasthan. Agric Sci Dig 30(2):94–97

Mehmood Y, Anjum MB, Sabir M, Arshad M (2011) Benefit cost ratio of organic and inorganic wheat production: a case study of district Sheikhupura. World Appl Sci J 13:175–180

Mukherjee A, Kapoor A, Dutta S (2018) Organic food business in India: a survey of companies. Res Econ Manage 3:72. https://doi.org/10.22158/rem.v3n2P72

Mzoughi N (2011) Farmers adoption of integrated crop protection and organic farming: do moral and social concerns matter? Ecol Econ 70(8):1536–1545

Oquist KA, Strock JS, Mulla DJ (2007) Influence of alternative and conventional farming practices on subsurface drainage and water quality. J Environ Qual 36:1194–1204

Pacini C, Wossink A, Giesen G, Vazzana C, Huirne R (2003) Evaluation of sustainability of organic, integrated and conventional farming systems: a farm and field-scale analysis. Agric Ecosyst Environ 95(1):273–288

Pandey J, Pandey U, Shubhashish K, Pandey R (2008) Inter-species variations in soil fertility stability in organic farming. Plant Arch 8:61–63

Pandey J, Singh A (2012) Opportunities and constraints in organic farming: an Indian perspective. J Sci Res 56:47–72. ISSN: 0447-9483

Petersen C, Drinkwater LE, Wagoner P (1999) The Rodale Institute farming System Trial: the first fifteen years. The Rodale Institute, Kutztown, PA

Pimental D, Hepperly P, Hanson J, Douds D, Seidel R (2005) Environmental, energetic and economic comparisons of organic and conventional farming systems. Bioscience 55:573–582

Ramesh P, Panwar NR, Singh AB, Ramana S, Yadav SK, Shrivastava R, Subba Rao A (2010) Status of organic farming in India. Curr Sci 98:1090–1194
Reganold J, Wachter J (2016) Organic agriculture in the twenty-first century. Nat Plants 2:15221. https://doi.org/10.1038/nplants.2015.221
Rezvanfar A, Olhan E (2011) Determine of factors associated with the adoption of organic agriculture among small farmers in Iran. Afr J Agric Res 6(13):2950–2956
Sadanandan AK, Hamza S (2006) Effect of organic farming on soil quality, nutrient uptake, yield and quality of Indian spice. In: 18th World Congress of Soil Sci. Philadelphia, PA, USA
Savage S (2016) USDA data on 370 crops: why organic farming has lower yields. Genetic Literacy Project. Available at https://geneticliteracyproject.org/2016/09/16/usda-data-370-crops-organic-farming-lower-yields
Scow KM, Somasco O, Ferris H, Gunapala N, Lau S, Venette R, Miller R, Shennan C (1994) Transition from conventional to low-input agriculture changes soil fertility and biology. Calif Agric 48:20–26
Singh RB (2000) Environmental consequences of agricultural development: a case study from the green revolution state of Haryana, India. Agric Ecosyst Environ 82(1):97–103
Smith P, Martino D, Cai Z, Gwary D, Janzen H, Kumar P et al (2007) Policy and technological constraints to implementation of greenhouse gas mitigation options in agriculture. Agric Ecosyst Environ 118:6–28
Soltani S, Azadi H, Mahmoudi H, Witlox F (2014) Organic agriculture in Iran: farmers' barriers to and factors influencing adoption. Renew Agric Food Syst 29(2):126–134
Soumya KM (2015) Organic Farming: an effective way to promote sustainable agriculture development in India. IOSR J Humanities Soc Sci 20:31–36, e-ISSN: 2279-0837, p-ISSN: 2279-0845. www.iosrjournals.org
Stolze M, Piorr A, Haring AM, Dabbert S (2000) Environmental impacts of organic farming in Europe, Organic farming in Europe: economics and policy, vol 6. University of Hohenheim, Stuttgart. Retrieved on 15 May 2011. http://orgprints.org/8400/1/Organic_Farming_in_Europe_Volume06_The_Environmental_Impacts_of_Organic_Farming_in_Europe.pdf
Suresh H, Kunnal LB (2004) Economics of Organic Farming of Rice. In: Veeresh GK (ed) Operational Methodologies and package of Practices in Organic farming. APOF, Banglore, pp 77–78
Sweezy SL, Werner MR, Buchanan M, Allison J (1998) Comparison of conventional and organic apple production systems during three years of conversion to organic management in coastal California. Am J Altern Agric 13(4):162–180
Tranter RB, Bennett RM, Costa L, Cowan C, Holt GC, Jones PJ, Miele M, Sottomayor M, Vestergaard J (2009) Consumers' willingness-to-pay for organic conversion-grade food: evidence from five EU countries. Food Policy 34(3):287–294
Tuck SL, Winqvist C, Mota F, Ahnstrom J, Turnbull LA, Bengstsson J (2014) Land-use intensity and the effects of organic farming on biodiversity: a hierarchical meta-analysis. J Appl Ecol 51:746–755
United States Environmental Protection Agency (2015) Understanding global warming potentials. Retrieved from http://epa.gov/climatechange/ghgemissions/gwps.html
Werner MR (1997) Soil quality characteristics during conversion to organic orchard management. Appl Soil Ecol 5:151–167
Willer H, Travnicek J, Meier C, Schlatter B (2021) The worlds of organic agriculture: statistics and emerging trends (Eds). FiBL and IFOAM—Organics International
Winter CK, Davis SF (2006) Organic food. J Food Sci 71:117–124
Woese K, Lange D, Boess C, Bogl KW (1997) A comparison of organically and conventionally grown foods- results of a review of the relevant literature. J Sci Food Agric 74:281–293

# Long-Term Fertilizer Experiments: Strategies for Efficient Carbon Sequestration

**Manoj Parihar, Rajendra Prasad Meena, Arjun Singh, Shalini, Saroj Choudhary, Kiran Rana, Jaideep Kumar Bisht, and Lakshmi Kant**

## Introduction

Soil is an important source or sink of carbon (C) and plays a major role in global C cycling (Stockmann et al. 2013; Parihar et al. 2020). Whether soil acts as a source or sink of C depends largely on management practices. Net C storage or sequestration depends on the transformation of applied C sources into soil organic matter or their mineralization through microbial decomposition (Baldock and Skjemstad 2000). Soil C sequestration can be enhanced through best management practices

M. Parihar (✉) · J. K. Bisht · L. Kant
ICAR-Vivekananda Parvatiya Krishi Anusandhan Sansthan (VPKAS), Almora, Uttarakhand, India
e-mail: manoj.parihar1@icar.gov.in

R. P. Meena
ICAR-Vivekananda Parvatiya Krishi Anusandhan Sansthan (VPKAS), Almora, Uttarakhand, India

ICAR-Indian Agricultural Research Institute, New Delhi, India

A. Singh
ICAR-National Research Centre for Banana, Tiruchirappalli, Tamil Nadu, India

Shalini
KrishiVigyan Kendra, Hamirpur, Banda University of Agriculture & Technology, Banda, Uttar Pradesh, India

S. Choudhary
Department of Soil Science and Agricultural Chemistry, Institute of Agricultural Sciences, Banaras Hindu University, Varanasi, Uttar Pradesh, India

K. Rana
Department of Agronomy, Institute of Agricultural Sciences, Banaras Hindu University, Varanasi, Uttar Pradesh, India



S. R. Mohanty, B. Kollah (eds.), *Greenhouse Gas Regulating Microorganisms in Soil Ecosystems*, https://doi.org/10.1007/978-3-031-70569-4_19

(BMPs) such as zero tillage, organic manure application, crop rotation, balanced fertilization, etc. (Benbi and Brar 2009; Minasny et al. 2017). The use of organic manure and chemical fertilizers is common in agriculture to increase soil and crop productivity (Shahid et al. 2013). Many long-term experiments have revealed that integrated use of organic and inorganic fertilizer improves labile soil carbon and productivity (Jha et al. 2014; Tang et al. 2018; Parihar et al. 2021). Similarly, long-term application of manure and compost enhances the soil carbon pool more than the application of the same amount of chemical fertilizers (Gregorich et al. 2001). However, the extent of C buildup is differentially influenced by the source of organic amendments, as their C:N ratio, lignin and cellulose content, and chemical composition determine the decomposition process (Cotrufo et al. 2013). To better understand how net C sequestration or loss occurs in the soil system, the total C stock is classified into active or labile and passive or recalcitrant pools (Oades 1984). Microbial biomass carbon, particulate organic carbon, KMnO4-oxidizable organic carbon, and mineralizable organic carbon are the major pools of labile SOC. The active or labile soil C pool has a rapid turnover rate and is considered very important from a crop production perspective, as it maintains the nutrient cycling process in soil (Chan et al. 2001; Majumder 2007). In contrast, the stable or recalcitrant C pool remains in the soil for a longer period and is important from a climate change perspective. To evaluate the various nutrient management practices for their relative potential for C sequestration and SOC dynamics, the carbon management index (CMI) can be used as a sensitive measure (Blair et al. 1995). The CMI helps in monitoring the long-term changes in C dynamics due to different nutrient supply options. A higher value of CMI indicates that management practices are sustainable in the long term, while a negative value means that system performance is declining (Gami et al. 2009).

Long-term fertilizer experiments (LTFEs) are important for predicting, quantifying, and assessing long-term changes in soil carbon content and evaluating the C sequestration potential of different soil and cropping systems (Powlson et al. 1986; Ladha et al. 2003). Therefore, LTFEs are more prominent in tropical and subtropical countries where inherent soil C content is very low due to high temperature and mineralization rate (Ladha et al. 2003; Mandal 2005; Pathak et al. 2011). In India, during 1970–1980, many LTFEs were imitated under different agro-climatic zones, covering various cropping systems. They were mostly designed to study the response of cropping systems under different inputs and management practices (Krause et al. 2020). Based on the management, these experiments are broadly classified into three main stages, such as static, semi-static, and dynamic (Fig. 1). Systematic analysis of long-term data received from various LTFEs will provide key information regarding the best nutrient supply option for increasing soil C sequestration and productivity. However, LTFEs must be managed carefully to ensure their relevance in sustainable agriculture production under the present scenario, and necessary changes can be incorporated accordingly. With this background information, the present chapter was prepared to cover the historical perspective of long-term studies

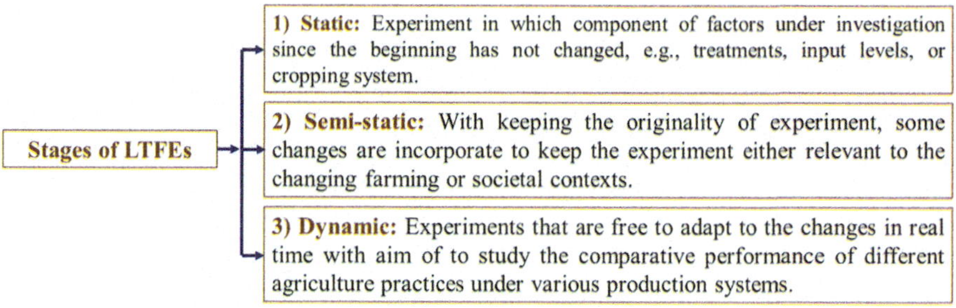


**Fig. 1** Various stages of long-term fertilizer experiments

with key findings, C dynamics and sequestration potential of various LTFEs, future challenges, and the way forward for selecting the best fertilization option for marginal farmers living in the Indian tropics.

## Historical Perspective

A large number of long-term fertilizer experiments in agriculture have been started in various parts of the world over the past two centuries. Most of these experiments were initiated in Europe and North America, such as at Rothamsted and Woburn (UK), Askov (Denmark), Bad Lauchstädt (Germany), Grignon (France), and the Sanborn Field (USA), with fewer in Asia, Australia, South America, and Africa. The oldest long-term fertilizer experiment was initiated by John Bennet Lawes and Joseph Henry Gilbert in 1843 (Johnston and Poulton 2018). In Indian literature, it is written that the application of manure is beneficial for crop production. During the mid-twentieth century (1930s–1940s), soil research was largely focused on composting various organic manures and green manuring. This was followed by research on chemical fertilizers, and in the early 1950s, permanent manure plot experiments were started. The results of these experiments revealed that the long-term use of chemical fertilizers does not affect soil productivity if applied judiciously. They also showed that declining soil fertility cannot be regained by crop rotation alone and that phosphate application plays a significant role in enriching soil fertility. In the 1960s, experimentation on soil fertility was strengthened with the introduction of input-responsive high-yield cultivars of major cereals. To study long-term changes in soil quality, crop productivity, and sustainability, the Indian Council of Agricultural Research (ICAR) started the "All India Coordinated Research Project on Long-term Fertilizer Experiments (AICRP-LTFE)" in September 1970 (Table 1). Subsequently, soil fertility experiments were more

**Table 1** Long-term fertilizer experiments in India

| Location | Est. year | Cropping system | Longitude/latitude |
|---|---|---|---|
| Almora, Uttarakhand | 1973 | Soybean–Wheat | 29.0° N; 79.0° E at 1250 m above mean sea level |
| Palampur, Himachal Pradesh | 1972 | Maize–Wheat | 32.0° N; 76.0° E at 1280 m above mean sea level |
| Umiam, Meghalaya | 2000 | Rice–Wheat | 25.6° N; 91.0° E at 968 m above mean sea level |
| Barackpore, West Bengal | 1972 | Rice–Wheat–Jute | 28.8° N; 88.4° E at 9.0 m above mean sea level |
| Mohanpur, West Bengal | 1986 | Rice–Wheat | 23.0° N; 89° E at 10.0 m above mean sea level |
| Gayeshpur, West Bengal | 1986 | Rice–Mustard–Sesame | 23° N; 89° E at 10.0 m above mean sea level |
| Kalyani, West Bengal | 1985 | Rice–Berseem | 22.90° N; 88.9° E at 10.0 m above mean sea level |
| Varanasi, Uttar Pradesh | 1985 | Rice–Wheat | 25.3° N; 83.0° E at 129.0 m above mean sea level |
| Faizabad, Uttar Pradesh | 1984 | Rice–Wheat | 26° N; 82° E at 113.0 m above mean sea level |
| Samastipur, Bihar | 1988 | Rice–Wheat | 25.5° N; 87.5° E at 52.0 m above mean sea level |
| Kanpur, Uttar Pradesh | 1984 | Rice–Wheat | 25.4° N; 84.6° E at 244.0 m above mean sea level |
| Pantnagar, Uttarakhand | 1983 | Rice–Wheat | 29.9° N; 79.1° E at 244.0 m above mean sea level |
| Ludhiana, Punjab | 1983 | Rice–Wheat | 30.9° N; 75.9° E at 247.0 m above mean sea level |
| New Delhi, Delhi | 1971 | Maize–Wheat–Cowpea (F) | 28.4° N; 77.2° E at 250.0 m above mean sea level |
| Karnal, Haryana | 1994 | Rice–Wheat | 29.1° N; 76.5° E at 252.0 m above mean sea level |
| Ranchi, Jharkhand | 1971 | Soybean–Wheat | 23.5° N; 85.0° E at 120.0 m above mean sea level |
| Nagpur, Maharashtra | 1986 | Cotton–Sorghum | 21.2° N; 79.1° E at 301.0 m above mean sea level |
| Jabalpur, Madhya Pradesh | 1972 | Soybean–Wheat–Maize (F) | 32.2° N; 79.9° E at 393.0 m above mean sea level |
| Indore, Madhya Pradesh | 1995 | Soybean–Wheat | 22.9° N; 75.9° E at 520.0 m above mean sea level |
| Akola, Maharashtra | 1988 | Sorghum–Wheat | 20.7° N; 77.0° E at 307.0 m above mean sea level |
| Bellary, Karnataka | 1978 | Maize–Chickpea | 16.5° N; 76.9° E at 448.0 m above mean sea level |

(continued)

**Table 1** (continued)

| Location | Est. year | Cropping system | Longitude/latitude |
|---|---|---|---|
| Hyderabad, Andhra Pradesh | 1995 | Sorghum–Castor | 17.3° N; 78.6° E at 220 m above mean sea level |
| Coimbatore, Tamil Nadu | 1972 | Finger millet–Maize–Cowpea (F) | 11.0° N; 76.0° E at 379.0 m above mean sea level |
| Cuttack, Orissa | 1984 | Rice–Rice | 20.0° N; 86.0° E at 36.0 m above mean sea level |
| Trivandrum, Kerala | 1977 | Cassava | 8.2° N; 76.6° E at 55.0 m above mean sea level |
| S.K. Nagar, Gujarat | 1988 | Pearl millet | 24.5° N; 72.7° E at 153.0 m above mean sea level |

focused on nutrient balance and soil fertility management in long-term multiple cropping systems. Experimentation on micronutrients received great attention in the 1980s after the discovery of zinc deficiency in soils of the Tarai region, foothills of the Indian Himalayas. Most LTFEs were started with the objective of selecting the right amount and type of fertilizer to maximize crop production. Over time, these experiments need to be modified to incorporate new research avenues without compromising their original aims.

## Importance and Major Challenges of LTFEs

The world's burgeoning population demands more food production from crops grown in soil. Therefore, fertile and productive soil is the most crucial resource for humankind's continued existence. However, globally, thousands of hectares of land are lost each year due to rapid urbanization and infrastructure development, or to soil pollution, erosion, and degradation. Moreover, the emphasis on producing food per unit area and per unit input at any cost is now being replaced by the concept of sustainability and environmental stewardship in agricultural land use for sustainable food production (Poulton and Johnston 2021). To meet the increasing demand for food, various changes are needed in the existing agricultural production system. However, it is difficult to predict the long-term effects of these changes on soil health and the ecosystem. The changes (e.g., nutrient, irrigation, and tillage management) may be beneficial, such as enhancing soil organic carbon for better soil health and crop growth, or they may be harmful, such as causing multi-nutrient deficiency, declining soil biological properties, and buildup of soil diseases and pests. We can show both the good and bad effects of changes in the existing agricultural system, but to identify both, we need to initiate long-term fertilizer experiments (LTFEs) to study the long-term changes in soil properties such as organic carbon and soil nutrient dynamics. LTFEs provide valuable information about the effects of long-term organic and inorganic nutrient management and various agricultural practices (irrigation, tillage) on agroecosystems. They are required to study the various long-term physical, chemical, and biological processes associated with the soil-plant production system. LTFEs are indispensable in soil and environmental research. Therefore, LTFEs are very important for answering the unsolved questions of the existing cropping system (Table 2).

Although long-term experiments have many advantages, they face many challenges in maintaining their importance and credibility in modern agriculture. LTFEs have set milestones in the past by answering the most complex puzzles of agricultural research. Concluding an LTFE can be the source of many unanswered questions. However, sometimes the reasons for ending an LTFE are more conclusive than the reasons for maintaining it. Sometimes, it is good to end LTFEs, especially when they have lost their importance over time or when all of the objectives of the experiments have been completed. The journey of long-term experiments is very challenging for the founders or project in-charge to keep the experiment alive without compromising the original aim. Here, we point out the most important challenges of LTFEs: (1) stay relevant to existing farming practices under rapidly changing agricultural practices, land use, and food habits; (2) difficulty in interpreting and presenting the data and findings in the context of changed practices (e.g., changes in crop varieties or crop rotation); (3) any changes to LTFE management cannot be made without first being standardized in separate field experiments and discussed with associated stakeholders, such as farmers and researchers; (4) trade-offs for the adoption of technologies may exist, as certain practices may not be adopted by farmers due to economic or other reasons; (5) transferring findings at the landscape level is also a big challenge for LTFEs, as the results of experiments may not be the same under different landscape conditions (such as soil erosion and biodiversity) and farm management practices (such as research fields vs. farmers' fields); (6) difficulty in obtaining financial support, as most of the LTFEs are funded by the government organizations with limited and fixed annual budgets. These budgets often become inadequate to maintain the experiments over time as the operational and maintenance costs of trials increase.

## Soil Carbon Pools, Importance, and Their Management

Total soil carbon content is primarily composed of soil organic carbon (SOC) and soil inorganic carbon (SIC) (Fig. 2). These pools are further divided into various sub-pools based on origin, resistance, and mean residence time (MRT). The MRT for soil inorganic carbon is about 85,000 years (ranges from 30,000 to 90,000), while the MRT for SOC is 35 years (Lal and Kimble 2000; Schlesinger 2006; Lorenz and Lal 2018).

**Table 2** Significant achievements of LTFEs situated in India

| S. No | LTFE center/ location | Significant achievements | Crops and soil |
|---|---|---|---|
| 1. | VPKAS, Almora | Soybean yield showed negative trend −22.50, −56.0, and −25.0 kg $ha^{-1}$ $year^{-1}$ under absolute control, N alone and NPK treatments, respectively, while application of organic manure showed positive yield trend. However, in wheat, NPK treatment showed positive yield trend (118 kg $ha^{-1}$ $year^{-1}$) | Silty clay loam, rice-wheat |
| 2. | BAU, Ranchi, India | Imbalanced fertilization with respect to K caused the largest depletions of K pools of soil.<br>Application of 150% NPK and NPK + FYM maintained higher K-supplying capacity compared with imbalanced fertilization.<br>K fertilizer recommendations based on NH4OAc-K need to be revised. | Kaolinitic red soil, soybean-wheat |
| 3. | Barrackpore, Pantnagar, and Raipur | Application of P in addition to N is essential to sustaining productivity of the rice-wheat cropping system.<br>Zn application is also inevitable in deficient areas (e.g., Pantnagar).<br>In situ green manuring with 50% NPK was found superior for sustaining the productivity of rice.<br>Highest sustainability yield index (SYI) was recorded with integrated application of organic and inorganic nutrients compared to inorganic alone (NPK + FYM > 150% NPK > 100% NPK > 100% NP > 100% N > control) | |
| 4. | Bhubaneshwar and Pattambi | Continuous use of FYM as soil amendment found more effective than lime to sustain productivity, especially in Alfisols | Alfisols |
| 5. | Palampur, Ranchi, and Bangalore | The lowest SYI was recorded with 100% NPK alone, probably due to decreasing soil health in terms of soil reaction and availability of P and K.<br>The SYI under 100% N was recorded near zero or negative in maize, wheat, soybean, and finger millet at all centers. Interestingly, absolute control (no manure/ no fertilizer) recorded little bit higher SYI value than 100% N. Therefore, the use of N alone should not be advisable to the farmers | Alfisols |

(continued)

**Table 2** (continued)

| S. No | LTFE center/ location | Significant achievements | Crops and soil |
|---|---|---|---|
| 6. | Jabalpur, Coimbatore, and Akola | Despite high organic carbon in Vertisols, the available nitrogen content always found lower compared to other soils having similar amounts of carbon. Thus, the study concluded that the availability of N relatively depends more on hydrolysable carbon. Therefore, it should be considered to predict the availability of N rather than total organic carbon.<br>P fractionation study revealed that Ca-P found most dominant, and the absence of P resulted in the decline of these fractions. Therefore, P fractions are in equilibrium and responsible for maintaining supply of P to plants | Vertisols |
| 7. | Delhi and Pantnagar | Despite the high status of non-exchangeable K and lattice K in Vertisols, crops need external supply of K through fertilizer.<br>However, in Inceptisols, the exchangeable K declining rate was very slow.<br>The 100% NPK + FYM treatment showed the highest soil enzymatic activity (FDA, acid and alkaline phosphatase), microbial biomass carbon, various carbon pools, and carbon management index.<br>Balance application of organic and inorganic nutrients increases soil quality index | Vertisols, Inceptisols |
| 8. | Gujarat and Varanasi | Use of chemical fertilizers alone leads to loss of SOC.<br>Long-term use of FYM along with fertilizers helps in maintaining microbial biomass and soil organic matter in both semi-arid and sub-humid regions of India | Entisols of Gujarat and Inceptisols of Varanasi |

## Soil Inorganic Carbon (SIC)

On the basis of origin, SIC is classified in two categories: (1) lithogenic inorganic carbon (LIC) derived primary carbonates from parent rocks/materials, and (2) pedogenic inorganic carbon (PIC). PICs are secondary carbonates derived from primary carbonates through a pedogenic process. Soil inorganic carbon is an important pool of soil carbon with a longer mean residence time; however, there are very few studies on the role of SIC in the global carbon cycle and soil carbon sequestration. Globally, the SIC stock amounts are estimated at around 700–1700 Pg C in the top 1 m of soil (Batjes 1996; Eswaran et al. 2000; Lorenz and Lal 2018). SIC is of more importance in arid and semi-arid regions, where it is found to be ten times higher than the organic carbon content (Eswaran et al. 2000). SIC is a relatively more stable form of soil carbon; however, it may change at a rate of 1 Mg $ha^{-1}$ $year^{-1}$ in

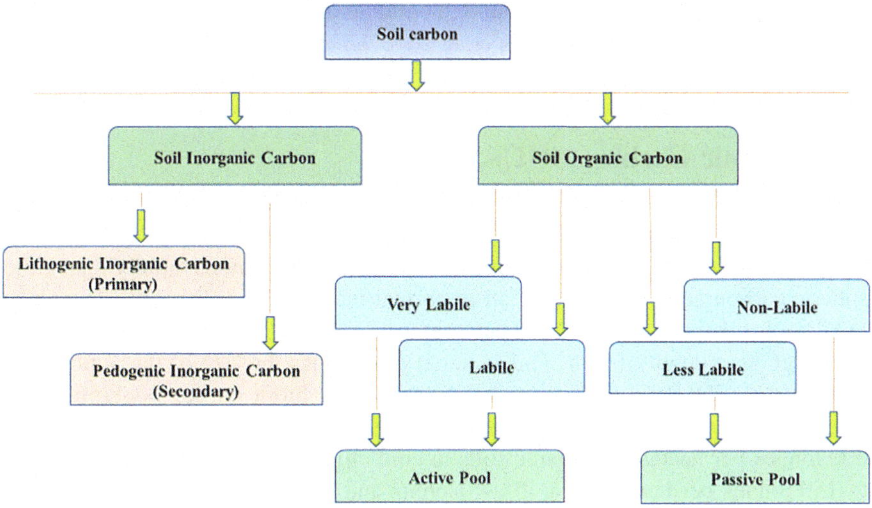


**Fig. 2** Major carbon pools in soil system

agricultural lands (Sanderman 2012). Between the two SICs, LIC components do not have much impact on the net C budget of an agroecosystem, given its origin from carbonate parent materials (Goddard et al. 2009). Pedogenic inorganic carbon (PIC) formation through reaction with $Ca^{2+}$ or $Mg^{2+}$ under arid conditions and with Na + under alkaline conditions is mainly considered as c sequestration process (Macías and Arbestain 2010). However, the role of SIC in C sequestration has remained doubtful and debatable. Some researchers believe precipitation of carbonates as sequestration, whereas others consider dissolution as the c sequestration process (Sanderman 2012; Monger et al. 2015). The formation of pedogenic carbonates can be enhanced through an increase in biogenic processes by soil application of organic amendments, materials rich in cations, lime, and amendments or conditioners that conserve water in the soil column (Lal and Follett 2009). By enhancing the above and belowground biomass, the formation of PIC may be increased in cultivated soils, as it enhances the partial pressure of carbon by promoting C production through enhanced decomposition rate (Lal 2004; Gao et al. 2017). The addition of organic manures or residues may increase the biogenic carbonate formation through higher activities of soil microfauna and macrofauna (Monger and Gallegos 2000). The application of soil amendments like Ca-rich minerals (Wollastonite, CaSiO3; Diopside, CaMgSi2O6) was found vital for improving the inorganic carbon sequestration by promoting precipitation of calcium carbonate (Haque et al. 2020; Kelland et al. 2020). Soil inorganic carbon role is more prominent in arid and semi-arid regions, and importantly in degraded land. Afforestation of such degraded lands has been found to build up the soil inorganic carbon by 50% in top 1 m soil (Gao et al. 2017; Su et al. 2010; Li et al. 2013).

## Soil Organic Carbon (SOC)

Soil organic carbon remains in equilibrium with atmospheric carbon under specific climate and management conditions. However, it changes more rapidly than atmospheric carbon and is considered an important indicator of soil health. The global SOC pool is estimated to be 1500 petagrams of carbon (PgC) to a depth of 1 m and 2300 PgC to a depth of 3 m. This is about three times the size of the atmospheric carbon pool (760 PgC) and four times the size of the biotic carbon pool (560 PgC). The SOC pool to a depth of 1 m varies from 30 tonnes per hectare in dry areas to 800 tonnes per hectare in organic soils in cold climates, with a typical range of 50 to 150 tonnes per hectare (Lal 2004; Lorenz and Lal 2018). Total organic carbon (TOC) can be divided into four different pools based on their degree of lability: very labile carbon (VLC), labile carbon (LC), less labile carbon (LLC), and non-labile/recalcitrant carbon (NL). These pools are generally further classified into an active carbon pool (comprising VLC and LC) and a passive carbon pool (comprising LLC and NL) (Sahoo et al. 2019). The active carbon pool is more vulnerable and has a shorter mean residence time (MRT) than the passive carbon pool. Changes in the active carbon pool occur frequently in response to external disturbances, especially land management practices. Therefore, it is considered a better indicator of soil health to account for changes in SOC due to different land-use patterns (Haynes 2005). The non-labile pool (passive pool) of SOC is a more stable and resistant component of SOC. It forms organic-mineral complexes with soil minerals and degrades slowly through microbial activity (Wiesenberg et al. 2010).

The critical threshold of soil organic carbon (SOC) in rhizospheric soil is 1.5 to 2.0%, which is essential for soil quality and health (Lal 2016). Maintaining SOC at or above this threshold is critical for improved soil structure, input use efficiency, climate regulation, and other provisioning services. For better management and restoration of SOC, it is crucial to identify the land-use changes that are primarily responsible for SOC degradation. The two main processes that affect SOC equilibrium are biomass input to the soil and decomposition of soil organic matter (SOM) (Sanderman et al. 2010). Technologies that encourage greater residue incorporation or retention in the soil, such as conservation agriculture, zero tillage, minimum tillage, and integrated and diverse cropping systems, are considered the most prominent SOC restoration strategies, in addition to the integrated use of manures and fertilizers (Lal 2016). However, long-term studies are important to predict the changes in SOC through various agricultural and land management practices and improve our understanding of their better and more sustainable management.

## Carbon Dynamics Under Various Long-Term Fertilized Experiment

The dynamics of soil organic carbon (SOC) primarily contribute to soil fertility and regulate its emission or sequestration for climate change mitigation and adaptation (Sainju et al. 2009). SOC constitutes 45–60% of soil organic matter and is defined as a heterogeneous mixture of simple and complex organic compounds, inert material, and pyrogenic compounds. Due to its greater charge density and surface area, SOC shows higher reactivity with clay matter and forms organo-mineral complexes in the soil system. The development of these organo-mineral complexes protects SOC from rapid degradation and stores it for millennia. However, the mean residence time or turnover period of SOC depends on the degree of protection within soil aggregates (Dungait et al. 2012).

Soil alone is considered to be a major carbon pool, even higher than the atmosphere and terrestrial vegetation combined (Lehmann and Kleber 2015). Certain agricultural practices, such as the adoption of high-yielding genotypes, residue retention, balanced fertilization, reduced or minimal tillage, optimal irrigation, crop rotation, and agricultural intensification, can improve SOC (Matson et al. 1997; Kucharik et al. 2001; Liao et al. 2015). All these practices improve agricultural productivity, which leads to greater biomass return to the soil system and thus enhances SOC. Among these practices, balanced fertilization with organic manure application is considered an important strategy for improving SOC. However, changes in the soil carbon pool or its dynamics are a very slow process and can only be assessed through long-term experiments (Richter et al. 2007). Therefore, to assess the fertilization effect on carbon accumulation and their dynamics, long-term fertilizer experiments running in various countries could be proven to be valuable resources. Many long-term studies had been conducted to predict the various organic and inorganic fertilizers on soil C sequestration potential, rate, efficiency, and dynamics in various C pools under diverse climatic and pedogenic conditions. Most of the studies reported that continuous combined application of organic (FYM) and chemical (NPK) fertilizers improved the soil C sequestration potential (Dou et al. 2016; Tang et al. 2018; Trivedi et al. 2020). To quantify the C storage under various LTFEs, Pathak et al. (2011) had reviewed 26 experiments conducting at various locations in India and observed that, in comparison to chemical fertilizer (NPK), combined application of NPK + FYM had provided higher SOC in all the LTFEs (Fig. 3). The average rate of C sequestration in NPK + FYM was 0.33 Mg C $ha^{-1}$ $yr^{-1}$, while in NPK it was only 0.16 Mg C $ha^{-1}$ $yr^{-1}$ (Fig. 4). Organic manure due to higher lignin content decomposes slowly and favor the accumulation of SOC in comparison to non-lignin material (Paustian et al. 1992; Mandal et al. 2007). Further, C sequestration efficiency (CSE) was estimated and found that average CSE was 12.24% and NPK + FYM in comparison to NPK recorded 6.53% CSE

Fig. 3 Mean SOC% in control, NPK, and FYM under 26 LTFEs in India

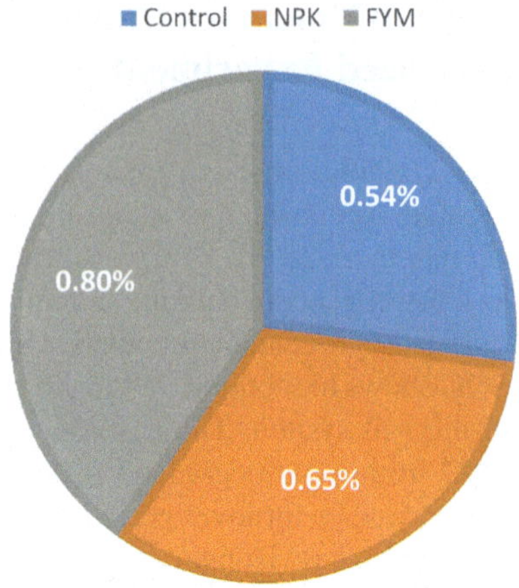


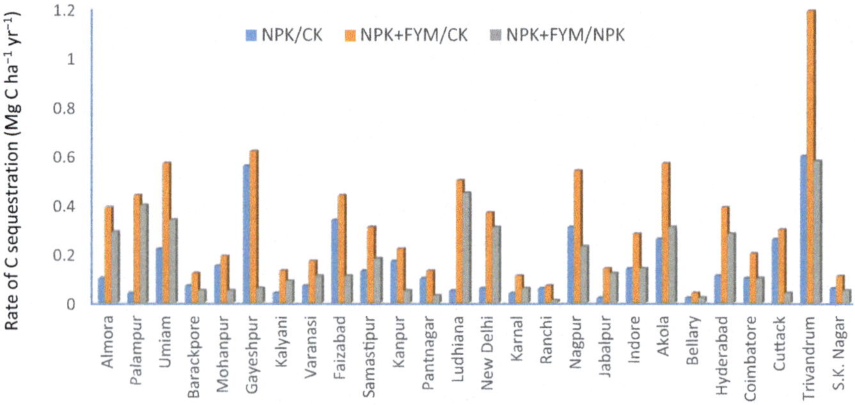


Fig. 4 Carbon sequestration rate of NPK and NPK + FYM under various LTFEs

(Fig. 5). Although NPK + FYM and NPK reported higher CSE in comparison to control but their magnitudes vary largely due to differences in climate, soil, agronomic management parameters, and quality of manure used (Fig. 6). Organic manure application improves soil C accumulation, as 62% of C applied through manure stabilizes into SOC, while NPK application only maintains the SOC (Fig. 7) (Majumder et al. 2008).

From a practical standpoint, Pathak et al. (2011) assessed the economic feasibility of combined NPK + FYM applications at nine locations and reported that it reduced net returns. They recommended urgent technological and policy interventions to support poor and marginal farmers. Regarding the feasibility of FYM, other options such as crop residue, urban waste, and sludge could be used to achieve the country's full C sequestration potential. Other organic sources could also be used,

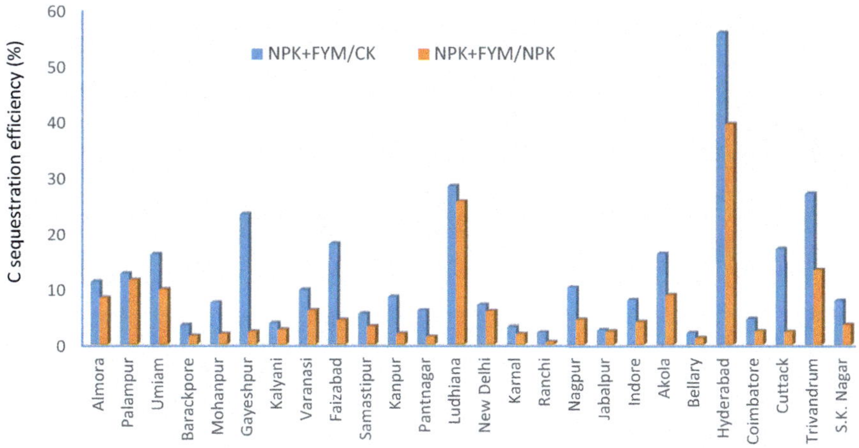


**Fig. 5** Carbon sequestration efficiency of NPK + FYM over control and NPK under various LTFEs

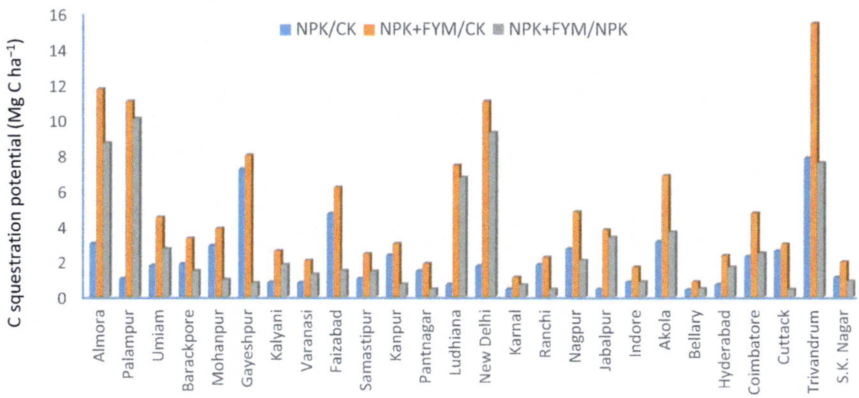


**Fig. 6** Carbon sequestration potential of NPK and NPK + FYM under various LTFEs

but their C sequestration potential and immediate effects on crop yield and soil quality need to be comprehensively assessed. In this regard, Ghosh et al. (2021) recently conducted a short-term study using various organic manures combined with NPK in a rice-rice cropping system. They reported that the rice straw + NPK treatment resulted in the highest total organic carbon (91.10 t $ha^{-1}$) and recalcitrant carbon (85.64 t $ha^{-1}$) due to its greater C:N ratio. The vermicompost + NPK treatment resulted in the highest labile carbon pool (10.24 t $ha^{-1}$), followed by the NPK + FYM treatment. Despite the high carbon sequestration through recalcitrant carbon pools, the rice straw + NPK treatment resulted in a small increment in crop yield (3.52 t $ha^{-1}$). Other organic sources such as vermicompost, FYM, and gliricidia improved the active C pool and showed a significant increment in yield. This suggests the importance of organic manure for changing soil C fractions and

**Fig. 7** Effect of long-term organic and inorganic fertilization on soybean crop at Almora, Uttarakhand

sequestration potential in the short term. Soil carbon content is influenced by multiple factors, including agricultural management practices, climatic conditions, and soil conditions. Zhang et al. (2010) conducted a long-term study at six locations in China representing different climatic and soil conditions, with different cropping patterns and management practices, to evaluate the impact of different cropping systems, climate, and soil conditions on SOC dynamics. The study confirmed that adding manures with inorganic fertilizers enhances total SOC. They further added that mineral fertilizers alone are not sufficient to maintain SOC in arid and semi-arid regions under monocropping systems and that adding organic manure is essential to maintaining SOC levels. Similar results have been reported previously, showing that organic manure application can sustain or improve soil organic carbon in humid and semi-arid to arid warm climates, even in monocropping systems (Cuvardic et al.

2004; Mando et al. 2005; Shen et al. 2007). The rate of organic matter conversion to SOC (carbon conversion rate) is significantly lower in warmer temperatures (6.8–7.7%) than in mild-temperate areas (15.8–31%). This is primarily due to the negative correlation between carbon conversion rate and annual precipitation and temperature, i.e., a higher rate of conversion occurs under lower precipitation and/or temperature and vice versa (Zhang et al. 2010). In summary, under Indian tropical conditions, cropping systems and agricultural practices are needed that can provide C inputs greater than the critical level (2.9 Mg C $ha^{-1}$ $yr^{-1}$) to sustain SOC levels and optimal soil health (Mandal et al. 2006).

## Key Lesson from LTFE for Improving Soil Carbon Sequestration

Several long-term field experiments (LTFEs) have been conducted worldwide to study the impact of inorganic and organic nutrient sources on crop yield and soil physicochemical properties (Table 3). The results of these studies provide key information for managing soil organic carbon (SOC) over the long term, as SOC would otherwise be transformed and released into the atmosphere as CO2, contributing to global warming. LTFEs are important for predicting long-term changes in soil carbon as a result of continuous use of various fertilizer practices. Some of the key findings of LTFEs regarding efficient SOC management are presented here, with suggestions for improving SOC levels:

Combined application of chemical fertilizer and organic manure is the most effective practice for improving crop productivity and SOC sequestration. Balanced and optimal use of NPK maintains SOC irrespective of cropping system and soil type. The source of organic matter, lignin content, C:N ratio, and their composition significantly influence C sequestration and soil carbon fractions. Integrated use of NPK and FYM improves SOC by increasing the formation and stability of macroaggregates. Excessive use of N fertilizers reduces yield and SOC in acid soils, while continuous applications of manure or manure plus mineral fertilizers are the most effective strategies for improving C sequestration. NPK + FYM increased SOC in both the labile and recalcitrant pools due to direct amendments of organic manure and greater return of belowground biomass, while chemical fertilizer showed no changes in SOC due to offsetting effects of enhanced recalcitrant and decreased labile carbon pools. Organic manure (OM) application arrests soil acidification and improves the productivity of red soils. However, OM addition increases the efflux of $CO_2$ and $N_2O$ more than inorganic fertilizers. Organic fertilization (M and MNPK) enhanced the aliphatic C content of humic acids (HA), which favored the increase in the organic C content and aggregate stability. Microaggregate-protected C is considered a sensitive measure for assessing the impact of long-term fertilization on SOC.

**Table 3** Examples of soil carbon management studies in various LTFEs experiments

| Location | Long./ Lat./Alt. | Study period | Cropping system | Results | Reference |
|---|---|---|---|---|---|
| Bhubaneswar, India | 20° 17′ N; 85° 49′ E; 30 m | 10 years | Rice–Rice | NPK+ FYM are adequate for maintaining SOC stock and sustainability of rice–rice cropping system | Mandal et al. (2020) |
| Xinjiang, China | 43° 57′ N; 87° 30′ E | 23 years | Seasonal or annual rotation of maize, spring wheat, and winter wheat | Omission of any primary nutrient (NPK) aggravates the rate of decrease in SOC<br>Highest decline was observed in K or P omitted fields<br>For sustainable yield and SOC maintenance in arid areas, integrated use of manure and NPK is advised | Lyu et al. (2017) |
| Almora, Uttarakhand | 29° 36′ N; 79° 40′ E; 1250 m | 30 years | Soybean–Wheat | SOC increased 40 and 70% in the NPK + FYM-treated plots as compared to NPK and unfertilized control plots, respectively<br>The estimated quantity of biomass C required to maintain the equilibrium SOM content was 321 kg $ha^{-1}$ $yr^{-1}$ | Kundu et al. (2007) |

(continued)

**Table 3** (continued)

| Location | Long./Lat./Alt. | Study period | Cropping system | Results | Reference |
|---|---|---|---|---|---|
| Mohanpur, West Bengal | 23° N; 89° E; 9.5 m | 20 years | Rice–Wheat | Minimum 2.9 Mg C ha-1 year-1 is required for sustenance of SOC level<br>FYM or compost per se could not appreciably improve SOC stock (~10.7%) but encouraged the rate of stabilization of crop residue C to SOC<br>Balanced fertilization with NPK increased SOC (9.3–51.8% over the control), its extent influenced by the cropping system | Mandal et al. (2007) |
| Wagga Wagga, Australia | 35° 05′ S; 147° 20′ E | 25 years | Pastures (Trifoliumsubterraneum L.; Lolium spp. and Hordeum spp.) | Improved soil nutrient and grazing management of permanent pasture can lead to an increase of 500–700 kg C ha $^{-1}$ year$^{-1}$ | Chan et al. (2011) |
| Agramunt, Spain | (41° 48′ N; 1° 07′ E; 330 m). | 13 years | Rainfed Barley | No-tillage (NT) and high fertilization rates (e.g., 60–120 kg N ha-1) could lead to significant SOC sequestration and associated CO2 emission offsetting | Álvaro-Fuentes et al. (2012) |

(continued)

**Table 3** (continued)

| Location | Long./ Lat./Alt. | Study period | Cropping system | Results | Reference |
|---|---|---|---|---|---|
| Shaanxi, China | 35° 12′ N; 107° 40′ E, 1220 m. | 25 years | Winter wheat | The SOC accumulation rate was 40.0, 48.0, 68.0, and 100.0 kg C $ha^{-1}$ $year^{-1}$ for the N45, N90, N135 and N180 treatments over 25 years, respectively<br>N fertilizer is a sustainable practice, especially in carbon sequestration and crop productivity | Guo et al. (2012) |
| Barrackpore, Ranchi, and Akola, India | 22° 45′ N, 88° 26′ E, 9 m; 23° 30′ N, 85° 15′ E, 120 m; 20° 42′ N, 77° 2′ E, 307.43 m | 29 year, 30 year, 14 year | Rice–wheat–jute, Soybean–wheat, and Sorghum–wheat | SOC in the unfertilized plot decreased by 41.5, 24.5, and 15.5% compared to initial values in Barrackpore, Ranchi, and Akola, respectively, wherein NPK and NPK + FYM either maintained or improved it over initial SOC content<br>100% recommended NPK are adequate for maintaining SOC, and yield declines are mostly location specific | Manna et al. (2005) |

(continued)

**Table 3** (continued)

| Location | Long./Lat./Alt. | Study period | Cropping system | Results | Reference |
|---|---|---|---|---|---|
| Shaanxi Province, China | 35° 12′ N, 107° 40′ E | 27 years | Pea, winter wheat, and millet rotation system | Fertilizers N + phosphorus (P) and manure had larger effects on the OC and N stocks in >2 mm aggregates in the legume-grain rotation system | Yajun et al. (2017) |
| Ludhiana, Punjab, India | 75° 48′ E, 30° 54′ N | 36 years | Maize–wheat rotation | The SOC pool was the lowest in control at 7.3 Mg $ha^{-1}$ and increased to 11.6 Mg $ha^{-1}$ with 100%NPK + FYM Balanced application of NPK fertilizers with FYM was best option for higher crop yields in maize–wheat rotation | Brar et al. (2017) |
| University of Alberta Breton, Canada | – | 28 years | Wheat–oat–barley–hay–hay rotation system | Application of NPK in combination with S improved SOC at 0.11 Mg C $ha^{-1}$ $yr^{-1}$ in the 0- to 15-cm depth over NPK alone | Giweta et al. (2014) |

(continued)

**Table 3** (continued)

| Location | Long./ Lat./Alt. | Study period | Cropping system | Results | Reference |
|---|---|---|---|---|---|
| Switzerland | (46° 23′ 55.72″ N, 06° 14′ 24.72″ E) | 37 years | Multi-crops | Increment in SOC content with organic amendments was only significant for very large inputs (70 t $ha^{-1}$) of fresh cattle manure<br>FYM provided greater benefits in comparison to other organic fertilizers on SOC management and soil quality preservation | Maltas et al. (2018) |
| North China Plain | 35° 00′ N, 114° 24′ E | 17 years | wheat–maize cropping system | The organic manure (OM and 1/2OMN) treatments were not a feasible option for farmers, but for sequestering soil carbon<br>The NPK treatment was important for increasing crop yields and sustainability of cropping systems | Gong et al. (2009) |

(continued)

**Table 3** (continued)

| Location | Long./ Lat./Alt. | Study period | Cropping system | Results | Reference |
|---|---|---|---|---|---|
| Punjab, India | 30°54′ 27″ N, 75°46′ 5 9″ | 15 years | Rice–wheat | Labile fractions of soil organic carbon were more sensitive to fertilizer management<br>C management index showed NPK + straw or NPK + FYM was statistically superior to the NPK + green manure for increasing the SOC status | Chaudhary et al. (2017) |
| IARI, New Delhi | 28° 37/–28° 39/N and 77° 9/–77° 11/E | 44 years | Maize–wheat cowpea (fodder) cropping sequence | NPK + FYM had the highest recalcitrant C pools within macro- and microaggregates and greater CMI (~26%) than NPK<br>50% of the accumulated C was observed in deep soil layers (30–90 cm) | Ghosh (2018) |
| IARI, New Delhi | 28° 37′–28° 39′ N and 77° 9′–77° 11′ E | 10 years | maize–wheat–cowpea cropping system | 100% NPK + FYM sequester 1.83 Tg C year$^{-1}$, which corresponds to ~1% of the fossil fuel emissions<br>POC and $KMnO_4$-C fractions had less potential to offset greenhouse gas emissions compared to SOC | Purakayastha et al. (2008) |

## Conclusion

This study showed that increasing soil organic carbon (SOC) improves soil productivity and crop yields. Long-term field experiments (LTFEs) have shown that continuous use of farmyard manure (FYM) improves SOC sequestration potential and can mitigate greenhouse gas emissions. However, the cost of organic manure is a major constraint at some locations, reducing farm profitability. Incentives, policy interventions, and technological advancements could be implemented to promote the use of organic manure on farmlands. Additionally, only 30–35% of the total FYM requirement is available for agricultural use, so the country needs an additional two-thirds of FYM to achieve its maximum C sequestration potential. In the absence of FYM, sludge from municipal solid waste, crop residues, and other manures could be used. In conclusion, under Indian conditions, where temperature and C mineralization rates are high, SOC content can be increased through the continuous use of balanced and integrated fertilization. However, the use of organic materials other than FYM needs to be validated and examined on a long-term basis using LTFEs.

## References

Álvaro-Fuentes J, Morell FJ, Plaza-Bonilla D, Arrúe JL, Cantero-Martínez C (2012) Modelling tillage and nitrogen fertilization effects on soil organic carbon dynamics. Soil Tillage Res 120:32–39

Baldock JA, Skjemstad JO (2000) Role of the soil matrix and minerals in protecting natural organic materials against biological attack. Org Geochem 31(7–8):697–710

Batjes NH (1996) Total carbon and nitrogen in the soils of the world. Eur J Soil Sci 47:151–163

Benbi DK, Brar JS (2009) A 25-year record of carbon sequestration and soil properties in intensive agriculture. Agron Sustain Dev 29(2):257–265. https://doi.org/10.1051/agro:2008070

Blair GJ, Lefroy RD, Lisle L (1995) Soil carbon fractions based on their degree of oxidation, and the development of a carbon management index for agricultural systems. Aust J Agric Res 46:1459–1466. https://doi.org/10.1071/AR9951459

Chan KY, Bowman A, Oates A (2001) Oxidizible organic carbon fractions and soil quality changes in an oxicpaleustalf under different pasture leys. Soil Sci 166:61–67

Chan KY, Conyers MK, Li GD, Helyar KR, Poile G, Oates A, Barchia IM (2011) Soil carbon dynamics under different cropping and pasture management in temperate Australia: Results of three long-term experiments. Soil Res 49:320–328

Chaudhary S, Dheri GS, Brar BS (2017) Long-term effects of NPK fertilizers and organic manures on carbon stabilization and management index under rice-wheat cropping system. Soil Tillage Res 166:59–66

Cotrufo MF, Wallenstein MD, Boot CM, Denef K (2013) The microbial efficiency-matrix stabilization (MEMS) framework integrates plant litter decomposition with soil organic matter stabilization: do labile plant inputs form stable soil organic matter? Glob Chang Biol 19(4):988–995. https://doi.org/10.1111/gcb.12113

Cuvardic M, Tveitnes S, Krogstad T, Lombnaes P (2004) Longterm effects of crop rotation and different fertilization systems on soil fertility and productivity. Acta Agric Scand 54:193–201. https://doi.org/10.1080/09064710410030258

Dou X, He P, Zhu P, Zhou W (2016) Soil organic carbon dynamics under long-term fertilization in a black soil of China: evidence from stable C isotopes. Sci Rep 6(1):1–10

Dungait JAJ, Hopkins DW, Gregory AS, Whitmore AP (2012) Soil organic matter turnover is governed by accessibility not recalcitrance. Glob Chang Biol 18:1781–1796

Eswaran H, Reich PF, Kimble JM (2000) Global carbon stocks. In: Lal R, Kimble JM, Eswaran H, Stewart BA (eds) Global climate change and pedogenic carbonates. CRC Press, Boca Raton, FL, pp 15–25

Gami SK, Lauren JG, Duxbury JM (2009) Influence of soil texture and cultivation on carbon and nitrogen levels in soils of the eastern Indo-Gangetic Plains. Geoderma 153:304–311

Gao Y, Tian J, Pang Y, Liu J (2017) Soil inorganic carbon sequestration following afforestation is probably induced by pedogenic carbonate formation in Northwest China. Front Plant Sci. https://doi.org/10.3389/fpls.2017.01282

Ghosh A, Bhattacharyya R, Meena MC, Dwivedi BS, Singh G, Agnihotri R, Sharma C (2018) Long-term fertilization effects on soil organic carbon sequestration in an Inceptisol. Soil Tillage Res 177:134–144

Ghosh M, Ashiq W, Bhogilal Vasava H, Gamage DNV, Patra PK, Biswas A (2021) Short-Term Carbon Sequestration and Changes of Soil Organic Carbon Pools in Rice under Integrated Nutrient Management in India. Agriculture 11(4):348

Giweta M, Dyck MF, Malhi SS, Puurveen D, Robertson JA (2014) Long-term S-fertilization increases carbon sequestration in a sulfur-deficient soil. Can J Soil Sci 94(3):295–301

Goddard MA, Mikhailova EA, Post CJ, Schlautman MA, Galbraith JM (2009) Continental United States atmospheric wet calcium deposition and soil inorganic carbon stocks. Soil Sci Soc Am J 73:989–994

Gong W, Yan XY, Wang JY, Hu TX, Gong YB (2009) Long-term manuring and fertilization effects on soil organic carbon pools under a wheat–maize cropping system in North China Plain. Plant Soil 314(1):67–76

Gregorich EG, Drury CF, Baldock JA (2001) Changes in soil carbon under long-term maize in monoculture and legume-based rotation. Can J Soil Sci 81:21–31

Guo S, Zhu H, Dang T, Wu J, Liu W, Hao M, Li Y, Syers JK (2012) Winter wheat grain yield associated with precipitation distribution under long-term nitrogen fertilization in the semiarid Loess Plateau in China. Geoderma 189:442–450

Haque F, Santos RM, Chiang YW (2020) Optimizing inorganic carbon sequestration and crop yield with wollastonite soil amendment in a microplot study. Front Plant Sci. https://doi.org/10.3389/fpls.2020.01012

Haynes RJ (2005) Labile organic matter fractions as central components of the quality of agricultural soils. Adv Agron 85:221–268

Jha P, Lakaria BL, Biswas AK, Saha R, Mahapatra P, Agrawal BK, Sahi DK, Wanjari RH, Lal R, Singh M, Rao AS (2014) Effects of carbon input on soil carbon stability and nitrogen dynamics. Agric Econ Environ 189:36–42. https://doi.org/10.1016/j.agee.2014.03.019

Johnston AE, Poulton PR (2018) The importance of long-term experiments in agriculture: their management to ensure continued crop production and soil fertility; the Rothamsted experience. Eur J Soil Sci 69:113–125

Kelland ME, Wade PW, Lewis AL, Taylor LL, Sarkar B, Andrews MG (2020) Increased yield and $CO_2$ sequestration potential with the $C_4$ cereal Sorghum bicolor cultivated in basaltic Kelland rock dust-amended agricultural soil. Glob Chang Biol 26(6):3658–3676. https://doi.org/10.1111/gcb.15089

Krause HM, Fliessbach A, Mayer J, Mäder P (2020) Implementation and management of the DOK long-term system comparison trial. In: Long-term farming systems research. Academic Press, pp 37–51

Kucharik CJ, Brye KR, Norman JM, Foley JA, Gower ST, Bundy LG (2001) Measurements and modeling of carbon and nitrogen cycling in agroecosystems of southern Wisconsin: potential for SOC sequestration during the next 50 years. Ecosystems 4:237–258

Kundu S, Bhattacharyya R, Prakash V, Ghosh BN, Gupta HS (2007) Carbon sequestration and relationship between carbon addition and storage under rainfed soybean–wheat rotation in a sandy loam soil of the Indian Himalayas. Soil Tillage Res 92(1–2):87–95

Ladha JK, Dawe D, Pathak H, Padre AT, Yadav RL, Singh B (2003) How extensive are yield declines in long-term rice-wheat experiments in Asia? Field Crop Res 81:159–180
Lal R (2004) Soil carbon sequestration impacts on global climate change and food security. Science 304:1623–1627
Lal R (2016) Soil health and carbon management. Food Energy Secur 5:212–222
Lal R, Follett RF (2009) Soils and climate change. In: Lal R, Follett RF (eds) Soil carbon sequestration and the greenhouse effect, SSSA Special Publication 57, second edition. Soil Science Society of America, Madison, pp xxi–xxviii
Lal R, Kimble JM (2000) Pedogenic carbonates and the global carbon cycle. In: Lal R, Kimble JM, Eswaran H, Stewart BA (eds) Global climate change and pedogeniccarbonates. CRC Press, Boca Raton, pp 291–302
Lehmann J, Kleber M (2015) The contentious nature of soil organic matter. Nature 528:60
Li YQ, Brandle J, Awada T, Chen YP, Han JJ, Zhang FX (2013) Accumulation of carbon and nitrogen in the plant–soil system after afforestation of active sand dunes in China's Horqin Sandy Land. Agric Ecosyst Environ 177:75–84. https://doi.org/10.1016/j.agee.2013.06.007
Liao Y, Wu WL, Meng FQ, Smith P, Lal R (2015) Increase in soil organic carbon by agricultural intensification in northern China. Biogeosciences 12(5):1403–1413
Lorenz K, Lal R (2018) Carbon sequestration in agricultural ecosystems. https://doi.org/10.1007/978-3-319-92318-5
Lyu J, Liu H, Wang X, Olave R, Tian C, Liu X (2017) Crop yields and soil organic carbon dynamics in a long-term fertilization experiment in an extremely arid region of northern Xinjiang, China. J Arid Land 9:345–354
Macías F, Arbestain MC (2010) Soil carbon sequestration in a changing global environment. Mitig Adapt Strateg Glob Chang 15:511–529
Majumder B (2007) Soil organic carbon pools and biomass productivity under agro-ecosystems of subtropical India. Ph.D. diss. Jadavpur University, Calcutta, India
Majumder B, Mandal B, Bandyopadhyay P (2008) Soil organic carbon pools and productivity in relation to nutrient management in a 20-year-old rice–berseem agroecosystem. Biol Fertil Soils 44:451–461
Maltas A, Kebli H, Oberholzer HR, Weisskopf P, Sinaj S (2018) The effects of organic and mineral fertilizers on carbon sequestration, soil properties, and crop yields from a long-term field experiment under a Swiss conventional farming system. Land Degrad Dev 29(4):926–938
Mandal B (2005) Assessment and improvement of soil quality and resilience for rainfed production systems. Completion Rep., Natl. Agric. Technol. Project. Indian Council of Agric Res. New Delhi, India
Mandal D, Dadhwal KS, Khola OP, Dhyani BL (2006) Adjusted T values for conservation planning in Northwest Himalayas of India. J Soil Water Conserv 61:391–397
Mandal B, Majumder B, Bandyopadhyay PK (2007) The potential of cropping systems and soil amendments for carbon sequestration in soils under long-term experiments in subtropical India. Glob Chang Biol 13:357–369
Mandal M, Kamp P, Singh M (2020) Effect of long term manuring on carbon sequestration potential and dynamics of soil organic carbon labile pool under tropical rice-rice agro-ecosystem. Commun Soil Sci Plant Anal 51(4):468–480
Mando A, Bonzi M, Wopereis MCS, Lompo F, Stroosnijder L (2005) Long-term effects of mineral and organic fertilization on soil organic matter fractions and sorghum yield under sudano-sahelian conditions. Soil Use Manag 21:396–401. https://doi.org/10.1079/sum2005339
Matson PA, Parton WJ, Power A, Swift M (1997) Agricultural intensification and ecosystem properties. Science 277:504–509
Minasny B, Malone BP, McBratney AB, Angers DA, Arrouays D, Chambers A (2017) Soil carbon 4 per mille. Geoderma 292:59–86. https://doi.org/10.1016/j.geoderma.2017.01.002
Monger HC, Gallegos RA (2000) Biotic and abiotic processes and rates of pedogenic carbonate accumulation in the southwestern United States. Relationship to atmospheric CO2 sequestration. In: Lal R, Kimble JM, Eswaran H, Stewart BA (eds) Global climate change and pedogenic carbonates. CRC/Lewis Publishers, Boca Raton, pp 273–290

Monger HC, Kraimer RA, Khresat S, Cole DR, Wang X, Wang J (2015) Sequestration of inorganic carbon in soil and groundwater. Geology 43:375–378

Oades JM (1984) Soil organic matter and structural stability: mechanisms and implications for management. Plant Soil 76:319–337. https://doi.org/10.1007/BF02205590

Parihar M, Rakshit A, Meena VS, Gupta VK, Rana K, Choudhary M, Tiwari G, Mishra PK, Pattanayak A, Bisht JK, Jatav SS (2020) The potential of arbuscularmycorrhizal fungi in C cycling: a review. Arch Microbiol 202:1581–1596

Parihar M, Panday SC, Meena RP, Kumar U, Meena VS, Choudhary M, Singh AK, Bisht JK, Kant L, Pattanayak A (2021) Long-term organic and inorganic fertilization on economics, energy budgeting and carbon footprint of soybean-wheat cropping system in the Indian mid-Himalayas. Arch Agron Soil Sci 30:1–5

Pathak H, Byjesh K, Chakrabarti B, Aggarwal PK (2011) Potential and cost of carbon sequestration in Indian agriculture: estimates from long-term field experiments. Field Crop Res 120:102–111

Paustian K, Parton WJ, Persson J (1992) Modeling soil organic matter in organic amendedand nitrogen fertilized long-term plots. Soil Sci Soc Am J 56:476–488

Poulton PR, Johnston AE (2021) Can long-term experiments help us understand, and manage, the wider landscape—Examples from Rothamsted, England. In: Exploring and optimizing agricultural landscapes. Cham: Springer International Publishing, pp 233–252

Powlson DS, Pruden G, Johnston AE, Jenkinson DS (1986) The nitrogen cycle in Broadbalk wheat experiment—recovery and losses of 15N-labelled fertility applied in springs and inputs of N from the atmosphere. J Agric Sci (Camb) 107:591–609

Purakayastha TJ, Rudrappa L, Singh D, Swarup A, Bhadraray S (2008) Long-term impact of fertilizers on soil organic carbon pools and sequestration rates in maize–wheat–cowpea cropping system. Geoderma 144(1–2):370–378

Richter DD, Callaham MA, Powlson DS, Smith P (2007) Long-term soil experiments: keys to managing earth's rapidly changing ecosystems. Soil Sci Soc Am J 71:266–279. https://doi.org/10.2136/sssaj2006.0181

Sahoo UK, Singh SL, Gogoi A, Kenye A, Sahoo SS (2019) Active and passive soil organic carbon pools as affected by different land use types in Mizoram, Northeast India. PLoS One 14(7):1–16

Sainju UM, Caesar-TonThat T, Jabro JD (2009) Carbon and nitrogen fractions in dryland soil aggregates affected by long-term tillage and cropping sequence. Soil Sci Soc Am J 73:1488–1495

Sanderman J (2012) Can management-induced changes in the carbonate system drive soil carbon sequestration? A review with particular focus on Australia. Agric Ecosyst Environ 155:70–77

Sanderman J, Farquharson R, Baldock J (2010) Soil carbon sequestration potential: a review for Australian agriculture. CSIRO

Schlesinger WH (2006) Inorganic carbon and the global C cycle. In: Lal R (ed) Encyclopedia of soil science. Taylor & Francis, London, pp 879–881

Shahid M, Nayak AK, Shukla AK, Tripathi R, Kumar A, Mohanty S, Bhattacharyya P, Raja R, Panda BB (2013) Long-term effects of fertilizer and manure applications on soil quality and yields in a sub-humid tropical rice rice system. Soil Use Manag 29:322–332. https://doi.org/10.1111/sum.2013.29.issue-3

Shen MX, Yang LZ, Yao YM, Wu DD, Wang JG, Guo RL, Yin SX (2007) Long-term effects of fertilizer managements on crop yields and organic carbon storage of a typical rice-wheat agroecosystem of China. Biol Fert Soils 44:187–200. https://doi.org/10.1007/s00374-007-0194-x

Stockmann U, Adams MA, Crawford JW, Field DJ, Henakaarchchi N, Jenkins M, Minasny B, McBratney AB, Courcelles VDR, Singh K, Wheeler I (2013) The knowns, known unknowns and unknowns of sequestration of soil organic carbon. Agric Ecosyst Environ 164:80–99

Su YZ, Wang XF, Yang R, Lee J (2010) Effects of sandy desertified land rehabilitation on soil carbon sequestration and aggregation in an arid region in China. J Environ Manag 91:2109–2116. https://doi.org/10.1016/j.jenvman.2009.12.014

Tang H, Xiao X, Tang W, Li C, Wang K, Li W (2018) Long-term effects of NPK fertilizers and organic manures on soil organic carbon and carbon management index under a double-cropping rice system in Southern China. Commun Soil Sci Plant Anal 49(16):1976–1989

Trivedi A, Bhattacharyya R, Biswas DR, Das S, Das TK, Mahapatra P, Shahi DK, Sharma C (2020) Long-term impacts of integrated nutrient management with equivalent nutrient doses to mineral fertilization on soil organic carbon sequestration in a sub-tropical Alfisol of India. Carbon Manage 11(5):483–497

Wiesenberg GLB, Dorodnikov M, Kuzyakov Y (2010) Source determination of lipids in bulk soil and soil density fractions after four years of wheat cropping. Geodarma 156:267–277

Yajun HAO, Yuhong WANG, Chang Q, Xiaorong WEI (2017) Effects of long-term fertilization on soil organic carbon and nitrogen in a highland agroecosystem. Pedosphere 27(4):725–736

Zhang WJ, Wang XJ, Xu MG, Huang SM, Liu H, Peng C (2010) Soil organic carbon dynamics under long-term fertilizations in arable land of Northern China. Biogeosciences 7:409–425

# Synthesis Report on Greenhouse Gas Emissions from Agroecosystems: GHG Cycling, Microbial Processes, and Strategies for GHG Mitigation

**Santosh R. Mohanty**

## GHG Emissions from Rice Paddy Fields

Chapter 1 reviewed methane and carbon dioxide emissions from rice paddies in Bangladesh under organic, inorganic, and integrated fertilizer regimes across multiple rice seasons. Amending soil with organic fertilizers like vermicompost increased methane emissions compared to chemical fertilizers, but vermicompost lowered emissions versus cow dung. Vermicompost also reduced carbon dioxide release and improved soil carbon sequestration and rice yields compared to cow dung. The results show vermicompost can mitigate greenhouse gases and boost productivity compared to conventional organic fertilizers like cow dung in intensive rice cropping. Key findings from the chapter were:

- Amending rice paddy soils with organic fertilizers like cow dung and vermicompost led to significantly higher seasonal methane ($CH_4$-C) emissions compared to recommended inorganic chemical fertilizers across rice growing seasons. This is attributed to the increased supply of organic carbon substrates for methanogenic archaea under flooded anaerobic conditions with organic inputs.
- However, amendment with vermicompost reduced $CH_4$-C emissions by 4–10% compared to the application of cow dung across major rice seasons like Boro (dry winter), T. Aman (monsoon), and Aus (pre-monsoon summer) in Bangladesh. The reduced emissions with vermicompost are likely due to its slower decomposition and nutrient release patterns.
- Carbon dioxide ($CO_2$-C) emissions were also substantially lowered by 15–17% under the vermicompost amendment versus the incorporation of cow dung. This highlights the potential of vermicompost to curb carbon losses through microbial respiration pathways.

S. R. Mohanty (✉)
ICAR Indian Institute of Soil Science, Bhopal, India



S. R. Mohanty, B. Kollah (eds.), *Greenhouse Gas Regulating Microorganisms in Soil Ecosystems*, https://doi.org/10.1007/978-3-031-70569-4_20

- The net ecosystem carbon balance (NECB) was improved by 11–26% with vermicompost application instead of cow dung, clearly indicating vermicompost's potential to enhance long-term soil organic carbon sequestration.
- Rice yields increased remarkably by 4–9% with vermicompost amendment compared to cow dung application. The higher yields verify the benefits of vermicompost in improving soil fertility and nutrient availability in intensive rice cropping systems.
- Specifically in the Boro season, vermicompost increased the net primary production input carbon (NPPIC) by 4–10% over cow dung amendment across studied years. Total carbon output was also 3–4% higher under cow dung than vermicompost, showing greater carbon losses.
- The NECB ranged from 733–1290 kg C/ha across Boro, T. Aman, and Aus cropping periods, conclusively demonstrating the potential for season-wise soil carbon sequestration under organic fertilizer regimes, especially vermicompost.

The results demonstrate that vermicompost can help mitigate GHG emissions and boost productivity compared to conventional organic fertilizers like cow dung in intensive triple rice cropping regimes.

## GHG Emissions from Diverse Agroecosystems

Chapter 2 provided an extensive overview of GHG emissions, particularly CO2, CH4, and N2O, from various agricultural ecosystems globally. Key points were:

- Agricultural soils contribute 60–80% of total anthropogenic N2O emissions, chiefly through chemical fertilizer use.
- Paddy crops generate substantially higher GHGs than other cereals, with maximal emissions of ~23.75 Tg CO2e/ha from rice paddies versus ~2.98 Tg CO2e/ha from pigeon pea.
- Global soil degassing accounts for ~35% of total $CO_2$ emissions, ~47% of CH4, ~53% of $N_2O$, and ~ 21% of NO emissions.
- Microorganisms like nitrifiers, denitrifiers, and methanogens play a vital role in GHG fluxes via processes such as nitrification, denitrification, and methane production.
- Land use changes like deforestation can increase GHG emissions by accelerating organic matter decomposition.
- Tillage, fertilization, irrigation practices, and crop choice significantly impact CO2 emissions from soils.
- Temperature, precipitation, soil moisture, texture, and aeration regulate the production and consumption of biogenic GHGs.
- Strategies like crop diversification, rotation, and incorporating legumes can help stabilize atmospheric GHG levels by improving soil conditions.

The extensive review of greenhouse gas emissions across diverse agricultural ecosystems worldwide unequivocally established agroecosystems as major contributors to rising atmospheric concentrations of potent greenhouse gases, including carbon dioxide, methane, and nitrous oxide. The significant emissions from activities like fertilizer use, livestock production, residue burning, and drainage of organic soils highlight the urgent need for targeted mitigation efforts to curb the climate impact of agriculture. Adoption of sustainable land management practices such as integrated nutrient management, conservation tillage, cover cropping, crop diversification, and agroforestry can help reduce greenhouse gas emissions through enhanced carbon storage, improved fertilizer efficiency, lowered methane production, and decreased carbon losses. However, the site-specific nature of agricultural emissions noted in the review emphasizes that mitigation strategies must be locally optimized based on soils, climate, and cropping patterns. Overall, the chapter clearly identifies agriculture as a priority sector for climate change mitigation and provides evidence-based guidance for transitioning to sustainable production systems with lower greenhouse gas footprints.

## Effect of Different Fertilizer Types on GHG Emissions

Chapter 3 focused on how various fertilizer types—organic, inorganic, and integrated nutrient management—affect GHG emissions and crop yields. Key aspects were:

- Organic fertilizers like vermicompost can elevate CH4 emissions but lower net GHG emissions and global warming potential (GWP) versus inorganic fertilizers by promoting soil C sequestration.
- Combined application of organic and inorganic fertilizers often leads to higher GHG emissions compared to sole organic or inorganic amendments.
- CH4 emissions are greater from organic inputs like cow dung with a high C:N ratio, while N2O dominates from inorganic nitrogenous fertilizers.
- Emissions also depend on factors like fertilizer amount, crop type, soil properties, and climate.
- Strategies like biochar addition or optimizing nitrogen application rates can mitigate GHG release while maintaining crop productivity under integrated nutrient management.
- Improving fertilizer use efficiency through precision application techniques is critical for reducing GHG emissions per unit crop yield.
- Organic fertilizers promote microbial growth and activity, enhancing soil health and quality.
- The GWP of CH4 is 25 times higher than CO2, while N2O is 298 times more potent than CO2 over a 100-year timescale.

The findings from this chapter underscore the need for judiciously integrating organic and inorganic nutrient sources to maximize agricultural yields while minimizing the environmental impacts of greenhouse gas emissions and reactive nitrogen pollution from farms. Strategies such as using vermicompost, optimizing fertilizer application rates, combining organic and chemical nutrients, and improving fertilizer use efficiency can help balance yields and emissions. However, the complex interactions between fertilizer type, soil properties, climate, and cropping patterns highlighted in the review indicate that integrated nutrient management plans must be tailored to local conditions. The potential tradeoffs between yield gains and elevated greenhouse gases also point to the need for holistic assessments to identify sustainable options for each production system and environment. Overall, the evidence synthesized in the chapter makes a compelling case for context-specific integrated nutrient management as a promising approach to sustainably enhance agricultural productivity while curbing greenhouse gas release and associated climate change impacts.

## GHG Emissions Under Conservation Agriculture

Chapter 4 extensively reviewed evidence on the impacts of conservation agriculture (CA) practices on GHG emissions from agricultural soils. Key aspects covered were:

- CA techniques like reduced or no-tillage, permanent soil cover through crop residues, and rotation cropping can lower CO2 emissions by 17–60% compared to intensive conventional tillage.
- CH4 and N2O fluxes vary considerably under CA depending on factors like soil texture, climate, crop choice, and management practices.
- CA may increase N2O loss in fine-textured or moist soils compared to conventional tillage due to higher denitrification rates.
- Strategies like residue retention, cover cropping, agroforestry integration, and fertilizer placement in CA systems can potentially offset elevated N2O by enhancing soil C sequestration.
- Incorporating livestock through well-managed rotational grazing and judicious manure application can curb reliance on inorganic fertilizers in CA systems.
- Overall, CA holds significant potential for mitigating GHGs, but optimal practices need to be identified based on local soil, climatic, and cropping conditions to maximize climate benefits.
- Long-term studies across diverse sites are essential to accurately quantify the GHG mitigation potentials of CA and related management strategies.

The extensive review highlighted the complex, context-specific nature of CA impacts on GHG fluxes, underscoring the need to tailor practices to local conditions for maximizing climate change mitigation.

## Key Soil Variables Influencing GHG Emissions

Chapter 5 summarized the current understanding of how different intrinsic soil properties and external environmental variables affect the production and consumption of major GHGs like CO2, CH4, and N2O in agroecosystems. Key points were:

- Soil moisture content and oxygen availability are primary drivers regulating the rates of microbial nitrification and denitrification and associated N2O emissions from soil.
- Temperature affects GHG fluxes by directly influencing microbial metabolic rates and soil respiration, contributing to CO2 emissions.
- Soil pH controls the N2O/N2 ratio emitted from denitrification, with lower pH favoring higher N2O production.
- Texture affects soil aeration, water retention, and gas diffusion, with fine-textured clayey soils having higher N2O and CH4 emissions compared to coarse sandy soils.
- Soil organic matter provides substrates like organic carbon and nitrogen for microbial GHG production but can also enhance nutrient immobilization and soil C storage.
- Land management practices, including fertilizer type and application rate, tillage techniques, irrigation regime, and residue retention, significantly impact GHG emission rates.
- Soil organisms ranging from microbes to macrofauna strongly regulate GHG dynamics via effects on C and N cycling.
- Environmental variables like temperature, rainfall, and water table depth affect GHG fluxes through impacts on moisture, aeration, and microbial activity.

The comprehensive review established the intricate complexity of soil biochemical and physicochemical processes governing GHG production and consumption but also identified potential management options for mitigation.

## Integrated Synthesis and Conclusions

This exhaustive synthesis report integrated detailed findings from multiple chapters examining diverse facets of GHG emissions from varied managed and natural agroecosystems worldwide. Several unifying themes emerged:

- Rice paddies and agricultural soils represent major anthropogenic sources of potent GHGs, including N2O, CH4, and CO2.
- Microbial activities like nitrification, denitrification, and methane production and consumption drive soil GHG fluxes.
- Environmental factors, including soil moisture, temperature, aeration, and pH, critically regulate biogenic GHG production and emission rates.

- Organic nutrient amendments generally elevate GHG emissions compared to inorganic fertilizers alone but enhance long-term soil C storage.
- Integrated nutrient management strategies and climate-smart conservation practices offer significant potential to curb GHGs while maintaining or improving agricultural productivity and profitability.
- Optimizing land management practices through site-specific combinations of organic residues, legume integration, fertilizer management, tillage reduction, and water management is essential for maximizing GHG mitigation in diverse agroecosystems and cropping systems.

The synthesized findings from multiple evidence streams clearly establish agriculture as a priority sector for climate change mitigation efforts. Widespread adoption of sustainable land management practices, including organic nutrient management, conservation tillage, crop diversification, and integrated farming systems, can help reduce GHG emissions, soil degradation, water pollution, and biodiversity losses while improving resilience and ecosystem service provisioning from managed landscapes. However, more interdisciplinary systems-level research is imperative to accurately quantify region-specific mitigation potentials, optimize adaptable practices, and incentivize farmer adoption. Overall, this exhaustive synthesis highlights the intricate biogeochemical and ecological interactions governing agricultural GHG dynamics but also demonstrates feasible options for transitioning to climate-smart sustainable farming systems across locales.

# Index




S. R. Mohanty, B. Kollah (eds.), *Greenhouse Gas Regulating Microorganisms in Soil Ecosystems*, https://doi.org/10.1007/978-3-031-70569-4

**R**



**S**



**T**

SPRINGER NATURE

# GPSR Compliance

*The European Union's (EU) General Product Safety Regulation (GPSR) is a set of rules that requires consumer products to be safe and our obligations to ensure this.*

*If you have any concerns about our products, you can contact us on ProductSafety@springernature.com*

In case Publisher is established outside the EU, the EU authorized representative is:

Springer Nature Customer Service Center GmbH
Europaplatz 3
69115 Heidelberg, Germany

The manufacturer's authorised representative in the EU is Springer Nature Customer Service Centre GmbH, Europaplatz 3, 69115 Heidelberg, Germany. If you have any concerns regarding our products, please contact ProductSafety@springernature.com

Printed and bound by CPI Group (UK) Ltd, Croydon, CR0 4YY
30/09/2026
02225586-0012